Space Science and Technologies

Series Editor

Peijian Ye, China Academy of Space Technology, Beijing, China

Space Science and Technologies publishes a host of recent advances and achievements in the field – quickly and informally. It covers a wide range of disciplines and specialties, with a focus on three main aspects: key theories, basic implementation methods, and practical engineering applications. It includes, but is not limited to, theoretical and applied overall system design, subsystem design, major space-vehicle supporting technologies, and the management of related engineering implementations.

Within the scopes of the series are monographs, professional books or graduate textbooks, edited volumes, and reference works purposely devoted to support education in related areas at the graduate and post-graduate levels.

More information about this series at http://www.springer.com/series/16385

Dayi Wang · Maodeng Li · Xiangyu Huang ·
Xiaowen Zhang

Spacecraft Autonomous Navigation Technologies Based on Multi-source Information Fusion

北京理工大学出版社
BEIJING INSTITUTE OF TECHNOLOGY PRESS

Springer

Dayi Wang
Beijing Institute of Spacecraft System Engineering
CAST
Beijing, China

Xiangyu Huang
Science and Technology on Space Intelligent Control Laboratory
Beijing Institute of Control Engineering
CAST
Beijing, China

Maodeng Li
Science and Technology on Space Intelligent Control Laboratory
Beijing Institute of Control Engineering
CAST
Beijing, China

Xiaowen Zhang
Science and Technology on Space Intelligent Control Laboratory
Beijing Institute of Control Engineering
CAST
Beijing, China

ISSN 2730-6410 ISSN 2730-6429 (electronic)
Space Science and Technologies
ISBN 978-981-15-4878-9 ISBN 978-981-15-4879-6 (eBook)
https://doi.org/10.1007/978-981-15-4879-6

Jointly published with Beijing Institute of Technology Press
The print edition is not for sale in China (Mainland). Customers from China (Mainland) please order the print book from: Beijing Institute of Technology Press.

This Springer imprint is published by the registered company Springer Nature Singapore Pte Ltd.
The registered company address is: 152 Beach Road, #21-01/04 Gateway East, Singapore 189721, Singapore

Series Editor's Preface

China's space technology and science research have earned a place in the world, but have not been compiled into a series of systematic publications yet. In 2018, the series *Space Science and Technology* edited mainly by me and co-authored by the leading figures in China's space industry was published in China, when China Academy of Space Technology was celebrating the 50th anniversary of its founding. This collection contains 23 volumes in Chinese, only 10 of which have been selected, recreated, and translated into English. In addition, each English volume has been recreated at the suggestion of the Springer, by deleting the contents similar to Springer's existing publications and adding the contents that are internationally advanced and even leading, and bear both Chinese characteristics and worldwide universality. This series fully reflects the knowledge and engineering experience recently accumulated by Chinese scientists and engineers in space technology and science research.

As Editor in Chief of this series, I always insist that this collection must be of high quality, either in Chinese version or in English version. First, the contents of this series must be condensed and sublimated based on the combination of theory and practice, so as to provide both a theoretical value and engineering guidance. Second, the relationships between past knowledge and state-of-the-art and between other people's work and our own new findings should be properly balanced in the book contents to ensure the knowledge systematicness and continuity and to highlight new achievements and insights. Each volume intends to introduce the readers something new. Third, the English version should be customized for international exposure and play a solid supporting role for China to contribute to the world's space field.

This collection consists of 10 volumes, including *Spacecraft Thermal Control Technologies, Spacecraft Power System Technologies, Spacecraft Electromagnetic Compatibility Technologies, Technologies for Spacecraft Antennas Engineering Design, Satellite Navigation Systems and Technologies, Satellite Remote Sensing Technologies, Spacecraft Autonomous Navigation Technologies Based on Multi-source Information Fusion, Technologies for Deep-Space Exploration, Space Robotics, and Manned Spacecraft Technologies.*

Spacecraft Autonomous Navigation Technologies Based on Multi-source Information Fusion takes deep-space spacecraft as application object and focuses on the method of multi-source information fusion and its application to autonomous navigation in the deep-space explorations. This volume focuses on estimation theory, fusion algorithm, performance analysis, autonomous navigation technique based on information fusion, and ground simulation test technique.

This volume is the fruit of the author's painstaking efforts and wisdom borne in his working practice. He has long been committed to the research of autonomous navigation technology of deep-space spacecrafts. He participated in the design of autonomous navigation scheme of Chang'e lunar exploration project. He has solved a series of key technical problems and made a number of breakthroughs. The autonomous navigation scheme for lunar exploration based on the research results of this volume has been successfully applied to the exploration missions of Chang'e series and will be applied to subsequent lunar and Mars exploration missions. Therefore, high practical value is one of the features of this volume.

Another feature of this volume pays close attention to the organic integration and later application of estimation theory, fusion theory, and autonomous navigation technology. The book follows the principles of integration and simplicity, gradually unveils the theory of estimation fusion, and builds a bridge between theory and engineering technology to help readers embark on the road leading to spacecraft autonomous navigation.

The publication of this series adds a new member to the international family of space technology and science publications, and intends to play an important role in promoting academic exchanges and space business cooperation. It provides comprehensive, authentic, and rich information for international space scientists and engineers, enterprises and institutions as well as government sectors to have a deeper understanding of China's space industry. Of course, I believe that this series will also be of great reference value to the researchers, engineers, graduate students, and university students in the related fields.

Peijian Ye
Academician
Chinese Academy of Sciences
Beijing, China

Preface

Spacecraft autonomous navigation refers to the technology of determining the spacecraft's position and velocity in real time using onboard equipment without relying on the ground support. Autonomous navigation based on multi-source information fusion is an autonomous navigation method that fuses multiple information sources (multiple observation targets, multi-sensors, prior knowledge, etc.). Autonomous navigation technology based on multi-source information fusion has great application prospects in the field of deep-space exploration. Compared with a single sensor and a single observation source, it can enhance information redundancy, extended time, and space coverage, and reduce information acquisition costs. This book focuses on the theory and methods of multi-source information fusion and its application in autonomous navigation for deep-space exploration. The content involves estimation theory, fusion algorithms, performance analysis, information fusion autonomous navigation technology, ground simulation test technology, etc.

The book is divided into four parts that consist of 12 chapters. The first part (Chap. 1) introduces basic concepts of autonomous navigation and summarizes main autonomous navigation methods for deep-space spacecraft. The second part (Chaps. 2–4) consists of estimation theory, fusion algorithm, and performance analysis. These chapters constitute the theoretical basis of multi-source information fusion autonomous navigation technology. The third part (Chaps. 5–11) presents time and coordinate systems, dynamic models and environment models, inertial navigation technology, optical autonomous navigation technology, optical/pulsar integrated autonomous navigation technology, altimeter- and velocimeter-/optical-aided inertial navigation technology, and simulation testing techniques of the autonomous navigation based on multi-source information fusion. Chapter 12 is the fourth part, which is the summary of the whole book and the prospect of technological development trends.

The main features of this book are as follows: (1) The basic estimation and fusion theory are emphasized in this book. Its main procedure is to consider a general situation first and then derive solutions for some then specific situations, which is easy for grasping the problem. In addition, when describing the basic theory of information fusion, this book will give as many different forms as possible, which will help researchers to consult and establish different forms of conceptual connections. (2) The autonomous navigation technology of the spacecraft involved in this book emphasizes practical application, some of which have been applied to actual flight missions such as Chang'e 3 and Chang'e 4 soft landing missions and will also be applied to subsequent Chinese planetary landing missions. (3) This book focuses on the combination of estimation, fusion theory, and autonomous navigation technology. Therefore, when narrating the application of the technology, a brief review of related theoretical methods will be made to facilitate the reader to establish the connection between theory and application.

Beijing, China Dayi Wang

Acknowledgements

Many thanks to the colleagues who helped prepare the material of this book: Yonghui Chu, Ji Li, Kai Xiong, and Chao Xu. We do wish to acknowledge the deep attention and strong support of Academician Peijian Ye, Prof. Liu Liangdong, and Prof. Wang Yongfu for the publication of this book. We acknowledge Academician Wu Hongxin, Academician Yang Mengfei, Prof. Li Guo, Prof. Sun Jing, Prof. He Yingzi, Prof. Wei Chunling, and Prof. Li Tieshou for their guidance and valuable advice. Mrs. Liang Xiaoheng (Secretary-General of Science and Technology Commission from Beijing Institute of Spacecraft System Engineering) and Beijing Institute of Technology Press have done a lot of work for publication, and we would like to thank them here.

Finally, we also would like to express our thanks for the support from National Outstanding Youth Science Foundation (61525301), National Defense Science and Technology Outstanding Young Talent Fund, National Natural Science Foundation (61690215, 61673057, 61503923), Beijing Institute of Spacecraft System Engineering, Beijing Institute of Control Engineering, and Science and Technology on Space Intelligent Control Laboratory.

Contents

About the Authors

Dayi Wang Dr. Wang Dayi, a Professor and doctoral supervisor, born in November 1973, is currently the deputy director and commander in field of space science and exploration of Beijing Institute of Spacecraft System Engineering. He is a winner of the National Science Fund for Distinguished Young Scholar and National Defense Fund for Outstanding Young Scholar. He is also selected by National Ten Thousand Talent Program, and serves as the Chief Technology Expert of 973 Project. He won the China Association for Science and Technology Qiu Shi Outstanding Young Scholar Award, was honored with the Prize for Scientific and Technological Innovation of Ho Leung Ho Lee Foundation, and was selected by National Million Talent Project, in 2015, 2016, and 2017, respectively. He is a national outstanding mid-aged expert entitled to the special government allowance of the State Council.

Dr. Dayi Wang has carried out innovative research works and solved a series of key technical problems in the field of spacecraft autonomous navigation and autonomous diagnosis and reconfiguration technology, making contributions to the successful flight test of Chang'e lunar probes and other spacecrafts.

Maodeng Li received the B.S. and Ph.D. degrees in aerospace engineering from Harbin Institute of Technology, Harbin, P.R. China, in 2006 and 2011, respectively. From September 2011 to August 2013, he was a postdoctoral research associate in Beijing Institute of Control Engineering (BICE). Since August 2013, he has been with Science and Technology on Space Intelligent Control Laboratory of BICE as a senior engineer. His current research areas include spacecraft autonomous navigation and GNC design for planetary landing missions.

Xiangyu Huang received the B.S. and Ph.D. degrees in aerospace engineering from Harbin Institute of Technology, Harbin, P.R. China, in 1999 and 2005, respectively. Since August 2005, he has been with Science and Technology on Space Intelligent Control Laboratory of Beijing Institute of Control Engineering (BICE), where he is currently Professor and Senior Research Specialist. He was GNC Engineer of the Chang'e-3 lunar landing mission, and his current research area is the GNC design for planetary landing missions. He won one National Technological Innovation prize (second class) and three ministerial-level prizes (first class).

Xiaowen Zhang received his B.S. degree from Harbin Engineering University, Harbin, China, in 2005, and his M.S. degree in control theory and control engineering from China Academy of Space Technology, Beijing, China, in 2009. He is currently Senior Engineer at Beijing Institute of Control Engineering. His research interest covers GNC system design of deep-space exploration missions and spacecraft autonomous navigation and guidance technology. He won one first-class ministerial-level prize.

Acronyms

APSR	Accretion-Powered Pulsars
ASG	Air Suspension Gyroscope
AXP	Anomalous X-ray Pulsar
BCRS	Barycentric Celestial Reference System
BLS	Batch Least Square
BLUE	Best Linear Unbiased Estimation
CDF	Cumulative Distribution Function
CF	Centralized Fusion
CGRO	Compton Gamma Ray Observatory
CI	Covariance Intersection
CKF	Cubature Kalman Filter
DARPA	Defense Advanced Research Projects Agency
DPS	Deep-Space Positioning System
DR	Dead Reckoning
DTG	Dynamically Tuned Gyroscope
EKF	Extended Kalman Filter
ESG	Electrostatically Suspended Gyroscope
ET	Ephemeris Time
FDIR	Fault Detection Isolation Recovery
FOG	Fiber Optical Gyroscope
FOM	Figure of Merit
GDOP	Geometric Dilution of Precision
GNC	Guidance, Navigation and Control
GNSS	Global Navigation Satellite System
HMXB	High-Mass X-ray Binary
HRG	Hemispherical Resonator Gyroscope
HRSC	High-Resolution Stereo Camera
HXMT	Hard X-ray Modulation Telescope
ICRF	International Celestial Reference Frame
IMM	Interacting Multiple Model

IMU	Inertial Measurement Unit
INS	Inertial Navigation System
ISAS	Institute of Space and Astronautical Science
ITS	Impactor Targeting Sensor
JD	Julian Date
JDL	Joint Directors of Laboratory
JPL	Jet Propulsion Laboratory
KF	Kalman Filter
LEC	Linear Equality Constraint
LFG	Liquid Floated Gyroscope
LIDAR	Light Detection and Ranging
LMMSE	Linear MMSE
LMXB	Low-Mass X-ray Binary
MAP	Maximum A Posteriori
MAPS	Multi-autonomous Positioning System
MDL	Minimum Description Length
MEMS	Microelectromechanical System
MF	Measurement Fusion
MICAS	Miniature Integrated Camera and Spectrometer
MJD	Modified Julian Date
ML	Maximum Likelihood
MMAE	Multiple Model Adaptive Estimation
MMSE	Minimum Mean Square Error
MOEMS	Microoptoelectromechanical Systems
MOLA	Mars Orbiter Laser Altimeter
MRI	Medium-Resolution Imager
MRO	Mars Reconnaissance Orbiter
MSE	Mean Square Error
MSL	Mars Science Laboratory
MSP	Millisecond Pulsar
NASA	National Aeronautics and Space Administration
NICER	Neutron star Interior Composition Explorer
NRL	Naval Research Laboratory
PDF	Probability Density Function
PDOP	Position Dilution of Precision
PF	Particle Filter
PINS	Platform Inertial Navigation System
PMF	Probability Mass Function
PSR	Pulsating Source of Radio
QEC	Quadratic Equality Constraint
RLG	Ring Laser Gyroscope
RPSR	Rotation-Powered Pulsar
SAW	Surface Acoustic Wave
SEXTANT	Station Explorer for X-ray Timing and Navigation Technology
SINS	Strapdown Inertial Navigation System

SSB	Solar System Barycenter
ST	Sidereal Time
STK	Systems Tool Kit
TAI	International Atomic Time
TCB	Barycentric Coordinate Time
TCG	Geocentric Coordinate Time
TCP/IP	Transmission Control Protocol/Internet Protocol
TDB	Barycentric Dynamical Time
TDOA	Time Difference of Arrival
TDOP	Time Dilution of Precision
TDT	Terrestrial Dynamical Time
TOA	Time of Arrival
TT	Terrestrial Time
UKF	Unscented Kalman Filter
USA	Unconventional Stellar Aspect
UT	Universal Time
UTC	Coordinated Universal Time
VSMM	Variable Structure Multiple Model
WLS	Weighted Least Square
XNAV	X-ray Pulsar-Based Navigation

Chapter 1
Introduction

Navigation generally refers to the determination of the orbit (position and velocity) and attitude parameters of a vehicle (or a moving body) relative to a coordinate system at a given time. In general, spacecraft navigation refers only to the determination of orbit parameters. In general, the navigation relying on only the onboard measuring and computing devices rather than ground support is referred to as spacecraft autonomous navigation, while the spacecraft autonomous navigation realized by the fusion processing of multiple information sources (multiple observed objects, multi-sensor measurements, priori knowledge, etc.) is called multi-source information fusion-based autonomous navigation.

Autonomous navigation is the core technology of autonomous spacecraft operation and is the premise for a spacecraft to autonomously control its orbital attitude and perform its space missions such as lunar soft landing and in-orbit services. Autonomous navigation can not only reduce the dependence of a spacecraft on ground control and improve its independent survivability, but also relieve the restriction on the layout of ground telemetry, track, and command (TT&C) stations posed by the limited land area and enhance the spacecraft capability of completing a task outside the TT&C area.

Compared with the autonomous navigation based on a single sensor and a single observed object, the autonomous navigation based on multi-source information fusion is an important development direction in the aerospace field, which can provide better time and space spreadability and more information redundancy, and enhance the autonomous capability of the system. A deep-space exploration spacecraft (hereinafter referred to as "probe" or "deep-space probe") generally flies a long distance and a long time in an almost unknown environment, with a particularly urgent need for a high degree of navigation autonomy. Taking deep-space exploration missions as the background, this book focuses on the theories, methods, and technical issues related to the autonomous navigation based on multi-source information fusion.

D. Wang et al., *Spacecraft Autonomous Navigation Technologies Based on Multi-source Information Fusion*, Space Science and Technologies, https://doi.org/10.1007/978-981-15-4879-6_1

1.1 Autonomous Navigation Technology

Spacecraft navigation can be divided into absolute navigation and relative navigation depending on the choice of the selected reference coordinate system. In general, the determination of motion parameters relative to the fixed coordinate system of a space target is called relative navigation, while the determination of motion parameters relative to an inertial system is called absolute navigation. Therefore, spacecraft autonomous navigation can also be divided into autonomous absolute navigation and autonomous relative navigation. It is in the research domain of relative navigation to determine the position and velocity of the tracking spacecraft relative to the target spacecraft in an in-orbit service for such a target. From the perspective of measurement principle, spacecraft autonomous navigation can be mainly divided into inertial navigation, autonomous optical navigation, autonomous navigation based on pulsars, and the navigation based on artificial beacons (global satellite navigation system, inter-satellite measurement, etc.). Strictly speaking, the navigation based on artificial beacon is not totally autonomous navigation and will not be discussed here. Next, only inertial, optical, and autonomous navigation based on pulsars will be introduced.

1.1.1 Inertial Navigation

Inertial navigation system (INS) is an autonomous navigation system based on dead reckoning (DR), which consists of a set of inertial devices and a navigation processor. Inertial devices, also called inertial measurement unit (IMU), mainly include three orthogonal gyroscopes and three orthogonal accelerometers. The gyroscopes are used for measuring the inertial angular velocity of a vehicle, and the accelerometers are used for measuring the specific force acting on the vehicle (also known as "non-gravitational acceleration"). In addition, through the numerical integration of IMU measurements, spacecraft position, velocity, and attitude can be determined.

Neither relying on external information nor radiating energy to the outside, INS has many advantages, such as high short-time accuracy, high bandwidth, complete navigation information, good concealment, and strong insusceptibility to disturbance. INS is not only used in spacecrafts, but also widely used in other moving vehicles. It is the most important navigation in aerospace, aviation, and sailing. However, the only weakness of INS is that its navigation accuracy will decrease over time.

1.1.1.1 Gyroscope

Gyroscope ("gyro" for short) is a device that uses the conservation of momentum to sense the direction. According to their working principle, the existing gyros fall into two categories [1]: those based on classical mechanics, such as mechanical gyros and vibratory gyros; and those based on modern physics, such as laser gyros and fiber-optic gyros.

1.1.1.2 Accelerometer

Accelerometers can be divided into mechanical accelerometers and solid-state accelerometers. Among them, the basic principle of a mechanical accelerometer is to connect the detected mass block with its shell through a spring. Under the steady state, the force acting on the mass block will balance with the spring tension, and the net elongation of the spring can be used to measure the received force proportional to the acceleration. Solid-state accelerometers include [2] surface acoustic wave (SAW) accelerometers, silicon accelerometers, quartz accelerometers, etc. With the development of MEMS technology, MEMS accelerometer emerged. There are two types of MEMS accelerometers: one works in the same way as mechanical accelerometers and the other is based on the principle of solid-state accelerometers. Like MEMS gyroscope, MEMS accelerometer is also small in size, light in weight, low in power consumption and cost, and fast in start-up, but poor in accuracy.

1.1.1.3 Inertial Navigation System

According to the different ways of installation of inertial devices in the vehicle, inertial navigation systems fall into two types, e.g., platform inertial navigation system (PINS) and strapdown inertial navigation system (SINS). The inertial devices of PINS are mounted on the platform and thus isolated from the external rotation. By installing universal joints, the platform can be aligned with the navigation coordinate system. The gyro on the platform detects the rotation signals, which are transmitted to the torque motor that rotates the universal joints to compensate the external rotation, thus keeping the platform in alignment with the navigation coordinate system. Once the included angle between universal joints is obtained, the information on vehicle attitude can be acquired. After compensating for the gravitational acceleration, the accelerometer signals can be integrated to obtain the velocity value and integrated again to obtain the position information.

The SINS system has no physical platform, with gyro and accelerometer directly mounted on the vehicle. Compared with PINS, this system is much smaller in size, weight, and cost. However, its inertial device has lower measurement accuracy due to the fixation on the vehicle, whose vibration and shock directly affect the component and the bad working environment. In addition, the accelerometer outputs in the SINS are the components in the vehicle coordinate system, which need to be converted into navigation coordinate system. As a result, the calculation becomes more complicated. However, with the rapid development of inertial devices and electronic computer technology, the above problems no longer restrict the development of SINS and this system has been widely used in the navigation field. Unless specially specified, all the inertial navigation systems in this book are SINS.

1.1.2 *Autonomous Optical Navigation*

As early as in the 1960s, institutes began to study the autonomous navigation technology for deep-space exploration, and gradually tested and applied it in missions. With continuous improvement in the performance and reliability of onboard computers, sensors, and actuators, the technology of autonomous navigation has been successfully applied to more and more deep-space missions, providing an important guarantee for improving the survivability of deep-space spacecrafts. So far, deep-space exploration relies mainly on autonomous optical navigation. The working principle of autonomous optical navigation is to identify the targeted celestial bodies with known ephemeris as navigation beacons, and then plan and process the observed optical images of the beacon, and finally derive the position, velocity, and attitude of the spacecraft from the known information [3, 4]. Key technologies of autonomous optical navigation include [3] the selection and planning of navigation beacons, the technology of optical navigation sensor, the technology of navigation information acquisition and processing, the navigation filtering algorithm, and other technologies. The main factors affecting the accuracy of autonomous optical navigation include state priori knowledge, unmodeled acceleration, sensor pointing error, measurement frequency, ephemeris errors of navigation beacons, random noise in measurement, and other systematic errors.

1.1.2.1 Deep-Space Autonomous Optical Navigation

The first application of autonomous optical navigation in deep-space dates back to 1960s. Battin et al. [5] proposed the autonomous navigation theory of interstellar spacecrafts, that is, to calculate the spacecraft position by measuring the angle between line of sight (LOS) of the known celestial body (such as the sun, earth, moon) and the star in distance, and using the ephemeris of that celestial body. The Apollo 8, launched in 1968, applied this theory and used sextant as the optical sensor of autonomous navigation system, so as to verify the feasibility of spacecraft autonomous navigation for the first time. The sextant on Apollo 8 was a dual-line narrow-field device used to measure the angle of LOS between the earth or the moon and a star. According to the geometric relationship, the spacecraft position in the space could be determined. Due to the technology restrictions at that time, the accuracy of position calculation obtained solely by angular measurement was not high enough. Therefore, autonomous navigation was only used as a supplement to ground TT&C to confirm the orbit safety and provide support for a spacecraft to return to earth when the navigation support from the ground was unavailable. In a later series of manned Apollo missions, autonomous celestial navigation was used as a backup of terrestrial navigation at the earth-moon transfer stage, and autonomous navigation and control was adopted at the lunar landing and ascending rendezvous stages.

The Mariner 9, the first Mars probe launched by the USA in May 1971, used photos of the natural satellites of Mars (Phobos and Deimos) with a star background taken by onboard optical system for autonomous navigation. Subsequent flight evaluations showed that the Mariner 9's optical observations were more accurate than previously expected and could have been used for navigation in the Mars orbiting phase without any other help [6].

In January 1994, the "Clementine" lunar probe launched by the USA used two star trackers to determine its own attitude and demonstrated the function of autonomous operation. It was originally intended to use the earth/moon images and inertial attitude information acquired by the imaging sensor for the earth-moon transfer and circum-lunar autonomous navigation. But due to the probe fault, navigation calculation based on the image data was only conducted on the ground [7].

On October 24, 1998, the "Deep Space 1" probe launched by the USA successfully demonstrated the onboard autonomous deep-space navigation and control system for the first time. At the cruise stage, the autonomous navigation based on the images of asteroids and background stars taken by the navigation camera was demonstrated with the positioning accuracy of about 250 km and the velocity accuracy of about 0.5 m/s, which are high enough to meet the requirements for cruising navigation accuracy. When approaching and flying over the asteroids or comets, the probe used the autonomous navigation technique based on the images of target asteroids or comets. Its autonomous navigation and control system can autonomously complete the photographing sequencing, image processing and analysis, orbit determination, ephemeris correction, orbit correction, and attitude maneuver [8, 9].

In January 2004, the "Stardust" probe launched by the USA in February 1999 flew by the "Wild-2" comet with the aid of autonomous optical navigation, and completed the mission of comet sampling. In 2006, it returned to the earth. The Stardust's closest approach to the "Wild-2" comet during its fly by the comet was expected to be 120–150 km, and the flyby time only lasted a few minutes. Because of the long distance from the earth, it could not rely on the ground-based navigation. Therefore, in the flyby process, the Stardust processed the comet images through center extraction, obtained the center of the comet, and with the help of attitude determination system and filtering technology, completed real-time autonomous navigation. [10]. The Stardust also used autonomous optical navigation during its flyby the "Tempel 1" comet on February 14, 2011.

In May 2003, the Hayabusa probe launched by Japan completed the first sample return from an asteroid. During the rendezvous and landing phases, it successfully accomplished autonomous attachment by using an optical navigation sensor, a radar rangefinder and a laser rangefinder, and communicating with pre-dropped navigation landmarks [11].

In September 2003, the ESA launched "SMART-1" to conduct an autonomous navigation experiment for future deep-space missions, and to return the in-orbit observations of autonomous navigation system to the ground for processing. In the experiment, a navigation camera was used to take photos of the identified navigation beacon. Those photos, combined with the image data and the probe attitude information, could determine the line-of-sight direction of the navigation beacon, which then was input to the navigation filter to estimate the orbit for the probe [12].

In March 2004, the "Rosetta" launched by ESA and the lander it carried adopted the autonomous control technique. Its autonomous control software could make autonomous decisions and controls far away from the earth (with a communication delay of 0.5 h) to ensure its correct operation. In order to ensure the probe safety, autonomous guidance, navigation, and control was implemented when landing on the comet nucleus. The orbit where the probe was located was determined by processing the measurements taken by a camera, a radar, and the orbit control was accomplished autonomously.

In January 2005, the "Deep Impact" launched by the USA completed its mission of rendezvous and impact with a comet and demonstrated autonomous navigation and control in the collision. Partially inheriting and developing the autonomous navigation and control system of Deep Space 1, an impactor can realize autonomous navigation and control within two hours before the impact with the target comet by using the images and attitude information of the comet taken by a navigation sensor [13].

In August 2005, when approaching the Mars, the "Mars Global Surveyor" launched by National Aeronautics and Space Administration (NASA) and equipped with an experimental navigation camera took pictures for two natural satellites of the Mars, and relied on their images and ephemeris information to realize autonomous navigation at an accuracy higher than that of ground TT&C, and demonstrated high-precision navigation necessary for future Mars-landing missions.

1.1.2.2 Analysis of Autonomous Optical Navigation at Different Flight Stages

The optical navigation sensors, image processing algorithms, and navigation methods selected for various probes at different mission stages will vary with the distance between the probe and the targeted celestial body, the position of the probe relative to a larger celestial body such as the sun or the earth, and the distribution of smaller celestial bodies around.

1. Autonomous Optical Navigation at the Stage of Free Flight

At the stage of free flight, the probe cannot be observed since it is far away from the targeted celestial body. Therefore, the navigation beacons for this stage are usually the celestial bodies with known ephemeris and near the flight orbit. Through continuous observation of the navigation beacon for a period of time, the orbit (and attitude) information of the probe can be estimated based on the ephemeris information of the navigation beacon. The Deep Space 1 relied on the optical measurements and ephemeris of asteroids and background stars to determine its own position and velocity. The "Clementine" probe extracted the direction vectors of the earth's center and the moon's center from the images of the earth and moon taken by the navigation sensor at the earth-moon transfer stage, and by combining the direction vectors with the attitude information and the earth/moon ephemeris information, estimated the orbit for the probe with the Kalman filter (KF) method. Due to the fault in the probe, its autonomous navigation experiment was not completed [14]. But this autonomous navigation method based on optical imaging was demonstrated by the "SMART-1" of ESA through a ground test.

2. Autonomous optical navigation at the approaching stage

When being approached by the probe, the size and brightness of targeted celestial body shall ensure that the navigation sensor of the probe is sensitive enough to image the target body on its focal plane. In this case, the targeted celestial body can be selected as the navigation beacon. By using the navigation sensor to continuously image the targeted celestial body and extracting its center vector information in combination with its inertial attitude and ephemeris, the orbit and attitude of the probe relative to the targeted celestial body can be determined. The center point extraction technique of a targeted celestial body was demonstrated when the Voyager probe launched by the USA encountered Neptune and Uranus. Thanks to this technique, optical navigation has been applied to the Stardust's approach to the comet [10]. This technique has also been used by Deep Space 1.

3. Autonomous optical navigation at the orbiting stage

When flying around the targeted celestial body, the probe is closer to it and thus obtaining its clear images by using the navigation camera. The navigation system uses the image processing algorithm to extract the edge image of the target and then calculates its apparent radius and center pointing vector as specially required for flying around spherical celestial bodies such as large planets. For an irregular small celestial body, its feature points are first identified by matching its edge image with its pre-processed model and then combined with the attitude information to determine the orbit of the probe through the filtering algorithm. The "Clementine" probe relied just on the optical navigation when flying around the moon. The apparent radius and centroid vector of the moon were extracted from the moon images acquired by the sensor and then combined with the attitude information to determine the orbit of the probe through the filtering algorithm.

To meet the need for autonomous navigation at the stage of flying around small celestial bodies, the JPL laboratory in the USA has developed an autonomous navigation method based on optical imaging [15]. In this method, the edge image of a small celestial body (the target) obtained by the wide-field camera is matched with the pre-processed target model, and then the position of the probe relative to the target center is determined in real time by using the weighted least-square (WLS) method, and at last, the position information is input into the navigation filter to calculate the orbit parameters of the probe.

4. Autonomous optical navigation at the landing or attachment stage

During autonomous landing or attachment, the probe is closer to the target. In order to accurately identify the surface conditions of the target, a safe landing site shall be selected, and the optical navigation camera shall have a small field of view and high resolution. The Institute of Space and Astronautical Science (ISAS) in Japan has developed a technique of autonomous navigation and guidance for asteroid landing [16]. To land on the asteroids where the state of the landing site is not known in detail, the institute proposed a method of autonomous landing navigation using a navigation camera and a laser rangefinder. According to that method, the navigation camera takes pictures of the landing target, the laser rangefinder measures the distance from the probe to the target surface, and then the filter acquires the information on probe position and velocity. Misu et al. in Japan invented an autonomous asteroid-landing navigation algorithm that determined the position and attitude of the probe relative to the target asteroid by extracting and tracking feature points [16]. Johnson et al. working in the JPL proposed an autonomous asteroid-landing navigation algorithm based on computer vision [17].

Through the above analysis, it is easy to see that, in the deep-space exploration, the method and technical means of autonomous control based on optical imaging navigation vary with stage and mission. At different operating stages, the different targets tracked by imaging navigation sensor have a different size, distance, brightness, and background stray light influence. This makes it very difficult to design the navigation sensor and operate the image processing algorithm. As a result, the research of optical sensor technology, navigation filtering algorithm, and image processing algorithm has become critical to the autonomous navigation of deep-space exploration spacecrafts. Table 1.1 summarizes the candidate solutions of autonomous navigation in deep-space exploration.

1.1.3 Autonomous Pulsar-Based Navigation

The autonomous navigation of deep-space probe based on X-ray pulsar is to determine the position unit vector of the pulsars in the barycentric celestial reference system (BCRS) and the standard arrival time of X-ray pulse through very-long-baseline interferometry (VLBI), to compare them with the pulsar LOS direction and actual arrival time measured by the X-ray detector installed on the deep-space probe, and to obtain the navigation information such as the position, velocity, and time of spacecraft through an appropriate filtering algorithm. Attitude determination can also be achieved through the navigation based on X-ray pulsar, in a way similar to the attitude determination based on optical camera. By imaging the pulsar, its coordinates in the probe coordinate system are obtained, so that the azimuth of LOS vector relative to the probe can be estimated.

Table 1.1 Candidate solutions of autonomous navigation in deep-space exploration [3]

Deep-space mission phase	Autonomous navigation task	Autonomous navigation scheme	Navigation sensors configuration	Navigation algorithm
Heliocentric transfer phase	Determine the heliocentric orbit	Based on the images of multiple asteroids and background stars	IMU, star sensor and high-precision NFOV navigation camera	Navigation asteroids selection and planning, image processing, navigation filtering, and celestial ephemeris calculation
Earth-moon transfer phase	Determine the geocentric orbit	Based on earth and moon images	IMU, star sensor and navigation camera	Image processing, navigation filtering, and celestial ephemeris calculation
Approach phase	Determine the attitude and orbit of the probe relative to the target body	Based on images of the target body (its center/edge/feature)	IMU, star sensor and navigation camera	Image processing, navigation filtering, and celestial ephemeris calculation
Impact phase	Determine the attitude and orbit of the probe relative to the target body, and the impact point information.	Autonomous navigation and predictive guidance scheme based on images of the target body (its center/surface feature)	IMU, High-precision WFOV navigation camera, star sensor and	Image processing, navigation filtering, celestial ephemeris calculation, and maneuvering guidance calculation
Orbiting phase	Determine the attitude and orbit of the probe relative to the orbited body	For a large orbited body, autonomous navigation scheme is based on the image/other optical information of the target body to acquire its orientation and apparent radius; for a small orbited body, autonomous navigation scheme is based on images of the target body (its center/edge/feature)	WFOV navigation camera, or other optical sensor, star sensor and IMU	Sensor data processing (image processing) and autonomous navigation filtering algorithm

(continued)

Table 1.1 (continued)

Deep-space mission phase	Autonomous navigation task	Autonomous navigation scheme	Navigation sensors configuration	Navigation algorithm
Landing or attachment phase	Determine the position and velocity of the probe relative to the landing site, and identify the obstacles in the landing area	Autonomous navigation and control scheme with the function of autonomous obstacle detection	Range finder and velocimeter, IMU, high-precision WFOV navigation camera, and 3D imager	Autonomous obstacle detection, prediction of safe landing zone, and high-precision autonomous navigation and control algorithm

1.1.3.1 Research History of Pulsar Navigation Theory

In 1974, the Jet Propulsion Laboratory (JPL) in the USA first proposed a pulsar-based method for determining the autonomous orbit of interplanetary probes at the accuracy of about 150 km that can be achieved by an antenna with a diameter of 25 m receiving radio signals from a pulsar [18]. However, since the radio signals are weak, a large antenna is required. This is not realistic for typical space missions. In addition, the visible light signals from pulsars are very weak and few [19]. To use visible light signals for navigation, a large optical telescope needs to be installed on the spacecraft, which is also very difficult for space missions.

In 1981, Chester and Butman from the Institute of Communication Systems in the USA proposed a navigation method based on the X-ray signals of pulsars, in order to achieve pulsar navigation and avoid the installation of large spaceborne equipment. Unlike the visible light and other radio signals emitted by pulsars, X-ray signals can be detected by small equipment, which is a good fit to spacecrafts. Simulation results show that an X-ray probe with an effective area of 1000 cm^2 can achieve a positioning accuracy of about 150 km [20].

In 1993, Dr. Wood [21] from the Naval Research Laboratory (NRL) in the USA designed a test for unconventional stars characterization, proposing the method of determining the orbit and attitude of spacecrafts with the help of X-ray sources and using X-ray pulsars for time keeping. Dr. Harson [22] from the Stanford University studied in depth this test and proposed the attitude determination algorithm based on X-ray source and the design scheme of time-keeping phase-locked loop.

In 2004, an ESA report analyzed the basic principle of pulsar navigation and the pulsar signal model and explained the engineering feasibility of the system [23]. Sheikh et al. [24] from the University of Maryland have done a lot of fruitful work. He established a navigation X-ray source database and used the existing pulsar timing model for reference to build a pulse time of arrival (TOA) conversion model considering parallax, Shapiro delay, and Romer delay. The results showed that the positioning accuracy of the spacecraft could reach 2 km when using an X-ray probe with an area of 0.1 m^2. Besides, by analyzing the time model of a certain pulsar, Sheikh pointed out that the positioning accuracy of about 300 m could be achieved theoretically by using that pulsar for navigation. In 2007, Graven analyzed the error in X-ray pulsar navigation system. He divided the error factors of that navigation system into sensor error, signal source measurement error, signal source model error, spacecraft systematic error, and other model errors, and initially analyzed the influence of pulse phase error on measurement information [25]. Emadzadeh studied the relative navigation method based on X-ray pulsars and analyzed the influence of velocity error on time delay estimation of signals [26].

Many other scholars [26–30] have also studied in depth and extensively the autonomous navigation by X-ray pulsars, discussed the basic framework of autonomous pulsar navigation, and analyzed the factors affecting key technologies and accuracy of pulsar navigation. The key technologies involved in X-ray pulsar navigation include: sky patrol and data processing of pulsars, pulse TOA measurement and error correction, X-ray probe and weak signal processing, establishment and maintenance of spatial and temporal references for navigation and positioning, and robust filtering of autonomous navigation information.

The main errors in pulsar navigation originate from the uncertainty of time measurement, including pulsar emitting noise, propagation noise, cosmic background noise, signal receiver noise, local time noise, time quantization noise, pulse shape uncertainty, pulse period uncertainty, and pulse phase error. The advantage of this method is that it can provide a time reference in addition to navigation information, but its disadvantage is that few X-ray pulsars are available and measurement accuracy cannot be guaranteed.

1.1.3.2 Science Research or Missions for Pulsar Navigation

In 1999, the ARGOS satellite was launched as an experimental platform for the USA. It was equipped with an X-ray telescope with an effective area of 1000 cm^2, which could accurately record the TOA of 1–15 keV photons.

In 2004, the Defense Advanced Research Projects Agency (DARPA) began to invite bids for the "Research on Autonomous Satellite Navigation Based on X-ray Sources (XNAV)" [31] to be funded in three phases. This program included four parts: X-ray pulsar characteristics research, X-ray probe development, pulsar navigation algorithm, and space system design. The aim of this program is to achieve the measured pulsar angular-position deviation less than 0.0001 arcsec, effective X-ray probe area less than 1.0 m^2, photon TOA measurement accuracy of 1–10 ns, and pulsar navigation and positioning accuracy better than 100 m, so as to provide highly reliable positioning, attitude measurement, and timing services for spacecrafts. In 2005, a project team led by Ball Aerospace received funding from the program.

To achieve the in-orbit application level of pulsar navigation, it is important to solve a series of key issues such as miniaturization and engineering of X-ray probes, accurate extraction of pulse TOA, high-fidelity ground simulation and in-orbit demonstration experiment, and to establish an accurate X-ray pulsar ephemeris, radiation signal outline and timing model through long-term in-orbit observation. In

2011, the NASA Goddard Space Flight Center and the Universities Space Research Association (USRA) launched the Station Explorer for X-ray Timing and Navigation Technology (SEXTANT) project [32]. The SEXTANT project would use nested focusing lenses and silicon-drifted imaging devices to demonstrate the orbit determination, with a millisecond pulsar as navigation signal source. The goal was to obtain the orbit determination accuracy better than 10 km in a highly dynamic low-earth orbit (LEO) through two weeks of observation. On June 3, 2017, the neutron star interior composition explorer (NICER) of SEXTANT was launched by SpaceX's "Falcon 9" rocket and installed on the space station. The data generated would be used to analyze the navigation performance of X-ray pulsars and tackle key problems in this technology.

At present, China is actively carrying out the in-orbit observation and navigation demonstration based on X-ray pulsars. On November 10, 2016, China launched its first pulsar or experiment satellite "XPNAV-1" to demonstrate the performance indexes and space adaptability of this onboard pulsar probe, and to accumulate the measured in-orbit pulsar data. On June 15, 2017, the hard X-ray modulation telescope (HXMT) was launched as China's first large X-ray astronomical satellite. HXMT can be used for X-ray sky survey and imaging in a large wide-band sky area and for accumulation of measured pulsar data in order to lay an experience foundation for realizing pulsar navigation.

1.2 Multi-source Information Fusion Technology

1.2.1 Definition of Multi-source Information Fusion

The research on multi-source information fusion dates back to the 1970s when the automatic data fusion of military modernization C^3I (Command, Control, Communication & Intelligence) system was studied in the USA and widely applied to various weapon platforms. In 1973, the research institutes in the USA, funded by the Department of Defense (DOD), took the lead in studying modern information fusion theory. In the early 1980s, the DARPA proposed the "Joint Tactical Synthesis Program" and successfully developed the All Source Analysis System/ENemy Situation Correlation Element (ASAS/ENSCE), the first practical information fusion system in intelligence and electronic warfare. In 1988, the DOD listed information fusion as one of the key technologies for focused research and development in the 1990s. Other countries have also strengthened their efforts in information fusion research and made fruitful achievements, such as battlefield utilization and target interception system, tactical command and control system and many other military information fusion systems. Information fusion technology has had a 40-year development history, but its definition is still controversial. Currently, the most accepted definition is that proposed by the Joint Directors of Laboratory (JDL) in 1991 [33], as described below:

Definition 1 information fusion is a *multi-level* process dealing with the association, correlation, and combination of data and information from single and multiple sources to achieve refined position and identity estimates, and complete and timely assessments of situations and threats, and their significance.

1.2.2 Classification of Multi-source Information Fusion Technologies

Multi-source information fusion technologies can be classified by many criteria.

If classified by fusion architecture, multi-source information fusion technologies can fall into three types: **centralized, distributed, and hybrid** [34].

In the centralized information fusion architecture, the observation data obtained by each sensor is uploaded to the higher-level information fusion center without being analyzed. The information fusion center combines, screens, correlates, and synthesizes all the initial data using certain criteria and algorithms and then outputs the information fusion conclusions once.

In the distributed information fusion architecture, every sensor will initially analyze the element observation data, make a local judgment conclusion, and report only this conclusion and its related information, or a potential but not entirely reliable conclusion (based on the preliminary analysis) and its related information to the information fusion center. Then the information fusion center centralizes the multifaceted data at a higher level for further synthesis and draws the final judgment conclusion.

The advantages of centralized information fusion are data comprehensiveness, no information loss, and high confidence of final judgment conclusion. However, it needs to transmit a large amount of data, which leads to demanding requirements for transmission network, a long time of information processing, and slow system response. In contrast, the distributed information fusion has much fewer data to transmit, so the requirements for transmission network can be relaxed, the time of information processing by information fusion center can be shortened, and the response speed can be increased.

As the combination of both centralized and distributed fusion architectures, the hybrid fusion architecture possesses the advantages of the two architectures, but pays a relatively high price in communication and computing. However, it also has the strong points incomparable to the above architectures, so it is widely applied in the practice.

If classified by fusion purpose, multi-source information fusion technologies can fall into three types: **detection fusion, estimation fusion, and attribute fusion**. Among them, detection fusion is to integrate the observation data or judgment results from different sensors into a more comprehensive and accurate judgment about a unified environment or event. The main purpose of detection fusion is to use multiple sensors for information fusion processing. This can eliminate the uncertainty in the detection by a single sensor or a single type of sensor, improve the reliability of detection system, and obtain more accurate knowledge of detected objects.

Estimation fusion is to study how to make best use of the useful information contained in multiple data sets when estimating the unknowns. These data sets typically come from multiple information sources, most often from multiple sensors.

Attribute fusion is to judge the attribute and type of the target from multi-sensor detection information.

1.2.3 Multi-source Information Fusion Methods

Typical multi-source information fusion methods include [35–37]: signal processing and estimation methods, statistical inference methods, information theory methods, and decision theory method.

Signal processing and estimation methods include linear estimation methods such as wavelet transform [38] for image enhancement and processing, weighted average, least squares and Kalman filter, and nonlinear estimation methods such as extended Kalman filter (EKF), particle filter [39], and unscented Kalman filter (UKF) [40], as well as multi-model adaptive estimation method [41].

Statistical inference methods include classical inference, Bayes inference, evidential inference [42, 43], as well as random set theory [44, 45] and support vector machine (SVM) theory.

Information theory method is to integrate multi-source data by means of information measure optimization, so as to effectively solve the problems. Typical algorithms include entropy method [46, 47] and minimum description length (MDL) method [48, 49].

Decision theory method [50, 51] is often applied to high-level decision fusion, such as fusing the decision theory method with visible light, infrared, and millimeter-wave radars, for alarm analysis [52].

1.3 Autonomous Navigation Technology Based on Multi-source Information Fusion

1.3.1 Research and Application Progress

Since autonomous navigation is essentially state estimation, the autonomous navigation based on multi-source information fusion belongs to estimation fusion if classified by fusion purpose.

The autonomous navigation based on multi-source information fusion has been applied in actual deep-space missions. A deep-space probe is generally equipped with both an IMU and an optical sensor. In fact, the autonomous navigation in deep-space exploration is based on multi-source information fusion, but still called autonomous optical navigation for customary reasons. In addition, the autonomous navigation [53] relying only on an optical sensor to observe different asteroids at the deep-space transfer stage can also be regarded as the fusion-based autonomous navigation that observes different optical landmarks at different stages. As far as China's deep-space exploration is concerned, the Chang'e-3 lander has achieved a soft landing on the lunar surface by autonomous navigation fusing with inertia and velocity/range measurement [54–56].

The US is developing a fusion navigation sensor for deep-space positioning system (DPS) [57]. This sensor is composed of an IMU, a wide angle camera and a narrow angle camera. Among them, the IMU can be used to measure acceleration and angular velocity at the powered flight stage. The wide angle camera can provide a wider field of vision. The narrow angle camera can provide high-resolution images when getting very close to the target. Through combination, their advantages are mutually complemented. China also developed a sun-earth-moon integrated sensor and a large-field wide-dynamic-range optical sensor is proposed, the former of which can observe the sun, earth, and the moon, and the latter of which can simultaneously observe central celestial bodies and stars.

Many scholars have also studied the fusion navigation methods [58–63] for deep-space exploration, such as the autonomous navigation based on inertial and pulsar fusion, the autonomous navigation based on optical and pulsar fusion, and the autonomous navigation fusing the information on multiple optical targets.

1.3.2 Necessity and Advantages

In the deep-space exploration, the targeted celestial bodies are very far away from the earth, so it is difficult for ground TT&C alone to meet the practical needs of deep-space probes, whether in terms of navigation accuracy or real-time performance. Therefore, it is necessary to develop the technology of autonomous navigation [64, 65]. The diversity and high autonomy of deep-space exploration missions determine that the autonomous navigation based on multi-source information fusion is an inevitable development trend. A typical deep-space mission is composed of at least three flight stages, namely earth departure, transfer, and approaching, possibly including orbiting, landing, and impact. A single way of autonomous navigation is obviously unable to meet the requirements for temporal and spatial coverage. Future deep-space exploration missions are bound to develop toward a high degree of autonomy, which means that fault detection, isolation, and recovery (FDIR) can be carried out by the satellites themselves. In order to realize FDIR, information redundancy is indispensable. This also determines the necessity of developing the autonomous navigation technology based on multi-source information fusion.

In conclusion, the autonomous navigation technology based on multi-source information fusion has great advantages over that based on a single sensor and a single observed target [35, 66, 67]:

1. Higher Information Redundancy

From multi-source information, redundant expression of the environment and of a certain feature can be obtained. In addition, when one of the multi-sensors fails, the redundant information provided by multi-sensors can be used for troubleshooting and thus improve the system robustness.

2. Wider Temporal Coverage

The synergy of multi-sensors improves the detection probability, because one sensor may detect the targets or events that other sensors have missed in a certain period. In addition, the parallel operations of multiple sensors can double the speed of information acquisition.

3. Wider Spatial Coverage

Multiple overlapping action zones of multi-sensors increase the spatial coverage, because one sensor may detect the areas that other sensors cannot. In addition, the observation of different targets in different spatial ranges also increases the spatial coverage to some extent.

4. Lower Information Acquisition Cost

A multi-sensor system may seem more expensive than a single-sensor system, but the acquisition of the same amount of information often costs more for a single sensor.

1.4 Outline

This book focuses on the estimation theory involved in the autonomous navigation based on multi-source information fusion and on its application in the transfer, approaching, orbiting, and landing missions of deep-space exploration. Figure 1.1 shows the organizational architecture and technology roadmap (excluding this chapter) of this book. It consists of 12 chapters grouped into four parts: theory (Chaps. 2–4), application (Chaps. 5–11), introduction, and perspective.

This chapter introduces the basic concept of the spacecraft autonomous navigation based on multi-source information fusion, reviews and summarizes the theories of autonomous deep-space navigation and multi-source information fusion, presents the progress, necessity and advantages of the autonomous navigation based on multi-source information fusion, and summarizes the main contents of this book.

Chapter 2 focuses on estimation theory, the basis of autonomous navigation algorithm. This chapter first gives the definition and evaluation criteria of estimation algorithm, introduces several optimal estimation algorithms in common use and their analytical formulas in special cases, and then based on this, expands and discusses the estimation algorithm for dynamic systems.

Chapter 3 focuses on fusion algorithm. This chapter first discusses several estimation fusion architectures in common use and linear fusion models/algorithms,

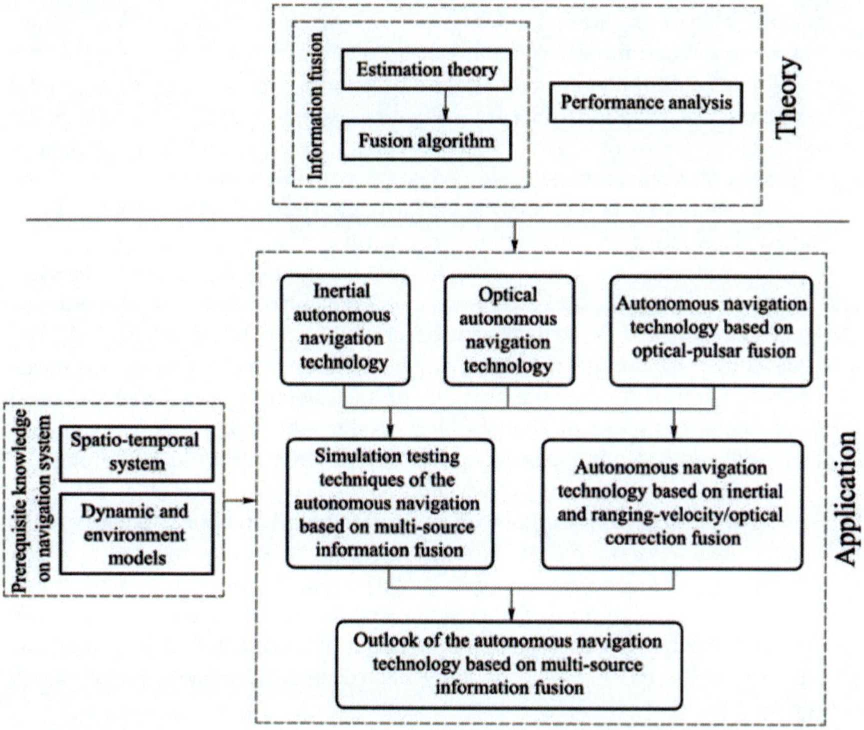

Fig. 1.1 Organizational structure and technology roadmap of this book

mainly including fusion algorithm in linear unified model and covariance intersection method in thc distributed fusion, and then elaborates on fusion navigation filtering algorithms for dynamic systems, which constitute the core algorithms of an autonomous navigation system based on multi-source information fusion.

Chapter 4 focuses on performance analysis. This chapter first presents the observability analysis methods for linear and nonlinear systems, describes the observability analysis methods for autonomous navigation system, and then introduces the Monte Carlo method and linear covariance analysis technique to evaluate the performance of navigation system quantitatively.

Chapter 5 focuses on time and space coordinate systems, the premise to describe the navigation state. Without the space-time concept, navigation descriptions are meaningless. This chapter first presents the definition of time, then provides the definition of reference coordinate system and the coordinate system transformation, and finally gives the ephemeris calculation method for navigation beacons.

Chapter 6 presents dynamic models and environment models, including spacecraft orbit dynamics model, spacecraft attitude kinematics model, celestial body surface model, and gravitational field model.

Chapter 7 focuses on inertial autonomous navigation technology. This chapter first presents the measurement equations of gyro and accelerometer, describes the INS equation in continuous-time and discrete-time forms, multiple sample algorithm and error model of strapdown inertial navigation system, and then introduces the calibration and error compensation methods of inertial devices, and finally illustrates the application of inertial autonomous navigation to spacecrafts.

Chapter 8 focuses on autonomous optical navigation technology. This chapter first introduces the principle of autonomous optical navigation as well as the selection criteria of optical imaging sensors and different types of optical navigation landmarks; and classifies and summarizes the measurement equations of autonomous optical navigation, analyzes the geometric observability of autonomous optical navigation, and studies the optimal route planning based on geometric observability analysis; and then presents the batch processing of autonomous optical navigation and the implementation process of Kalman filtering algorithms; and finally introduces the simulation results of autonomous optical navigation against the background of deep-space transfer, approaching, and orbiting.

Chapter 9 focuses on the autonomous navigation based on optical-pulsar fusion. This chapter first reviews the autonomous optical navigation and pulsar-based autonomous navigation; and then analyzes geometric and dynamic observability, and presents the planning algorithm of navigation landmarks; and studies the algorithm of the autonomous navigation based on optical-pulsar fusion; and finally simulates and demonstrates this navigation technology in the deep-space transfer, approaching, and orbiting phases.

Chapter 10 focuses on the autonomous navigation technology based on inertial and ranging-velocity/optical fusion and its application to lunar soft landing. This chapter first introduces the flight process of soft landing and the composition and workflow of soft landing autonomous navigation system, and presents the observation equations of altimeter, velocimeter, and optical sensor; and then analyzes the observability of the autonomous navigation system based on IMU + velocity and ranging correction and of the autonomous navigation system based on IMU + optical correction, and studies and presents the inertial-fusion autonomous navigation method based on Kalman filter framework; and finally introduces the simulation results of inertial-fusion autonomous navigation for lunar soft landing and the flight results of Chang'e-3 lander.

Chapter 11 focuses on the experiment techniques of the autonomous navigation based on multi-source information fusion. This chapter first presents the simulation testing techniques of the autonomous navigation based on optics-pulsar fusion against the background of deep-space transfer, approaching and orbiting; and then introduces the simulation testing techniques of the autonomous navigation based on inertial and ranging-velocity fusion against the background of lunar soft landing hovering, obstacle avoidance and slow descent.

Chapter 12 summarizes the whole book and presents the perspective. It suggests the future research direction and development strategy in three areas: the autonomous navigation based on multi-source information fusion, information fusion technology, and fusion-based autonomous navigation sensor.

References

1. Guangqu, Y. 1987. *Principles of inertial navigation.* Beijing: Aviation Industry Publishing House.
2. Titterton, D. and J.L. Weston, *Strapdown inertial navigation technology.* Vol. 17. The Institution of Electrical Engineers.
3. Wang, D.Y., and X. Huang. 2009. Survey of autonomous navigation and control for deep space exploration. *Aerospace Control and Application* 35 (3): 6–12.
4. Li, J.F., W. Cui, and H. Baoyin. 2012. A survey of autonomous navigation for deep space exploration. *Mechanics in Engineering* 34 (2): 1–9.
5. Battin, R.H. 1999. *An introduction to the mathematics and methods of astrodynamics.* New York: American Institute of Aeronautics Astronautics.
6. Duxbury, T.C., G. Born, and N. Jerath. 1972. Viewing Phobos and Deimos for navigating mariner 9. *Journal of Spacecraft and Rockets* 11 (4): 215–222.
7. Regeon, P., R. Chapman, and R. Baugh. 1995. CLEMENTINE: the deep space program science experiment. *Acta Astronautica* 35 (1): 307–321.
8. Riedel, J., et al. 2000. *Deep space 1 technology validation report: autonomous optical navigation.* Jet Propulsion Laboratory: Pasadena.
9. Bhaskaran, S., et al. 1998. Orbit determination performance evaluation of the deep space 1 autonomous navigation system. In *AAS/AIAA spaceflight mechanics meeting.* Monterrey, CA.
10. Bhaskaran, S., J.E. Riedel, and S.P. Synnott. 1998. Autonomous nucleus tracking for comet/asteroid encounters: the STARDUST example. In IEEE aerospace conference. Snowmass at Aspen, CO, USA: IEEE.
11. Kubota, T., et al. 2000. Navigation, guidance and control of asteroid sample return spacecraft: MUSES-C. In *Proceedings of the 4th ESA international conference.* ESTEC, Noordwijk.
12. Marini, A., G. Racca, and B. Foing. 2002. SMART-1 technology preparation for future planetary missions. *Advances in Space Research* 30 (8): 1895–1900.
13. Mastrodemos, N., D.G. Kubitschek, and S.P. Synnott. 2005. Autonomous navigation for the deep impact mission encounter with comet Tempel 1. *Space Science Reviews* 117 (1–2): 95–121.
14. Bernard, K.A.J., M. and D. Robert. 1993. The deep space program science experiment mission (DSPSE) astrodynamics mission planning. In *The 44th international astronautical federation congress.* Graz, Austria.
15. Bhaskaran, S., J.E. Riedel, and S.P. Synnott. 1995. Demonstration of autonomous orbit determination around small bodies. In *AAS/AIAA astrodynamics specialist conference.* Halifax, Nova Scotia, Canada.
16. Kubota, T., et al. 2003. An autonomous navigation and guidance system for MUSES-C asteroid landing. *Acta Astronautica* 52 (2): 125–131.
17. Johnson, E., Y. Cheng, and L.H. Matthies. 2000. Machine vision for autonomous small body navigation. In *Aerospace conference proceedings, 2000 IEEE.* IEEE.
18. Downs, G. 1974. *Interplanetary navigation using pulsating radio sources.* NASA Technique Reports. 1–12. NASA N74-34150.
19. Shearer, A., and A. Golden. 2001. Implications of the optical observations of isolated neutron stars. *The Astrophysical Journal* 547 (2): 967.
20. Chester, T. and S. Butman. 1981. Navigation using X-ray pulsars. Jet propulsion laboratory, Pasadena, CA, NASA Tech. Rep. 81N27129.
21. Wood, K.S. 1993. Navigation studies utilizing the NRL-801 experiment and the ARGOS satellite. In *Small Satellite Technology and Applications III.* International Society for Optics and Photonics.
22. Hanson, J.E. 2006. *Principles of X-ray navigation.* Stanford: Stanford Linear Accelerator Center (SLAC).
23. Sala Álvarez, J., et al. 2004. *Feasibility study for a spacecraft navigation system relying on pulsar timing information*, vol. 3. Paris: European Space Agency.

24. Sheikh, S.I. 2005. The use of variable celestial X-ray sources for spacecraft navigation, in *Phd thesis*. University of Maryland: Maryland.
25. Graven, P., et al. 2007. XNAV beyond the moon. In *Proceeding ION 63rd annual meeting*. Cambridge, Massachusetts.
26. Emadzadeh, A.A., and J.L. Speyer. 2011. X-ray pulsar-based relative navigation using epoch folding. *IEEE Transactions on Aerospace and Electronic Systems* 47 (4): 2317–2328.
27. Shuai P., et al. 2009. *Principle and method of X ray pulsar navigation system*. Beijing: China Astronautic Publishing House (in Chinese).
28. Zheng, W., et al. 2015. *X-ray pulsar-based navigation: theory and application*. Beijing: Science Press.
29. Liu, J., et al. 2010. X-ray pulsar navigation method for spacecraft with pulsar direction error. *Advances in Space Research* 46 (11): 1409–1417.
30. Wang, Y., et al. 2014. X-ray pulsar-based navigation using time-differenced measurement. *Aerospace Science and Technology* 36: 27–35.
31. Pines, D. 2004. *ARPA/DARPA space programs*, 1–15 Arlington: XNAV Industry Day.
32. Winternitz, L.M., et al. 2015. X-ray pulsar navigation algorithms and testbed for SEXTANT. In *IEEE aerospace conference*. IEEE.
33. White, F.E. 1991. Data fusion lexicon. In *Joint directors of laboratories, technical panel for C3, data fusion sub-panel, naval ocean systems center*. San Diego.
34. Castanedo, F. 2013. A review of data fusion techniques. *The Scientific World Journal*.
35. Han, C., H. Zhu, and Z. Duan. 2009. *Multi-source information fusion*. Beijing: Tsinghua University Press.
36. Pan, Q., et al. 2012. Basic methods and progress of information fusion]. *Control Theory & Applications* 29 (10): 1233–1244.
37. Khaleghi, B., et al. 2011. Multisensor data fusion: A review of the state-of-the-art. *Information Fusion* 14 (1): 28–44.
38. Nunez, J., et al. 1999. Multiresolution-based image fusion with additive wavelet decomposition. *IEEE Transactions on Geoscience and Remote Sensing* 37 (3): 1204–1211.
39. Van Der Merwe, R., et al. 2001. The unscented particle filter. In *Advances in neural information processing systems*
40. Julier, S., J. Uhlmann, and H.F. Durrant-Whyte. 2000. A new method for the nonlinear transformation of means and covariances in filters and estimators. *IEEE Transactions on Automatic Control* 45 (3): 477–482.
41. Li, X.-R., and Y. Bar-Shalom. 1996. Multiple-model estimation with variable structure. *IEEE Transactions on Automatic Control* 41 (4): 478–493.
42. Murphy, R.R. 1998. Dempster-Shafer theory for sensor fusion in autonomous mobile robots. *IEEE Transactions on Robotics and Automation* 14 (2): 197–206.
43. Bloch, I. 1996. Information combination operators for data fusion: a comparative review with classification. *IEEE Transactions on Systems, Man, and Cybernetics-Part A: Systems and Humans* 26 (1): 52–67.
44. Goutsias, J., R.P. Mahler, and H.T. Nguyen. 2012. *Random sets: theory and applications*, vol. 97. Berlin: Springer Science & Business Media.
45. Mori, S. 1997. Random sets in data fusion problems. In *Signal and Data Processing of Small Targets 1997*. International Society for Optics and Photonics.
46. Manyika, J., and H. Durrant-Whyte. 1995. *Data fusion and Sensor Management: a decentralized information-theoretic approach*. New Jersey: Prentice.
47. Zhou, Y. and H. Leung. 1997. Minimum entropy approach for multisensor data fusion. In *Proceedings of the IEEE signal processing workshop on higher-order statistics*.
48. Barron, A., J. Rissanen, and B. Yu. 1998. The minimum description length principle in coding and modeling. *IEEE Transactions on Information Theory* 44 (6): 2743–2760.
49. Joshi, R., and A.C. Sanderson. 1999. Minimal representation multisensor fusion using differential evolution. *IEEE Transactions on Systems, Man, and Cybernetics-Part A: Systems and Humans* 29 (1): 63–76.

50. Goodman, I.R., R.P. Mahler, and H.T. Nguyen. 2013. *Mathematics of data fusion*, vol. 37. Berlin: Springer Science & Business Media.
51. Berger, J.O. 2013. *Statistical decision theory and Bayesian analysis*. Berlin: Springer Science & Business Media.
52. Nelson, C.L. and D.S. Fitzgerald. 1996. Sensor fusion for intelligent alarm analysis. In *30th annual 1996 international carnahan conference*. IEEE.
53. Riedel, J., et al. 1997. Navigation for the new millennium: Autonomous navigation for Deep-Space 1. In *Space flight dynamics*.
54. Zhang, H.H., et al. 2014. Guidance navigation and control for Chang'E-3 powered descent. *Sci Sin Tech* 44 (4): 377–384.
55. Huang, X.Y., et al. 2014. Autonomous navigation and guidance for chang'e-3 soft landing. *Journal of Deep Space Exploration* 1 (1): 52–59.
56. Zhang, H.H., et al. 2014. Autonomous navigation for powered descent phase of Chang'E-3 lunar Lander. *Control Theory & Applications* 31 (12): 1686–1694.
57. Vaughan, A.T. and J.E. Riedel. 2016. *Deep space positioning system*. Google Patents.
58. Wang, Y., et al. 2013. XNAV/CNS integrated navigation based on improved kinematic and static filter. *The Journal of Navigation* 66 (6): 899–918.
59. Wang, Y., W. Zheng, and S. Sun. 2015. X-ray pulsar-based navigation system/Sun measurement integrated navigation method for deep space explorer. *Proceedings of the Institution of Mechanical Engineers, Part G: Journal of Aerospace Engineering* 229 (10): 1843–1852.
60. Wang, Y., et al. 2015. Autonomous navigation method for low-thrust interplanetary vehicles. *Journal of Aerospace Engineering* 29 (1): 04015009.
61. Ning, X., et al. 2017. Impact of the pulsar's direction on CNS/XNAV integrated navigation. *IEEE Transactions on Aerospace and Electronic Systems* 53 (6): 3043–3055.
62. Liu, J., E. Wei, and S. Jin. 2017. Mars cruise orbit determination from combined optical celestial techniques and X-ray pulsars. *The Journal of Navigation* 70: 719–734.
63. Kai, X., W. Chunling, and L. Liangdong. 2016. Performance enhancement of X-ray pulsar navigation using autonomous optical sensor. *Acta Astronautica* 128: 473–484.
64. Wu, W.R., D. Wang, and X. Ning. 2011. *Autonomous navigation principle and technology for deep space spacecraft.*, Beijing China Astronautic Publishing House.
65. Han, H., and J. Chen. 2008. 21st century foreign deep space exploration development plans and their progresses. *Spacecraft Engineering* 17 (3): 1–22.
66. Raol, J.R. 2009. *Multi-sensor data fusion with MATLAB*. CRC press.
67. He, Y. 2010. *Information fusion theory with applications*. Beijing: Publishing House of Electronics Industry.

Chapter 2
Point Estimation Theory

Estimation theory is a branch of statistics that deals with estimating interested parameters from random observations (or sampled data). An estimator (also called estimation rules), defining the corresponding rules of inference, can be classified into point and interval estimators. A point estimator gives an approximate value of unknown parameters, whereas an interval estimator calculates an interval that covers the unknown parameters with a high probability. Navigation estimation is often the issue of point estimation, so this chapter only focuses on the relevant contents of point estimation.

2.1 Basic Concepts

Denote $\boldsymbol{x} \in \mathbb{R}^n$ as an unknown parameter to be estimated, $z \in \mathbb{R}^m$ as measurement, and $\boldsymbol{Z}$ as a set of measurements, i.e., $\boldsymbol{Z} = \{z_1, z_2, \ldots, z_N\}$. Then an estimate of $\boldsymbol{x}$, denoted by $\hat{\boldsymbol{x}}$, can be written as a function of $\boldsymbol{Z}$:

$$\hat{\boldsymbol{x}} = \varphi(\boldsymbol{Z}) \tag{2.1}$$

where φ is called the estimator [1].

Depending on whether to treat $\boldsymbol{x}$ as a random variable or not, estimation algorithm can be classified into Bayesian estimation and non-Bayesian estimation [1]. In Bayesian estimation, $\boldsymbol{x}$ is treated as a random variable and assumed to have a known prior probability density function (PDF) $p(\boldsymbol{x})$. According to the Bayes' rule (see Eq. C.27 of Appendix C3.2), the following posterior PDF is obtained

$$p(\boldsymbol{x}|\boldsymbol{Z}) = \frac{p(\boldsymbol{Z}|\boldsymbol{x})p(\boldsymbol{x})}{p(\boldsymbol{Z})} \tag{2.2}$$

D. Wang et al., *Spacecraft Autonomous Navigation Technologies Based on Multi-source Information Fusion*, Space Science and Technologies, https://doi.org/10.1007/978-981-15-4879-6_2

Because **Z** is the available, the denominator in Eq. (2.2) is a normalized constant defined by

$$p(\mathbf{Z}) = \int p(\mathbf{Z}|\boldsymbol{x})p(\boldsymbol{x})d\boldsymbol{x} \tag{2.3}$$

The issue of Bayesian estimation is to obtain the posterior probability density function of Eq. (2.2). Once $p(\boldsymbol{x}|\mathbf{Z})$ is computed, a point estimate of $\boldsymbol{x}$ can be obtained using some loss function. Typical Bayesian estimators include minimum mean square error (MMSE) estimator and maximum a posteriori (MAP) estimator.

In non-Bayesian estimation, $\boldsymbol{x}$ is treated as an unknown constant, and thus, its prior PDF cannot be defined. In this case, the likelihood function is defined as the conditional probability density function of **Z** given by $\boldsymbol{x}$, i.e.,

$$\mathcal{L}(\boldsymbol{x}) \triangleq p(\mathbf{Z}|\boldsymbol{x}) \tag{2.4}$$

A non-Bayesian estimate can be obtained by applying some criterion to function (2.4). Non-Bayesian estimators in common use include maximum likelihood (ML) estimator and weighted least squares (WLS) estimator. Given an estimator, its properties can be evaluated according to the following definitions.

Definition 2.1 The estimation error of an estimate $\hat{\boldsymbol{x}}$ for a random or non-random variable $\boldsymbol{x}$, denoted by $\tilde{\boldsymbol{x}}$, is defined $\tilde{\boldsymbol{x}} \triangleq \boldsymbol{x} - \hat{\boldsymbol{x}}$.

Definition 2.2 If $\mathrm{E}\{\tilde{\boldsymbol{x}}\} = 0$, then the estimator (2.1) is an **unbiased** estimator of $\boldsymbol{x}$.

Definition 2.3 If $\lim\limits_{N\to\infty} \mathrm{E}\{\tilde{\boldsymbol{x}}\} = \mathbf{0}$, then the estimator (2.1) is an **asymptotic unbiased** estimator.

Definition 2.4 If $\lim\limits_{N\to\infty} \hat{\boldsymbol{x}} = \boldsymbol{x}$, then the estimator (2.1) is a **consistent** estimator [2]. (Another definition is $\operatorname{p}\lim\limits_{N\to\infty} \hat{\boldsymbol{x}} = \boldsymbol{x}$ [3], where p lim is the probability limit.)

Definition 2.5 The mean square error (MSE) of an estimate $\hat{\boldsymbol{x}}$ is defined as

$$\begin{aligned}\mathrm{MSE}(\hat{\boldsymbol{x}}) &\triangleq \mathrm{E}\{(\boldsymbol{x} - \hat{\boldsymbol{x}})^{\mathrm{T}}(\boldsymbol{x} - \hat{\boldsymbol{x}})\} \\ &= \mathrm{E}\{(\hat{\boldsymbol{x}} - \boldsymbol{x})^{\mathrm{T}}(\hat{\boldsymbol{x}} - \boldsymbol{x})\}\end{aligned} \tag{2.5}$$

Definition 2.6 The estimator (2.1) is an **efficient** estimator if its MSE is the smallest among all possible estimators of $\boldsymbol{x}$ [2].

Theorem 2.1 (Cramer–Rao Low Bound Theorem [4]) *If the joint PDF* $p(\boldsymbol{x}; \mathbf{Z})$ *satisfies the regularity condition*

$$\mathrm{E}\left\{\frac{\partial p(\boldsymbol{x}; \mathbf{Z})}{\partial \boldsymbol{x}}\right\} = \mathbf{0}, \quad \forall \boldsymbol{x} \tag{2.6}$$

Then the error covariance for any unbiased estimator $\hat{x}$ satisfies the following Cramer–Rao inequality:

$$\begin{aligned}\operatorname{var}(\tilde{x}) &\triangleq \mathrm{E}\left\{(\tilde{x}-\bar{\tilde{x}})(\tilde{x}-\bar{\tilde{x}})^{\mathrm{T}}\right\} \\ &= \mathrm{E}\left\{\tilde{x}\tilde{x}^{\mathrm{T}}\right\} \geq \mathcal{I}^{-1}\end{aligned} \tag{2.7}$$

where $\bar{\tilde{x}} \triangleq \mathrm{E}\{\tilde{x}\}$ and $\mathcal{I}$ is the Fisher information matrix defined by

$$\mathcal{I} \triangleq \mathrm{E}\left\{\left[\frac{\partial \log p(\mathbf{Z}|x)}{\partial x}\right]^{\mathrm{T}}\left[\frac{\partial \log p(\mathbf{Z}|x)}{\partial x}\right]\right\} \tag{2.8}$$

where $p(\mathbf{Z}|x)$ is the conditional probability density function of $\mathbf{Z}$ given x.

Definition 2.7 The estimator (2.1) is **asymptotically Gaussian distributed** if its limiting distribution is Gaussian distributed.

Definition 2.8 The estimator (2.1) is **asymptotically efficient** if it is consistent, asymptotically normally distributed, and has an asymptotic covariance matrix that is not large than the asymptotic covariance matrix of any other consistent, asymptotically Gaussian distributed estimator [5].

2.2 Common Parameter Estimators

2.2.1 MMSE Estimation

The MMSE estimator of x is defined by

$$\hat{x}^{\mathrm{MMSE}} \triangleq \arg\min_{x} J^{\mathrm{MMSE}} \tag{2.9}$$

where $J^{\mathrm{MMSE}} \triangleq \mathrm{E}\{(\hat{x}-x)^{\mathrm{T}}(\hat{x}-x)|\mathbf{Z}\}$ is the corresponding loss function. Using the optimal necessary condition $\frac{\partial J^{\mathrm{MMSE}}}{\partial \hat{x}} = 0$, $\hat{x}^{\mathrm{MMSE}}$ is given by

$$\hat{x}^{\mathrm{MMSE}} = \mathrm{E}\{x|z\} \tag{2.10}$$

Properties of the MMSE Estimator

1. $\hat{x}^{\mathrm{MMSE}}$ is an unbiased estimate of x, i.e., $\mathrm{E}\{\hat{x}^{\mathrm{MMSE}}\} = \mathrm{E}\{x\}$. (see Eq. (C.26) of Appendix C3.2).

2. Denote $\tilde{\boldsymbol{x}}^{\text{MMSE}} = \boldsymbol{x} - \hat{\boldsymbol{x}}^{\text{MMSE}}$. Then [6]

$$\text{var}\{\tilde{\boldsymbol{x}}^{\text{MMSE}}\} = \text{E}\{\text{var}\{\boldsymbol{x}|\boldsymbol{z}\}\} \tag{2.11}$$

(Hint: this property can be derived from the fact of $\text{E}\{\tilde{\boldsymbol{x}}^{\text{MMSE}}\} = \mathbf{0}$).
3. $\hat{\boldsymbol{x}}^{\text{MMSE}}$ has the smallest variance among all possible estimates of $\boldsymbol{x}$:

$$\text{var}\{\tilde{\boldsymbol{x}}\} \geq \text{E}\{\text{var}\{\boldsymbol{x}|\boldsymbol{z}\}\} = \text{var}\{\tilde{\boldsymbol{x}}^{\text{MMSE}}\} \tag{2.12}$$

where $\tilde{\boldsymbol{x}}$ is the error of any estimate of $\boldsymbol{x}$.
(Hint: this property can be derived using Eq. (2.11) and the definition of the MMSE estimator).

2.2.2 ML Estimator

The maximum likelihood estimate (MLE) of $\boldsymbol{x}$, denoted by $\hat{\boldsymbol{x}}^{\text{MLE}}$, maximizes the likelihood function (2.4):

$$\hat{\boldsymbol{x}}^{\text{MLE}} \triangleq \arg\max_{\boldsymbol{x}} \ \mathcal{L}(\boldsymbol{x}) \triangleq \arg\max_{\boldsymbol{x}} p(\boldsymbol{Z}|\boldsymbol{x}) \tag{2.13}$$

Because $p(\boldsymbol{Z}|\boldsymbol{x}) > 0$ and monotonic aspect of natural logarithm function, it is often to define $\ln \mathcal{L}(\boldsymbol{x})$ as the log-likelihood function. Then

$$\hat{\boldsymbol{x}}^{\text{MLE}} = \arg\max_{\boldsymbol{x}} \ln \mathcal{L}(\boldsymbol{x}) \tag{2.14}$$

Obviously, $\hat{\boldsymbol{x}}^{\text{MLE}}$ satisfies the optimal necessary condition

$$\frac{\partial}{\partial \boldsymbol{x}} \ln \mathcal{L}(\boldsymbol{x})|_{\boldsymbol{x}=\hat{\boldsymbol{x}}^{\text{MLE}}} = 0 \tag{2.15}$$

and the second sufficient condition

$$\frac{\partial^2}{\partial \boldsymbol{x}^2} \ln \mathcal{L}(\boldsymbol{x})|_{\boldsymbol{x}=\hat{\boldsymbol{x}}^{\text{MLE}}} < 0 \tag{2.16}$$

Properties of the ML estimator

1. Invariance property: suppose that $\boldsymbol{g}(\boldsymbol{x})$ is a vector (or scalar) function of $\boldsymbol{x}$. Then, $\boldsymbol{g}(\hat{\boldsymbol{x}}^{\text{MLE}})$ is the maximum likelihood estimate of $\boldsymbol{g}(\boldsymbol{x})$ (see [7] for the proof).
2. The ML estimator is asymptotically unbiased [1].
3. The ML estimator is consistent.
4. An advantage of MLE is that it is not limited to the measurement noise of Gaussian distribution.

5. In the case of zero-mean Gaussian measurement noise, the ML estimate is the same as that of the MMSE estimate.
6. The ML estimator is asymptotically efficient [8]:

$$\lim_{N\to\infty} \frac{\text{var}(\hat{\boldsymbol{x}}^{\text{MLE}})}{\mathcal{I}^{-1}} = 1 \tag{2.17}$$

where $\mathcal{I}$ is the Fisher information matrix defined by Eq. (2.8)
7. The ML estimator is asymptotically Gaussian [8]: for large n, $\sqrt{n}(\hat{\boldsymbol{x}}^{\text{MLE}} - \boldsymbol{x})$ has a Gaussian distribution with zero mean and variance $\mathcal{I}^{-1}$, where $\mathcal{I}$ is the Fisher information matrix given by Eq. (2.8).

2.2.3 Maximum a Posteriori (MAP) Estimator

The MAP estimator is also a Bayesian estimator. The MAP estimate of $\boldsymbol{x}$, denoted as $\hat{\boldsymbol{x}}^{\text{MAP}}$, maximizes the posterior PDF of $\boldsymbol{x}$ given available measurements $\boldsymbol{Z}$

$$\begin{aligned}\hat{\boldsymbol{x}}^{\text{MAP}} &= \arg\max_{\boldsymbol{x}} p(\boldsymbol{x}|\boldsymbol{Z}) \\ &= \arg\max_{\boldsymbol{x}} \ln p(\boldsymbol{x}|\boldsymbol{Z})\end{aligned} \tag{2.18}$$

Certainly, $\hat{\boldsymbol{x}}^{\text{MAP}}$ satisfies

$$\frac{\partial}{\partial \boldsymbol{x}} \ln p(\boldsymbol{x}|\boldsymbol{z})|_{\boldsymbol{x}=\hat{\boldsymbol{x}}^{\text{MAP}}} = 0 \tag{2.19}$$

Remark: for large samples, the MAP estimate is convergent to the ML estimate.

2.2.4 Weight Least-Square (WLS) Estimator

The MMSE, ML, and MAP estimators are derived from the probability, whereas the WLS estimator is derived from the measurement equation directly. Consider the measurement equation with the form of

$$\boldsymbol{z} = \boldsymbol{h}(\boldsymbol{x}) + \boldsymbol{v} \tag{2.20}$$

where $\boldsymbol{v}$ is the measurement noise. Then the WLS estimate of $\boldsymbol{x}$, denoted by $\hat{\boldsymbol{x}}^{\text{WLS}}$, is given by

$$\hat{\boldsymbol{x}}^{\text{WLS}} = \arg\min_{\boldsymbol{x}} [\boldsymbol{z} - \boldsymbol{h}(\boldsymbol{x})]^{\text{T}} \boldsymbol{W} [\boldsymbol{z} - \boldsymbol{h}(\boldsymbol{x})] \tag{2.21}$$

where $\boldsymbol{W} = \boldsymbol{W}^{\mathrm{T}} > \mathbf{0}$. Specially, if $\boldsymbol{W} = \boldsymbol{I}$, then the WLS estimator is called least-square (LS) estimator and $\hat{\boldsymbol{x}}^{\mathrm{WLS}}$ denoted by $\hat{\boldsymbol{x}}^{\mathrm{LS}}$.

2.3 Closed Form Parameter Estimators

Section 2.2 presents several common estimators which can be used for theoretical analysis. To apply them in practice, the distributional information or the form of measurement equations should be specified. However, even if they are provided, the calculation of these estimators may be prohibitively complicated. In view of this, this section introduces some closed form parameters estimators under some assumptions. The closed form allows them to be applied in practice conveniently.

2.3.1 Linear Estimator

A linear estimator assumes that the estimate $\hat{\boldsymbol{x}}$ is a linear function of measurement $\boldsymbol{z}$, i.e.,

$$\hat{\boldsymbol{x}} = \boldsymbol{A}\boldsymbol{z} + \boldsymbol{b} \tag{2.22}$$

where $\boldsymbol{A}$ and $\boldsymbol{b}$ are the matrix and the vector to be determined, respectively.

A typical linear estimator used in practice is the linear MMSE (LMMSE) estimator. Given the measurement $\boldsymbol{z}$, the LMMSE estimate of $\boldsymbol{x}$ is denoted as $\hat{\boldsymbol{x}}^{\mathrm{LMMSE}}(\boldsymbol{z})$. The goal of LMMSE estimator is to find an unbiased estimator with the form of

$$\hat{\boldsymbol{x}}^{\mathrm{LMMSE}} = \boldsymbol{A}\boldsymbol{z} + \boldsymbol{b} \tag{2.23}$$

that minimizes the following MSE loss function

$$J^{\mathrm{MSE}} = \mathrm{E}\{(\hat{\boldsymbol{x}} - \boldsymbol{x})^{\mathrm{T}}(\hat{\boldsymbol{x}} - \boldsymbol{x})\} \tag{2.24}$$

The unbiasedness implies that

$$\boldsymbol{b} = \bar{\boldsymbol{x}} - \boldsymbol{A}\bar{\boldsymbol{z}} \tag{2.25}$$

where $\bar{\boldsymbol{x}} \triangleq \mathrm{E}(\boldsymbol{x})$ and $\bar{\boldsymbol{z}} \triangleq \mathrm{E}(\boldsymbol{z})$. Then Eq. (2.23) can be rewritten as

$$\hat{\boldsymbol{x}}^{\mathrm{LMMSE}} = \bar{\boldsymbol{x}} + \boldsymbol{A}(\boldsymbol{z} - \bar{\boldsymbol{z}}) \tag{2.26}$$

Substituting Eq. (2.26) into Eq. (2.24), gives

$$J^{\text{MSE}} = \text{E}\{[A(z-\bar{z})+\bar{x}-x]^{\text{T}}[A(z-\bar{z})+\bar{x}-x]\} \tag{2.27}$$

Using the fact of $\boldsymbol{y}^{\text{T}}\boldsymbol{x} = \text{Tr}[\boldsymbol{x}\boldsymbol{y}^{\text{T}}]$, Eq. (2.27) can be rewritten as

$$\begin{aligned} J^{\text{MSE}} &= \text{Tr}[\boldsymbol{A}\boldsymbol{P}_{zz}\boldsymbol{A}^{\text{T}}] - \text{Tr}[\boldsymbol{A}\boldsymbol{P}_{zx}] - \text{Tr}[\boldsymbol{A}^{\text{T}}\boldsymbol{P}_{xz}] + \text{Tr}[\boldsymbol{P}_{xx}] \\ &= \text{Tr}[\boldsymbol{A}\boldsymbol{P}_{zz}\boldsymbol{A}^{\text{T}}] - \text{Tr}[\boldsymbol{A}\boldsymbol{P}_{zx}] - \text{Tr}[\boldsymbol{P}_{zx}\boldsymbol{A}] + \text{Tr}[\boldsymbol{P}_{xx}] \end{aligned} \tag{2.28}$$

where

$$\begin{aligned} \boldsymbol{P}_{zz} &\triangleq \text{var}(z) \triangleq \text{E}\{(z-\bar{z})(z-\bar{z})^{\text{T}}\} \\ \boldsymbol{P}_{zx} &\triangleq \text{cov}(z,\boldsymbol{x}) \triangleq \text{E}\{(z-\bar{z})(\boldsymbol{x}-\bar{\boldsymbol{x}})^{\text{T}}\} \\ \boldsymbol{P}_{xz} &\triangleq \text{cov}(\boldsymbol{x},z) \triangleq \text{E}\{(\boldsymbol{x}-\bar{\boldsymbol{x}})(z-\bar{z})^{\text{T}}\} = \boldsymbol{P}_{zx}^{\text{T}} \\ \boldsymbol{P}_{xx} &\triangleq \text{var}(\boldsymbol{x}) \triangleq \text{E}\{(\boldsymbol{x}-\bar{\boldsymbol{x}})(\boldsymbol{x}-\bar{\boldsymbol{x}})^{\text{T}}\} \end{aligned} \tag{2.29}$$

The optimal necessary condition $\frac{\partial J^{\text{MSE}}}{\partial A} = 0$ leads to

$$\boldsymbol{A} = \boldsymbol{P}_{xz}\boldsymbol{P}_{zz}^{-1} \tag{2.30}$$

Substitution of Eq. (2.30) into Eq. (2.26) gives the LMMSE estimate

$$\hat{\boldsymbol{x}}^{\text{LMMSE}} = \bar{\boldsymbol{x}} + \boldsymbol{P}_{xz}\boldsymbol{P}_{zz}^{-1}(z-\bar{z}) \tag{2.31}$$

Denote $\tilde{\boldsymbol{x}}^{\text{LMMSE}} = \boldsymbol{x} - \hat{\boldsymbol{x}}^{\text{LMMSE}}$. Then

$$\begin{aligned} \boldsymbol{P}_{\tilde{x}^{\text{LMMSE}}} &= \text{var}(\tilde{\boldsymbol{x}}^{\text{LMMSE}}) \\ &\triangleq \text{E}\left\{\left(\tilde{\boldsymbol{x}}^{\text{LMMSE}} - \text{E}\{\tilde{\boldsymbol{x}}^{\text{LMMSE}}\}\right)\left(\tilde{\boldsymbol{x}}^{\text{LMMSE}} - \text{E}\{\tilde{\boldsymbol{x}}^{\text{LMMSE}}\}\right)^{\text{T}}\right\} \\ &= \text{E}\{\tilde{\boldsymbol{x}}^{\text{LMMSE}}(\tilde{\boldsymbol{x}}^{\text{LMMSE}})^{\text{T}}\} \\ &= \boldsymbol{P}_{xx} - \boldsymbol{P}_{xz}\boldsymbol{P}_{zz}^{-1}\boldsymbol{P}_{zx} \end{aligned} \tag{2.32}$$

Properties of the LMMSE Estimator

1. Principle of orthogonality.

   $\tilde{\boldsymbol{x}}^{\text{LMMSE}}$ is orthogonal with the affine transformation of z:

$$\text{E}\{\tilde{\boldsymbol{x}}^{\text{LMMSE}}(\boldsymbol{L}z+\boldsymbol{b})^{T}\} = \boldsymbol{0} \tag{2.33}$$

where $\boldsymbol{L}$ and $\boldsymbol{b}$ are constants. Equation (2.33) is derived as follows

$$\begin{aligned}
\mathrm{E}\{\tilde{\boldsymbol{x}}^{\mathrm{LMMSE}}(\boldsymbol{L}\boldsymbol{z}+\boldsymbol{b})^T\} &= \mathrm{E}\{\tilde{\boldsymbol{x}}^{\mathrm{LMMSE}}\boldsymbol{z}\}\boldsymbol{L}^T \\
&= \mathrm{E}\{\tilde{\boldsymbol{x}}^{\mathrm{LMMSE}}(\boldsymbol{z}-\bar{\boldsymbol{z}})^T\}\boldsymbol{L}^T \\
&= \mathrm{E}\big\{\big(\boldsymbol{x}-\bar{\boldsymbol{x}}-\boldsymbol{P}_{xz}\boldsymbol{P}_{zz}^{-1}(\boldsymbol{z}-\bar{\boldsymbol{z}})\big)(\boldsymbol{z}-\bar{\boldsymbol{z}})^T\big\}\boldsymbol{L}^T \\
&= (\boldsymbol{P}_{xz}-\boldsymbol{P}_{xz}\boldsymbol{P}_{zz}^{-1}\boldsymbol{P}_{zz})\boldsymbol{L}^T = \boldsymbol{0}
\end{aligned} \tag{2.34}$$

where the identity of $\mathrm{E}\{\tilde{\boldsymbol{x}}^{\mathrm{LMMSE}}\} = \boldsymbol{0}$ is used.

From Eq. (2.33) we have

$$\mathrm{E}\{\tilde{\boldsymbol{x}}^{\mathrm{LMMSE}}\tilde{\boldsymbol{z}}^T\} = \boldsymbol{0} \tag{2.35}$$

$$\mathrm{E}\{\tilde{\boldsymbol{x}}^{\mathrm{LMMSE}}\boldsymbol{z}^T\} = \boldsymbol{0} \tag{2.36}$$

$$\mathrm{E}\{\tilde{\boldsymbol{x}}^{\mathrm{LMMSE}}(\hat{\boldsymbol{x}}^{\mathrm{LMMSE}})^T\} = \boldsymbol{0} \tag{2.37}$$

where $\tilde{\boldsymbol{z}} = \boldsymbol{z} - \bar{\boldsymbol{z}}$.

Lemma 2.1 *Consider the LMMSE estimate (2.31) and the following linear model*

$$\boldsymbol{y} = \boldsymbol{H}\boldsymbol{x} + \boldsymbol{v} \tag{2.38}$$

where $\boldsymbol{v}$ is the noise with

$$\begin{aligned}
\mathrm{E}\{\boldsymbol{v}\} &= \boldsymbol{0} \\
\mathrm{E}\{\boldsymbol{v}\boldsymbol{v}^T\} &= \boldsymbol{R} \\
\mathrm{E}\{\boldsymbol{v}\boldsymbol{z}^T\} &= \boldsymbol{0}
\end{aligned} \tag{2.39}$$

Then the LMMSE estimate of $\boldsymbol{y}$ is given by

$$\hat{\boldsymbol{y}}^{\mathrm{LMMSE}}(\boldsymbol{z}) = \boldsymbol{H}\hat{\boldsymbol{x}}^{\mathrm{LMMSE}}(\boldsymbol{z}) \tag{2.40}$$

If $\mathrm{E}\{\boldsymbol{x}\boldsymbol{v}^T\} = \boldsymbol{0}$, then

$$\boldsymbol{P}_{\tilde{y}^{\mathrm{LMMSE}}} = \boldsymbol{H}\boldsymbol{P}_{\tilde{x}^{\mathrm{LMMSE}}}\boldsymbol{H}^T + \boldsymbol{R} \tag{2.41}$$

where $\boldsymbol{P}_{\tilde{y}^{\mathrm{LMMSE}}} = \mathrm{var}(\tilde{\boldsymbol{y}}^{\mathrm{LMMSE}})$ and $\tilde{\boldsymbol{y}}^{\mathrm{LMMSE}} = \boldsymbol{y} - \hat{\boldsymbol{y}}^{\mathrm{LMMSE}}$

Proof According to Eqs. (2.31) and (2.32), we have

$$\begin{aligned}
\hat{\boldsymbol{y}}^{\mathrm{LMMSE}} &= \bar{\boldsymbol{y}} + \boldsymbol{P}_{yz}\boldsymbol{P}_{zz}^{-1}(\boldsymbol{z}-\bar{\boldsymbol{z}}) \\
\boldsymbol{P}_{\tilde{y}^{\mathrm{LMMSE}}} &= \boldsymbol{P}_{yy} - \boldsymbol{P}_{yz}\boldsymbol{P}_{zz}^{-1}\boldsymbol{P}_{zy}
\end{aligned} \tag{2.42}$$

From Eqs. (2.38) and (2.39), we have

$$\begin{aligned}\bar{\boldsymbol{y}} &= \boldsymbol{H}\bar{\boldsymbol{x}} \\ \boldsymbol{P}_{yz} &= \boldsymbol{H}\boldsymbol{P}_{xz}\end{aligned} \tag{2.43}$$

Substituting Eq. (2.43) into the first equation of Eq. (2.42) gives Eq. (2.40). Because $\mathrm{E}\{\boldsymbol{x}\boldsymbol{v}^T\} = \boldsymbol{0}$, we have

$$\boldsymbol{P}_{yy} = \boldsymbol{H}\boldsymbol{P}_{xx}\boldsymbol{H}^T + \boldsymbol{R} \tag{2.44}$$

where $\boldsymbol{P}_{yy} = \mathrm{var}(\boldsymbol{y})$. Substituting Eq. (2.44) and the second equation of Eq. (2.43) into the second equation of (2.42) gives (2.41).

2. Suppose $\boldsymbol{z}$ can be written as $\boldsymbol{z} = (\boldsymbol{z}_1; \boldsymbol{z}_2)$ and $\mathrm{cov}(\boldsymbol{z}_1, \boldsymbol{z}_2) = \boldsymbol{0}$. Then

$$\hat{\boldsymbol{x}}^{\mathrm{LMMSE}}(\boldsymbol{z}) = \hat{\boldsymbol{x}}^{\mathrm{LMMSE}}(\boldsymbol{z}_1) + \hat{\boldsymbol{x}}^{\mathrm{LMMSE}}(\boldsymbol{z}_2) - \bar{\boldsymbol{x}} \tag{2.45}$$

and

$$\begin{aligned}\mathrm{var}\{\tilde{\boldsymbol{x}}^{\mathrm{LMMSE}}(\boldsymbol{z})\} &= \mathrm{var}\{\boldsymbol{x}\} - \mathrm{cov}(\boldsymbol{x}, \boldsymbol{z}_1)\mathrm{var}(\boldsymbol{z}_1)^{-1}\mathrm{cov}(\boldsymbol{x}, \boldsymbol{z}_1)^T \\ &\quad - \mathrm{cov}(\boldsymbol{x}, \boldsymbol{z}_2)\mathrm{var}(\boldsymbol{z}_2)^{-1}\mathrm{cov}(\boldsymbol{x}, \boldsymbol{z}_2)^T \\ &= \mathrm{var}\{\tilde{\boldsymbol{x}}^{\mathrm{LMMSE}}(\boldsymbol{z}_1)\} - \mathrm{cov}(\boldsymbol{x}, \boldsymbol{z}_2)\mathrm{var}(\boldsymbol{z}_2)^{-1}\mathrm{cov}(\boldsymbol{x}, \boldsymbol{z}_2)^T\end{aligned} \tag{2.46}$$

Equation (2.45) is derived as follows

$$\begin{aligned}\hat{\boldsymbol{x}}^{\mathrm{LMMSE}}(\boldsymbol{z}) &= \bar{\boldsymbol{x}} + \boldsymbol{P}_{xz}\boldsymbol{P}_{zz}^{-1}(\boldsymbol{z} - \bar{\boldsymbol{z}}) \\ &= \bar{\boldsymbol{x}} + \begin{bmatrix}\boldsymbol{P}_{xz_1} & \boldsymbol{P}_{xz_2}\end{bmatrix}\begin{bmatrix}\boldsymbol{P}_{z_1z_1}^{-1} & \\ & \boldsymbol{P}_{z_2z_2}^{-1}\end{bmatrix}\begin{bmatrix}\boldsymbol{z}_1 - \bar{\boldsymbol{z}}_1 \\ \boldsymbol{z}_2 - \bar{\boldsymbol{z}}_2\end{bmatrix} \\ &= \bar{\boldsymbol{x}} + \boldsymbol{P}_{xz_1}\boldsymbol{P}_{z_1z_1}^{-1}(\boldsymbol{z}_1 - \bar{\boldsymbol{z}}_1) + \bar{\boldsymbol{x}} + \boldsymbol{P}_{xz_2}\boldsymbol{P}_{z_2z_2}^{-1}(\boldsymbol{z}_2 - \bar{\boldsymbol{z}}_2) - \bar{\boldsymbol{x}} \\ &= \hat{\boldsymbol{x}}^{\mathrm{LMMSE}}(\boldsymbol{z}_1) + \hat{\boldsymbol{x}}^{\mathrm{LMMSE}}(\boldsymbol{z}_2) - \bar{\boldsymbol{x}}\end{aligned} \tag{2.47}$$

and thus

$$\tilde{\boldsymbol{x}}^{\mathrm{LMMSE}}(\boldsymbol{z}) = \boldsymbol{x} - \bar{\boldsymbol{x}} - \boldsymbol{P}_{xz_1}\boldsymbol{P}_{z_1z_1}^{-1}(\boldsymbol{z}_1 - \bar{\boldsymbol{z}}_1) - \boldsymbol{P}_{xz_2}\boldsymbol{P}_{z_2z_2}^{-1}(\boldsymbol{z}_2 - \bar{\boldsymbol{z}}_2) \tag{2.48}$$

After proper calculations, Eq. (2.46) can be derived.

3. The LMMSE estimator is easy to implement since it only depends on the first and second moment of PDF.

2.3.2 MMSE Estimator for Jointly Gaussian Distribution

If the parameter $\boldsymbol{x}$ to be estimated and the observation $\boldsymbol{z}$ are jointly Gaussian (see Appendix C5.2 for the definition of jointly Gaussian distribution), then the MMSE estimate of $\boldsymbol{x}$ is given by

$$\hat{\boldsymbol{x}}^{\text{MMSE}} \triangleq \text{E}\{\boldsymbol{x}|\boldsymbol{z}\} = \bar{\boldsymbol{x}} + \boldsymbol{P}_{xz}\boldsymbol{P}_{zz}^{-1}(\boldsymbol{z} - \bar{\boldsymbol{z}}) \tag{2.49}$$

where $\boldsymbol{P}_{xz}$ and $\boldsymbol{P}_{zz}$ are defined by Eq. (2.29).

The corresponding conditional covariance matrix for the MMSE estimator is

$$\begin{aligned}\boldsymbol{P}_{xx|z} &\triangleq \text{E}\{(\boldsymbol{x} - \text{E}\{\boldsymbol{x}|\boldsymbol{z}\})(\boldsymbol{x} - \text{E}\{\boldsymbol{x}|\boldsymbol{z}\})^{\text{T}}|\boldsymbol{z}\} \\ &= \text{E}\{(\boldsymbol{x} - \hat{\boldsymbol{x}}^{\text{MMSE}})(\boldsymbol{x} - \hat{\boldsymbol{x}}^{\text{MMSE}})^{\text{T}}|\boldsymbol{z}\} \\ &= \boldsymbol{P}_{xx} - \boldsymbol{P}_{xz}\boldsymbol{P}_{zz}^{-1}\boldsymbol{P}_{zx}\end{aligned} \tag{2.50}$$

where the definitions of $\boldsymbol{P}_{xx}$, $\boldsymbol{P}_{xz}$, $\boldsymbol{P}_{zz}$ and $\boldsymbol{P}_{zx}$ are given by Eq. (2.29).

2.3.3 Estimation Algorithms for Linear Measurement Equation

This section considers estimators for linear measurement equation with the form of

$$\boldsymbol{z} = \boldsymbol{H}\boldsymbol{x} + \boldsymbol{v} \tag{2.51}$$

where $\boldsymbol{H}$ is measurement sensitivity matrix and $\boldsymbol{v}$ is measurement noise.

1. **LMMSE estimator**

If the measurement noise in Eq. (2.51) satisfies

$$\begin{aligned}&\text{E}\{\boldsymbol{v}\} = \boldsymbol{0}, \text{E}\{\boldsymbol{v}\boldsymbol{v}^{\text{T}}\} = \boldsymbol{R}, \\ &\text{E}\{\boldsymbol{x}\boldsymbol{v}^{\text{T}}\} = \boldsymbol{0}\end{aligned} \tag{2.52}$$

, then $\hat{\boldsymbol{x}}^{\text{LMMSE}}$ in Eq. (2.31) and $\boldsymbol{P}_{\tilde{x}^{\text{LMMSE}}}$ in Eq. (2.32) can be written as

$$\begin{aligned}\hat{\boldsymbol{x}}^{\text{LMMSE}} &= (\boldsymbol{P}_{xx}^{-1} + \boldsymbol{H}^{\text{T}}\boldsymbol{R}^{-1}\boldsymbol{H})^{-1}(\boldsymbol{H}^{\text{T}}\boldsymbol{R}^{-1}\boldsymbol{z} + \boldsymbol{P}_{xx}^{-1}\bar{\boldsymbol{x}}) \\ \boldsymbol{P}_{\tilde{x}^{\text{LMMSE}}} &= (\boldsymbol{P}_{xx}^{-1} + \boldsymbol{H}^{\text{T}}\boldsymbol{R}^{-1}\boldsymbol{H})^{-1}\end{aligned} \tag{2.53}$$

Proof From Eqs. (2.51) and (2.52), the following equation can be derived

$$\bar{\boldsymbol{z}} = \boldsymbol{H}\bar{\boldsymbol{x}}$$

$$
\begin{aligned}
\boldsymbol{P}_{xz} &= \boldsymbol{P}_{xx}\boldsymbol{H}^{\mathrm{T}} \\
\boldsymbol{P}_{zz} &= \boldsymbol{H}\boldsymbol{P}_{xx}\boldsymbol{H}^{\mathrm{T}} + \boldsymbol{R}
\end{aligned} \tag{2.54}
$$

Substituting Eq. (2.54) into Eq. (2.31) gives

$$
\begin{aligned}
\hat{\boldsymbol{x}}^{\mathrm{LMMSE}} &= \bar{\boldsymbol{x}} + \boldsymbol{P}_{xx}\boldsymbol{H}^{\mathrm{T}}(\boldsymbol{H}\boldsymbol{P}_{xx}\boldsymbol{H}^{\mathrm{T}} + \boldsymbol{R})^{-1}(\boldsymbol{z} - \bar{\boldsymbol{z}}) \\
&= \bar{\boldsymbol{x}} + (\boldsymbol{P}_{xx}^{-1} + \boldsymbol{H}^{\mathrm{T}}\boldsymbol{R}^{-1}\boldsymbol{H})^{-1}\boldsymbol{H}^{\mathrm{T}}\boldsymbol{R}^{-1}(\boldsymbol{z} - \bar{\boldsymbol{z}}) \\
&= (\boldsymbol{P}_{xx}^{-1} + \boldsymbol{H}^{\mathrm{T}}\boldsymbol{R}^{-1}\boldsymbol{H})^{-1}\boldsymbol{H}^{\mathrm{T}}\boldsymbol{R}^{-1}\boldsymbol{z} \\
&\quad + (\boldsymbol{I} - (\boldsymbol{P}_{xx}^{-1} + \boldsymbol{H}^{\mathrm{T}}\boldsymbol{R}^{-1}\boldsymbol{H})^{-1}\boldsymbol{H}^{\mathrm{T}}\boldsymbol{R}^{-1}\boldsymbol{H})\bar{\boldsymbol{x}} \\
&= (\boldsymbol{P}_{xx}^{-1} + \boldsymbol{H}^{\mathrm{T}}\boldsymbol{R}^{-1}\boldsymbol{H})^{-1}(\boldsymbol{H}^{\mathrm{T}}\boldsymbol{R}^{-1}\boldsymbol{z} + \boldsymbol{P}_{xx}^{-1}\bar{\boldsymbol{x}})
\end{aligned} \tag{2.55}
$$

where Eq. (B.24) in Appendix B.6:

$$
\boldsymbol{P}_{xx}\boldsymbol{H}^{\mathrm{T}}(\boldsymbol{H}\boldsymbol{P}_{xx}\boldsymbol{H}^{\mathrm{T}} + \boldsymbol{R})^{-1} = (\boldsymbol{P}_{xx}^{-1} + \boldsymbol{H}^{\mathrm{T}}\boldsymbol{R}^{-1}\boldsymbol{H})^{-1}\boldsymbol{H}^{\mathrm{T}}\boldsymbol{R}^{-1} \tag{2.56}
$$

and

$$
\begin{aligned}
&\boldsymbol{I} - (\boldsymbol{P}^{-1} + \boldsymbol{H}^{\mathrm{T}}\boldsymbol{R}^{-1}\boldsymbol{H})^{-1}\boldsymbol{H}^{\mathrm{T}}\boldsymbol{R}^{-1}\boldsymbol{H} \\
&\quad = (\boldsymbol{P}^{-1} + \boldsymbol{H}^{\mathrm{T}}\boldsymbol{R}^{-1}\boldsymbol{H})^{-1}(\boldsymbol{P}^{-1} + \boldsymbol{H}^{\mathrm{T}}\boldsymbol{R}^{-1}\boldsymbol{H} - \boldsymbol{H}^{\mathrm{T}}\boldsymbol{R}^{-1}\boldsymbol{H}) \\
&\quad = (\boldsymbol{P}^{-1} + \boldsymbol{H}^{\mathrm{T}}\boldsymbol{R}^{-1}\boldsymbol{H})^{-1}\boldsymbol{P}^{-1}
\end{aligned} \tag{2.57}
$$

are used.

Substituting Eq. (2.55) into Eq. (2.32) gives the second equation of Eq. (2.53), which completes the proof.

As shown in Table 2.1, Eq. (2.53) has several different forms

Table 2.1 Different forms of the LMMSE estimator

	$\hat{\boldsymbol{x}}^{\mathrm{LMMSE}}$ and $\boldsymbol{P}_{\tilde{x}^{\mathrm{LMMSE}}}$	Gain matrix K
Form 1	$\hat{\boldsymbol{x}}^{\mathrm{LMMSE}} = (\boldsymbol{P}_{xx}^{-1} + \boldsymbol{H}^{\mathrm{T}}\boldsymbol{R}^{-1}\boldsymbol{H})^{-1}(\boldsymbol{H}^{\mathrm{T}}\boldsymbol{R}^{-1}\boldsymbol{z} + \boldsymbol{P}_{xx}^{-1}\bar{\boldsymbol{x}})$ $\boldsymbol{P}_{\tilde{x}^{\mathrm{LMMSE}}} = (\boldsymbol{P}_{xx}^{-1} + \boldsymbol{H}^{\mathrm{T}}\boldsymbol{R}^{-1}\boldsymbol{H})^{-1}$	
Form 2	$\hat{\boldsymbol{x}}^{\mathrm{LMMSE}} = \bar{\boldsymbol{x}} + \boldsymbol{K}(\boldsymbol{z} - \boldsymbol{H}\bar{\boldsymbol{x}})$ $\boldsymbol{P}_{\tilde{x}^{\mathrm{LMMSE}}} = (\boldsymbol{P}_{xx}^{-1} + \boldsymbol{H}^{\mathrm{T}}\boldsymbol{R}^{-1}\boldsymbol{H})^{-1}$	$\boldsymbol{K} = \boldsymbol{P}_{\tilde{x}^{\mathrm{LMMSE}}}\boldsymbol{H}^{\mathrm{T}}\boldsymbol{R}^{-1}$
Form 3	$\hat{\boldsymbol{x}}^{\mathrm{LMMSE}} = \bar{\boldsymbol{x}} + \boldsymbol{K}(\boldsymbol{z} - \boldsymbol{H}\bar{\boldsymbol{x}})$ $\boldsymbol{P}_{\tilde{x}^{\mathrm{LMMSE}}} = (\boldsymbol{I} - \boldsymbol{K}\boldsymbol{H})\boldsymbol{P}_{xx}(\boldsymbol{I} - \boldsymbol{K}\boldsymbol{H})^{\mathrm{T}} + \boldsymbol{K}\boldsymbol{R}\boldsymbol{K}^{\mathrm{T}}$	$\boldsymbol{K} = \boldsymbol{P}_{xx}\boldsymbol{H}^{\mathrm{T}}(\boldsymbol{H}\boldsymbol{P}_{xx}\boldsymbol{H}^{\mathrm{T}} + \boldsymbol{R})^{-1}$
Form 4	$\hat{\boldsymbol{x}}^{\mathrm{LMMSE}} = \bar{\boldsymbol{x}} + \boldsymbol{K}(\boldsymbol{z} - \boldsymbol{H}\bar{\boldsymbol{x}})$ $\boldsymbol{P}_{\tilde{x}^{\mathrm{LMMSE}}} = (\boldsymbol{I} - \boldsymbol{K}\boldsymbol{H})\boldsymbol{P}_{xx}$	$\boldsymbol{K} = \boldsymbol{P}_{xx}\boldsymbol{H}^{\mathrm{T}}(\boldsymbol{H}\boldsymbol{P}_{xx}\boldsymbol{H}^{\mathrm{T}} + \boldsymbol{R})^{-1}$

Theorem 2.2 *For the linear measurement* Eq. (2.51)*, if the measurement noise satisfies*

$$\mathrm{E}\{\boldsymbol{v}\} = \mathbf{0}, \mathrm{E}\{\boldsymbol{v}\boldsymbol{v}^{\mathrm{T}}\} = \boldsymbol{R} \tag{2.58}$$

and there is no priori knowledge about $\bar{\boldsymbol{x}}$*, then the LMMSE estimate of* $\boldsymbol{x}$ *is given by*

$$\begin{aligned}\hat{\boldsymbol{x}}^{\mathrm{LMMSE}} &= (\boldsymbol{H}^{\mathrm{T}}\boldsymbol{R}^{-1}\boldsymbol{H})^{-1}\boldsymbol{H}^{\mathrm{T}}\boldsymbol{R}^{-1}\boldsymbol{z} \\ \boldsymbol{P}_{\tilde{\boldsymbol{x}}^{\mathrm{LMMSE}}} &= (\boldsymbol{H}^{\mathrm{T}}\boldsymbol{R}^{-1}\boldsymbol{H})^{-1}\end{aligned} \tag{2.59}$$

Detail derivation of Theorem 2.2 is given in [7, 9]. Intuitively, no priori knowledge about $\bar{\boldsymbol{x}}$ implies that $\boldsymbol{P}_{xx}^{-1} \to 0$. Then Theorem 2.2 can be derived from Eq. (2.53) directly.

2. **WLS estimator**

The WLS estimate in Eq. (2.21) for the linear measurement equation Eq. (2.51) can be written as

$$\hat{\boldsymbol{x}}^{\mathrm{WLS}} = \arg\min(\boldsymbol{z} - \boldsymbol{H}\boldsymbol{x})^{\mathrm{T}}\boldsymbol{W}(\boldsymbol{z} - \boldsymbol{H}\boldsymbol{x}) \tag{2.60}$$

The optimal necessary condition leads to

$$\hat{\boldsymbol{x}}^{\mathrm{WLS}} = (\boldsymbol{H}^{\mathrm{T}}\boldsymbol{W}\boldsymbol{H})^{-1}\boldsymbol{H}^{\mathrm{T}}\boldsymbol{W}\boldsymbol{z} \tag{2.61}$$

Remark

1. If $\mathrm{E}\{\boldsymbol{v}\} = \mathbf{0}$, then $\hat{\boldsymbol{x}}^{\mathrm{WLS}}$ is an unbiased estimate, i.e., $\mathrm{E}\{\hat{\boldsymbol{x}}^{\mathrm{WLS}}\} = \mathrm{E}\{\boldsymbol{x}\}$.
2. Let $\tilde{\boldsymbol{x}}^{\mathrm{WLS}} = \boldsymbol{x} - \hat{\boldsymbol{x}}^{\mathrm{WLS}}$. If $\mathrm{E}\{\boldsymbol{v}\} = \mathbf{0}$, $\mathrm{E}\{\boldsymbol{v}\boldsymbol{v}^{\mathrm{T}}\} = \boldsymbol{R}$, then

$$\begin{aligned}\boldsymbol{P}_{\tilde{\boldsymbol{x}}^{\mathrm{WLS}}} &\triangleq \mathrm{var}(\tilde{\boldsymbol{x}}^{\mathrm{WLS}}) \\ &\triangleq \mathrm{E}\left\{\left(\tilde{\boldsymbol{x}}^{\mathrm{WLS}} - \mathrm{E}\{\tilde{\boldsymbol{x}}^{\mathrm{WLS}}\}\right)\left(\tilde{\boldsymbol{x}}^{\mathrm{WLS}} - \mathrm{E}\{\tilde{\boldsymbol{x}}^{\mathrm{WLS}}\}\right)^{\mathrm{T}}\right\} \\ &= \mathrm{E}\{(\tilde{\boldsymbol{x}}^{\mathrm{WLS}})(\tilde{\boldsymbol{x}}^{\mathrm{WLS}})^{\mathrm{T}}\} \\ &= (\boldsymbol{H}^{\mathrm{T}}\boldsymbol{W}\boldsymbol{H})^{-1}\boldsymbol{H}^{\mathrm{T}}\boldsymbol{W}\boldsymbol{R}\boldsymbol{W}\boldsymbol{H}(\boldsymbol{H}^{\mathrm{T}}\boldsymbol{W}\boldsymbol{H})^{-1}\end{aligned} \tag{2.62}$$

Compare Eq. (2.62) with Theorem 2.1, we see that and if $\boldsymbol{P}_{\tilde{\mathrm{x}}^{\mathrm{WLS}}}$ has a minimum value of

$$\boldsymbol{P}_{\tilde{\mathrm{x}}^{\mathrm{WLS}}} = (\boldsymbol{H}^{\mathrm{T}}\boldsymbol{R}^{-1}\boldsymbol{H})^{-1} \tag{2.63}$$

when $\boldsymbol{W} = \boldsymbol{R}^{-1}$.

2.4 State Estimation Algorithms in Dynamic Systems

Sections 2.2 and 2.3 address general parameter estimation, whereas in the spacecraft autonomous navigation system, the interested parameters are states of dynamic system. Their time evolution equation is governed by orbit dynamics. The state estimator can either be batch form or recursive form. The batch form can be seen as a general parameter estimator. However, for real-time application, it is preferred to use recursive form estimators. Therefore, we only focus on the recursive estimators for dynamic systems. We begin with the recursive Bayesian estimation. Most of different types Kalman filters can be derived in this framework. In Sect. 2.4.2, the well-known Kalman filter is presented and is derived in two ways. One is deriving the KF as the LMMSE estimator and the other one is deriving it as the MMSE estimator using the recursive Bayesian estimation algorithm. In Sects. 2.4.3 and 2.4.4, two nonlinear Kalman filters, extended Kalman filter and unscented Kalman filter, are introduced. Finally, the recent developed constrained Kalman filter is presented in Sect. 2.4.5.

2.4.1 Recursive Bayesian Estimation

A discrete-time stochastic dynamic system can be written in the form of

$$\boldsymbol{x}_k = f(\boldsymbol{x}_{k-1}, \boldsymbol{w}_{k-1}, \boldsymbol{x}_{k-2}, \boldsymbol{w}_{k-2} \cdots) \tag{2.64}$$

where $\boldsymbol{x}$ is the discrete-time state, $\boldsymbol{w}$ is the process noise, subscripts $k, k-1$, and $k-2$ correspond to quantities at time t_k, t_{k-1}, and t_{k-2}, respectively. The dynamical model (2.64) can be written in a probabilistic form of

$$\boldsymbol{x}_k \sim p(\boldsymbol{x}_k|\boldsymbol{x}_{k-1}, \boldsymbol{x}_{k-2} \cdots) \tag{2.65}$$

where $p(\boldsymbol{x}_k|\boldsymbol{x}_{k-1}, \boldsymbol{x}_{k-2} \cdots)$ is the conditional PDF.

The dynamical model (2.65) is a **Markov model of order** p if the current state only depends on the past p states:

$$p(\boldsymbol{x}_k|\boldsymbol{x}_{k-1}, \boldsymbol{x}_{k-2} \cdots) = p(\boldsymbol{x}_k|\boldsymbol{x}_{k-1}, \boldsymbol{x}_{k-2} \cdots \boldsymbol{x}_{k-p}) \quad k \geq p \tag{2.66}$$

Therefore, a first-order Markov model can be written as

$$\boldsymbol{x}_k \sim p(\boldsymbol{x}_k|\boldsymbol{x}_{k-1}) \tag{2.67}$$

In a **hidden Markov model** (HMM), the states cannot be directly observed, but the random variable z, which gives indirect information about $\boldsymbol{x}$, can be observed [10]. Figure 2.1 illustrates a first-order HMM.

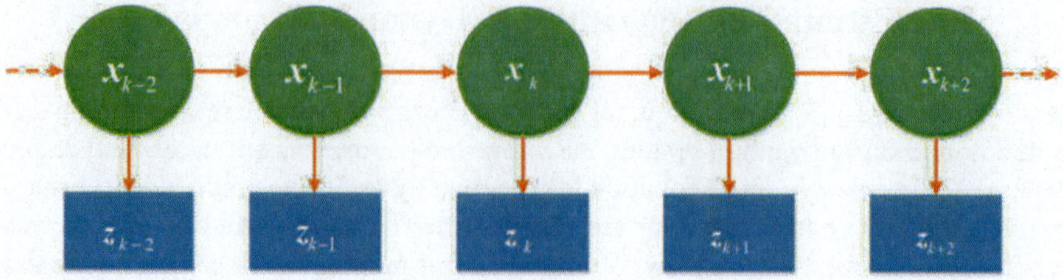

Fig. 2.1 A first-order hidden Markov model

The first-order HMM has two properties

$$p(\boldsymbol{x}_k|\boldsymbol{x}_{1:k-1}, z_{1:k-1}) = p(\boldsymbol{x}_k|\boldsymbol{x}_{k-1}) \tag{2.68}$$

$$p(z_k|\boldsymbol{x}_{1:k}, z_{1:k-1}) = p(z_k|\boldsymbol{x}_k) \tag{2.69}$$

where $\boldsymbol{x}_{1:k-1} \triangleq \{\boldsymbol{x}_1, \ldots, \boldsymbol{x}_{k-1}\}$ and $z_{1:k} \triangleq \{z_1, \ldots, z_k\}$. Given these two properties, the recursive Bayesian filtering equation can be derived.

Firstly, the Bayes' rule gives

$$p(\boldsymbol{x}_k|z_{1:k}) = \frac{p(z_{1:k}|\boldsymbol{x}_k)p(\boldsymbol{x}_k)}{p(z_{1:k})} \tag{2.70}$$

The numerator of the right-hand side of Eq. (2.70) can be written as

$$\begin{aligned} p(z_{1:k}|\boldsymbol{x}_k)p(\boldsymbol{x}_k) &= p(z_k, z_{1:k-1}|\boldsymbol{x}_k,)p(\boldsymbol{x}_k) \\ &= p(z_k|z_{1:k-1}, \boldsymbol{x}_k,)p(z_{1:k-1}|\boldsymbol{x}_k,)p(\boldsymbol{x}_k) \end{aligned} \tag{2.71}$$

where the following identity is used

$$p(\boldsymbol{x}, \boldsymbol{y}|z) = p(\boldsymbol{x}|\boldsymbol{y}, z)p(\boldsymbol{y}|z) \tag{2.72}$$

Using Eq. (2.69), Eq. (2.71) can be rewritten as

$$\begin{aligned} p(z_{1:k}|\boldsymbol{x}_k)p(\boldsymbol{x}_k) &= p(z_k|\boldsymbol{x}_k)p(z_{1:k-1}|\boldsymbol{x}_k,)p(\boldsymbol{x}_k) \\ &= p(z_k|\boldsymbol{x}_k,)p(z_{1:k-1}, \boldsymbol{x}_k) \\ &= p(z_k|\boldsymbol{x}_k,)p(\boldsymbol{x}_k|z_{1:k-1})p(z_{1:k-1}) \end{aligned} \tag{2.73}$$

Note that the denominator of Eq. (2.70) can be written as

$$p(z_{1:k}) = p(z_k|z_{1:k-1})p(z_{1:k-1}) \tag{2.74}$$

Substituting Eqs. (2.73) and (2.74) into Eq. (2.70) gives

$$p(\boldsymbol{x}_k|z_{1:k}) = \frac{p(z_k|\boldsymbol{x}_k)p(\boldsymbol{x}_k|z_{1:k-1})}{p(z_k|z_{1:k-1})} \tag{2.75}$$

where $p(z_k|z_{1:k-1})$ is a normalized constant, $p(z_k|\boldsymbol{x}_k)$ corresponds to the measurement update equation, and $p(\boldsymbol{x}_k|z_{1:k-1})$ is given by

$$\begin{aligned} p(\boldsymbol{x}_k|z_{1:k-1}) &= \int p(\boldsymbol{x}_k, \boldsymbol{x}_{k-1}|z_{1:k-1})\mathrm{d}\boldsymbol{x}_{k-1} \\ &= \int p(\boldsymbol{x}_k|\boldsymbol{x}_{k-1}, z_{1:k-1})p(\boldsymbol{x}_{k-1}|z_{1:k-1})\mathrm{d}\boldsymbol{x}_{k-1} \\ &= \int p(\boldsymbol{x}_k|\boldsymbol{x}_{k-1})p(\boldsymbol{x}_{k-1}|z_{1:k-1})\mathrm{d}\boldsymbol{x}_{k-1} \end{aligned} \tag{2.76}$$

Equation (2.76) is the well-known Chapman–Kolmogorov equation and $p(\boldsymbol{x}_k|\boldsymbol{x}_{k-1})$ corresponds to the state predict equation (also called the time update equation). In the derivation of Eq. (2.76), the identity of Eq. (2.68) has been used.

Substituting Eq. (2.76) into Eq. (2.75) gives the general Bayesian filtering equation for a first-order HMM. Note that $p(\boldsymbol{x}_k|z_{1:k})$ depends on $p(\boldsymbol{x}_{k-1}|z_{1:k-1})$. Given the prior distribution $p(\boldsymbol{x}_0)$, the filtering equation can be recursively implemented.

2.4.2 Kalman Filtering

Kalman filter (KF), or Kalman–Bucy filter [11, 12], is an optimal recursive estimator. It can be proved that, under certain assumptions, the KF is optimal no matter which optimal criterion is used. KF combines all available information about measurements and system and minimizes the error in some statistical sense. Nevertheless, KF does not need to store all the measurement data from the previous time and reprocess it when there is new measurement data. This feature is very important for the actual implementation of the filtering algorithm.

Compared with the Wiener filtering algorithm designed in frequency domain, KF uses the state-space method in the time domain space. The recursive form of KF is easy to be implemented in a computer. Due to above characteristics, the KF immediately attracts much attentions by engineering communities after being proposed. The design of the Apollo spacecraft navigation system is a successful case of the KF algorithm in its early application. At present, KF is widely used in various fields, especially in the design of an integrated navigation system based on multi-source measurement information fusion, such as inertial navigation system/global positioning system (INS/GPS) integrated navigation system. This section will derive the KF equations for linear systems in two ways. We first show that it is the LMMSE estimator for arbitrary distribution white random process and then show that it is the MMSE estimator under Gaussian assumptions.

Consider the following dynamic system

$$\begin{aligned} \boldsymbol{x}_k &= \boldsymbol{F}_{k,k-1}\boldsymbol{x}_{k-1} + \boldsymbol{w}_{k-1} \\ \boldsymbol{z}_k &= \boldsymbol{H}_k\boldsymbol{x}_k + \boldsymbol{v}_k \end{aligned} \tag{2.77}$$

where $\boldsymbol{x}$ and $\boldsymbol{z}$ are state and measurements, respectively, $\boldsymbol{F}_{k,k-1}$ is the state transition matrix from t_{k-1} to t_k, $\boldsymbol{H}_k$ is the measurement sensitive matrix, $\boldsymbol{w}_k$ is the process noise, and $\boldsymbol{v}_k$ is the measurement noise. Here, $\boldsymbol{w}_k$ and $\boldsymbol{v}_k$ are uncorrelated, zero-mean white-noise processes with known covariance. Then

$$\mathrm{E}(\boldsymbol{w}_k\boldsymbol{w}_j^T) = \begin{cases} \boldsymbol{Q}_k, & k = j \\ \boldsymbol{0}, & k \neq j \end{cases}, \quad \mathrm{E}(\boldsymbol{v}_k\boldsymbol{v}_j^T) = \begin{cases} \boldsymbol{R}_k, & k = j \\ \boldsymbol{0}, & k \neq j \end{cases}, \quad \mathrm{E}\{\boldsymbol{w}_k\boldsymbol{v}_j^{\mathrm{T}}\} = \boldsymbol{0}\ \forall k, j \tag{2.78}$$

In addition, the initial state $\boldsymbol{x}_0$, uncorrelated with both process and measurement noises, has a known mean and covariance matrix

$$\begin{aligned} \hat{\boldsymbol{x}}_0 &= \mathrm{E}\{\boldsymbol{x}_0\} \\ \boldsymbol{P}_0 &= \mathrm{E}\{(\hat{\boldsymbol{x}}_0 - \boldsymbol{x}_0)(\hat{\boldsymbol{x}}_0 - \boldsymbol{x}_0)^{\mathrm{T}}\} \end{aligned} \tag{2.79}$$

The KF can be summarized as follows:

1. Prediction (Time update)

$$\begin{aligned} \hat{\boldsymbol{x}}_{k|k-1} &= \boldsymbol{F}_{k,k-1}\hat{\boldsymbol{x}}_{k-1} \\ \boldsymbol{P}_{k|k-1} &= \boldsymbol{F}_{k,k-1}\boldsymbol{P}_{k-1}\boldsymbol{F}_{k,k-1}^{\mathrm{T}} + \boldsymbol{Q}_{k-1} \end{aligned} \tag{2.80}$$

where $\hat{\boldsymbol{x}}_{k-1}$ denotes the posterior estimate of $\boldsymbol{x}_{k-1}$ after $\boldsymbol{z}_{k-1}$ has been taken into account, $\hat{\boldsymbol{x}}_{k|k-1}$ denotes the prior estimate of $\boldsymbol{x}_k$ before $\boldsymbol{z}_k$ has been taken into account, $\boldsymbol{P}_{k-1}$ and $\boldsymbol{P}_{k|k-1}$ are the error covariance matrices of $\hat{\boldsymbol{x}}_{k-1}$ and $\hat{\boldsymbol{x}}_{k|k-1}$, respectively.

2. Measurement Update

Innovations:

$$\tilde{\boldsymbol{z}}_k = \boldsymbol{z}_k - \boldsymbol{H}_k\hat{\boldsymbol{x}}_{k|k-1} \tag{2.81}$$

Innovations covariance:

$$\boldsymbol{P}_{\tilde{z}_k\tilde{z}_k} = \boldsymbol{H}_k\boldsymbol{P}_{k|k-1}\boldsymbol{H}_k^{\mathrm{T}} + \boldsymbol{R}_k \tag{2.82}$$

Kalman gain:

$$\boldsymbol{K}_k = \boldsymbol{P}_{x_k\tilde{z}_k}\boldsymbol{P}_{\tilde{z}_k\tilde{z}_k}^{-1} \tag{2.83}$$

State measurement update:

$$\hat{\boldsymbol{x}}_k = \hat{\boldsymbol{x}}_{k|k-1} + \boldsymbol{K}_k \tilde{\boldsymbol{z}}_k \tag{2.84}$$

Covariance measurement update:

$$\boldsymbol{P}_k = \boldsymbol{P}_{k|k-1} - \boldsymbol{K}_k \boldsymbol{P}_{\tilde{z}_k \tilde{z}_k} \boldsymbol{K}_k^{\mathrm{T}} \tag{2.85}$$

2.4.2.1 Deriving the KF as the LMMSE Estimator

This subsection will derive the KF as the LMMSE estimator. The original derivation was proposed by Kalman [12] using properties of the LMMSE estimator.

For the linear model (2.77), the uncorrelatedness between $\boldsymbol{x}_0$ and process and measurement noises leads to

$$\begin{aligned} &\mathrm{E}\{\boldsymbol{w}_{k-1}\boldsymbol{z}_{1:k}^{\mathrm{T}}\} = \boldsymbol{0} \\ &\mathrm{E}\{\boldsymbol{x}_{k-1}\boldsymbol{w}_{k-1}^{\mathrm{T}}\} = \boldsymbol{0} \\ &\mathrm{E}\{\boldsymbol{v}_k\boldsymbol{z}_{1:k-1}^{\mathrm{T}}\} = \boldsymbol{0} \\ &\mathrm{E}\{\boldsymbol{x}_k\boldsymbol{v}_k^{\mathrm{T}}\} = \boldsymbol{0} \end{aligned} \tag{2.86}$$

Given the measurements $z_{1:k-1}$, suppose we have obtained the LMMSE estimate of the state and use the following notations

$$\begin{aligned} \hat{\boldsymbol{x}}_{k-1} &\triangleq \hat{\boldsymbol{x}}_{k-1}^{\mathrm{LMMSE}}(z_{1:k-1}) \\ \tilde{\boldsymbol{x}}_{k-1} &\triangleq \boldsymbol{x}_{k-1} - \hat{\boldsymbol{x}}_{k-1}^{\mathrm{LMMSE}}(z_{1:k-1}) \\ \boldsymbol{P}_{k-1} &\triangleq \mathrm{var}(\tilde{\boldsymbol{x}}_{k-1}) \end{aligned} \tag{2.87}$$

Note that the first equation of Eq. (2.77) is in the form of Eq. (2.38). According to Lemma 2.1 and Eq. (2.78), we have

$$\hat{\boldsymbol{x}}_{k|k-1} = \boldsymbol{F}_{k,k-1}\hat{\boldsymbol{x}}_{k-1} \tag{2.88}$$

where

$$\hat{\boldsymbol{x}}_{k|k-1} \triangleq \hat{\boldsymbol{x}}_{k}^{\mathrm{LMMSE}}(z_{1:k-1}) \tag{2.89}$$

Denote

$$\tilde{\boldsymbol{x}}_{k|k-1} \triangleq \boldsymbol{x}_k - \hat{\boldsymbol{x}}_{k-1}^{\mathrm{LMMSE}}(z_{1:k-1}) \tag{2.90}$$

Using the fact of $\mathrm{E}\{\boldsymbol{x}_{k-1}\boldsymbol{w}_{k-1}^{T}\} = \boldsymbol{0}$ and applying Eq. (2.41) to $\tilde{\boldsymbol{x}}_{k|k-1}$ gives

$$\begin{aligned}\boldsymbol{P}_{k|k-1} &\triangleq \mathrm{var}(\tilde{\boldsymbol{x}}_{k|k-1}) \\ &= \boldsymbol{F}_{k,k-1}\boldsymbol{P}_{k-1}\boldsymbol{F}_{k,k-1}^{\mathrm{T}} + \boldsymbol{Q}_{k-1}\end{aligned} \tag{2.91}$$

Equations (2.88) and (2.91) are the time update equations of the KF.

Next, the measurement update for the KF will be derived. Using the fact of $\mathrm{E}\{\boldsymbol{v}_k \boldsymbol{z}_{1:k-1}^T\} = \boldsymbol{0}$ and Lemma 2.1 gives

$$\hat{\boldsymbol{z}}_k = \boldsymbol{H}_k \hat{\boldsymbol{x}}_{k|k-1} \tag{2.92}$$

where

$$\hat{\boldsymbol{z}}_k \triangleq \hat{\boldsymbol{z}}_k^{\mathrm{LMMSE}}(\boldsymbol{z}_{1:k-1}) \tag{2.93}$$

Define the innovation as

$$\begin{aligned}\tilde{\boldsymbol{z}}_k &\triangleq \boldsymbol{z}_k - \hat{\boldsymbol{z}}_k \\ &= \boldsymbol{H}_k \tilde{\boldsymbol{x}}_{k|k-1} + \boldsymbol{v}_k\end{aligned} \tag{2.94}$$

Using the fact of $\mathrm{E}\{\boldsymbol{x}_k \boldsymbol{v}_k^T\} = \boldsymbol{0}$ and applying Eq. (2.41) to $\tilde{\boldsymbol{z}}_k$ gives

$$\begin{aligned}\boldsymbol{P}_{\tilde{z}_k \tilde{z}_k} &\triangleq \mathrm{var}(\tilde{\boldsymbol{z}}_k) \\ &= \boldsymbol{H}_k \boldsymbol{P}_{k|k-1} \boldsymbol{H}_k^{\mathrm{T}} + \boldsymbol{R}_k\end{aligned} \tag{2.95}$$

Denote

$$\begin{aligned}\hat{\boldsymbol{x}}_k &\triangleq \hat{\boldsymbol{x}}_k^{\mathrm{LMMSE}}(\boldsymbol{z}_{1:k}) \\ \tilde{\boldsymbol{x}}_k &\triangleq \boldsymbol{x}_k - \hat{\boldsymbol{x}}_k^{\mathrm{LMMSE}}(\boldsymbol{z}_{1:k}) \\ \boldsymbol{P}_k &\triangleq \mathrm{var}(\tilde{\boldsymbol{x}}_k)\end{aligned} \tag{2.96}$$

Because $\boldsymbol{z}_{1:k}$ and $\tilde{\boldsymbol{z}}_{1:k}$ are linear functions of each other, which implies that

$$\hat{\boldsymbol{x}}_k^{\mathrm{LMMSE}}(\boldsymbol{z}_{1:k}) = \hat{\boldsymbol{x}}_k^{\mathrm{LMMSE}}(\tilde{\boldsymbol{z}}_{1:k}) \tag{2.97}$$

For the innovation $\tilde{\boldsymbol{z}}_k$, using properties of the LMMSE estimator, we have

$$\begin{aligned}&\mathrm{E}\{\tilde{\boldsymbol{z}}_k\} = \boldsymbol{0} \\ &\mathrm{cov}(\tilde{\boldsymbol{z}}_{1:k-1}, \tilde{\boldsymbol{z}}_k) = \boldsymbol{0}\end{aligned} \tag{2.98}$$

Then applying Eq. (2.45) to (2.97) gives

$$\hat{\boldsymbol{x}}_k^{\mathrm{LMMSE}}(\boldsymbol{z}_{1:k}) = \hat{\boldsymbol{x}}_k^{\mathrm{LMMSE}}(\tilde{\boldsymbol{z}}_{1:k-1}) + \hat{\boldsymbol{x}}_k^{\mathrm{LMMSE}}(\tilde{\boldsymbol{z}}_k) - \bar{\boldsymbol{x}}_k \tag{2.99}$$

From Eq. (2.31), we have

$$\hat{\boldsymbol{x}}_k^{\text{LMMSE}}(\tilde{z}_k) = \bar{\boldsymbol{x}}_k + \boldsymbol{P}_{x_k\tilde{z}_k}\boldsymbol{P}_{\tilde{z}_k\tilde{z}_k}^{-1}\tilde{z}_k \tag{2.100}$$

where

$$\begin{aligned}
\boldsymbol{P}_{x_k\tilde{z}_k} &\triangleq \text{cov}(\boldsymbol{x}_k, \tilde{z}_k) \\
&= \text{cov}(\boldsymbol{x}_k, \boldsymbol{H}_k\tilde{\boldsymbol{x}}_{k|k-1} + \boldsymbol{v}_k) \\
&= \text{cov}(\boldsymbol{x}_k, \boldsymbol{H}_k\tilde{\boldsymbol{x}}_{k|k-1}) \\
&= \text{E}\{(\tilde{\boldsymbol{x}}_{k|k-1} + \hat{\boldsymbol{x}}_{k|k-1})\tilde{\boldsymbol{x}}_{k|k-1}^{\text{T}}\boldsymbol{H}_k^{\text{T}}\} \\
&= \text{E}\{\tilde{\boldsymbol{x}}_{k|k-1}\tilde{\boldsymbol{x}}_{k|k-1}^{\text{T}}\}\boldsymbol{H}_k^{\text{T}} = \boldsymbol{P}_{k|k-1}\boldsymbol{H}_k^{\text{T}}
\end{aligned} \tag{2.101}$$

where the identities of $\text{E}\{\boldsymbol{x}_k\boldsymbol{v}_k^T\} = \boldsymbol{0}$ and $\text{E}\{\hat{\boldsymbol{x}}_{k|k-1}\tilde{\boldsymbol{x}}_{k|k-1}^T\} = \boldsymbol{0}$ (see Eq. (2.37)) are used. Substituting Eq. (2.100) into Eq. (2.99) and using the notations of (2.96) gives

$$\hat{\boldsymbol{x}}_k = \hat{\boldsymbol{x}}_{k|k-1} + \boldsymbol{P}_{x_k\tilde{z}_k}\boldsymbol{P}_{\tilde{z}_k\tilde{z}_k}^{-1}\tilde{z}_k \tag{2.102}$$

where $\boldsymbol{P}_{x_k\tilde{z}_k}$ is given by Eq. (2.101), $\boldsymbol{P}_{\tilde{z}_k\tilde{z}_k}$ is given by Eq. (2.95), and $\tilde{z}_k$ is given by Eq. (2.94).

According to Eq. (2.46), we have

$$\boldsymbol{P}_k = \boldsymbol{P}_{k|k-1} - \boldsymbol{P}_{x_k\tilde{z}_k}\boldsymbol{P}_{\tilde{z}_k\tilde{z}_k}^{-1}\boldsymbol{P}_{x_k\tilde{z}_k}^{\text{T}} \tag{2.103}$$

where the notations of (2.96) is used.

Equations (2.102) and (2.103) are the measurement update equations for the KF. Usually, the Kalman gain is defined as

$$\boldsymbol{K}_k \triangleq \boldsymbol{P}_{x_k\tilde{z}_k}\boldsymbol{P}_{\tilde{z}_k\tilde{z}_k}^{-1} \tag{2.104}$$

Equations (2.102) and (2.103) can be rewritten as

$$\begin{aligned}
\hat{\boldsymbol{x}}_k &= \hat{\boldsymbol{x}}_{k|k-1} + \boldsymbol{K}_k\tilde{z}_k \\
\boldsymbol{P}_k &= \boldsymbol{P}_{k|k-1} - \boldsymbol{K}_k\boldsymbol{P}_{\tilde{z}_k\tilde{z}_k}\boldsymbol{K}_k^{\text{T}}
\end{aligned} \tag{2.105}$$

Remark

1. Similar with Table 2.1, KF has several different forms of measurement update equation, as shown in Table 2.2.

Table 2.2 Different forms of KF measurement updating equation

Form 1	$\boldsymbol{P}_k^{-1} = \boldsymbol{P}_{k\vert k-1}^{-1} + \boldsymbol{H}_k^{\mathrm{T}}\boldsymbol{R}_k^{-1}\boldsymbol{H}_k\hat{\boldsymbol{x}}_k = \boldsymbol{P}_k(\boldsymbol{H}_k^{\mathrm{T}}\boldsymbol{R}_k^{-1}z_k + \boldsymbol{P}_{k\vert k-1}^{-1}\hat{\boldsymbol{x}}_{k\vert k-1})$
Form 2 (information filter)	$\boldsymbol{P}_k^{-1} = \boldsymbol{P}_{k\vert k-1}^{-1} + \boldsymbol{H}_k^{\mathrm{T}}\boldsymbol{R}_k^{-1}\boldsymbol{H}_k\boldsymbol{K}_k = \boldsymbol{P}_k\boldsymbol{H}_k^{\mathrm{T}}\boldsymbol{R}_k^{-1}\hat{\boldsymbol{x}}_k = \hat{\boldsymbol{x}}_{k\vert k-1} + \boldsymbol{K}_k(z_k - \boldsymbol{H}_k\hat{\boldsymbol{x}}_{k\vert k-1})$
Form 3	$\boldsymbol{K} = \boldsymbol{P}_{k\vert k-1}\boldsymbol{H}^{\mathrm{T}}(\boldsymbol{H}\boldsymbol{P}_{k\vert k-1}\boldsymbol{H}\mathrm{T} + \boldsymbol{R})^{-1}\hat{\boldsymbol{x}}_k = \hat{\boldsymbol{x}}_{k\vert k-1} + \boldsymbol{K}_k(z_k - \boldsymbol{H}_k\hat{\boldsymbol{x}}_{k\vert k-1})\boldsymbol{P}_{k\vert k} = (\boldsymbol{I} - \boldsymbol{K}\boldsymbol{H})\boldsymbol{P}_{k\vert k-1}(\boldsymbol{I} - \boldsymbol{K}\boldsymbol{H})^{\mathrm{T}} + \boldsymbol{K}\boldsymbol{R}\boldsymbol{K}^{\mathrm{T}}$
Form 4	$\boldsymbol{K} = \boldsymbol{P}_{k\vert k-1}\boldsymbol{H}^{\mathrm{T}}(\boldsymbol{H}\boldsymbol{P}_{k\vert k-1}\boldsymbol{H}^{\mathrm{T}} + \boldsymbol{R})^{-1}\hat{\boldsymbol{x}}_k = \hat{\boldsymbol{x}}_{k\vert k-1} + \boldsymbol{K}_k(z_k - \boldsymbol{H}_k\hat{\boldsymbol{x}}_{k\vert k-1})\boldsymbol{P}_{k\vert k} = (\boldsymbol{I} - \boldsymbol{K}\boldsymbol{H})\boldsymbol{P}_{k\vert k-1}$

2.4.2.2 Deriving the KF as the MMSE Estimator Under Gaussian Assumptions

This subsection derivates the KF using the recursive Bayesian estimator and shows that the KF is indeed the MMSE estimator for Gaussian assumption.

Consider the model described by Eq. (2.77), (2.78) and (2.79) again. Here, we assume that $\boldsymbol{w}_k$ and $\boldsymbol{v}_k$ are Gaussian noises. In addition, the prior PDF of state is assumed to be $p(\boldsymbol{x}_0) = \mathcal{N}(\boldsymbol{x}; \hat{\boldsymbol{x}}_0^-, \boldsymbol{P}_0)$. Then the probabilistic form of Eq. (2.77) can be written as (see Appendix C5.2 for proof)

$$\begin{aligned} p(\boldsymbol{x}_k|\boldsymbol{x}_{k-1}) &= \mathcal{N}(\boldsymbol{x}_k; \boldsymbol{F}_k\boldsymbol{x}_{k-1}, \boldsymbol{Q}_{k-1}) \\ p(z_k|\boldsymbol{x}_k) &= \mathcal{N}(z_k; \boldsymbol{H}_k\boldsymbol{x}_k, \boldsymbol{R}_k) \end{aligned} \tag{2.106}$$

The Bayesian filtering equation Eq. (2.75) has a closed form with [13]

$$\begin{aligned} p(\boldsymbol{x}_k|z_{1:k-1}) &= \mathcal{N}(\boldsymbol{x}_k; \hat{\boldsymbol{x}}_{k|k-1}, \boldsymbol{P}_{k|k-1}) \\ p(\boldsymbol{x}_k|z_{1:k}) &= \mathcal{N}(\boldsymbol{x}_k; \hat{\boldsymbol{x}}_k, \boldsymbol{P}_k) \\ p(z_k|z_{1:k-1}) &= \mathcal{N}(\boldsymbol{x}_k; \boldsymbol{H}_k\hat{\boldsymbol{x}}_{k|k-1}, \boldsymbol{S}_k) \end{aligned} \tag{2.107}$$

The parameters in Eq. (2.107) are given by the KF equations:

$$\begin{aligned} \hat{\boldsymbol{x}}_{k|k-1} &= \boldsymbol{F}_{k,k-1}\hat{\boldsymbol{x}}_{k-1} \\ \boldsymbol{P}_{k|k-1} &= \boldsymbol{F}_{k,k-1}\boldsymbol{P}_{k-1}\boldsymbol{F}_{k,k-1}^{\mathrm{T}} + \boldsymbol{Q}_{k-1} \\ \hat{\boldsymbol{x}}_k &= \hat{\boldsymbol{x}}_{k|k-1} + \boldsymbol{K}_k(z_k - \boldsymbol{H}_k\hat{\boldsymbol{x}}_{k|k-1}) \\ \boldsymbol{P}_k &= \boldsymbol{P}_{k|k-1} - \boldsymbol{K}_k\boldsymbol{S}_k\boldsymbol{K}_k^{\mathrm{T}} \end{aligned} \tag{2.108}$$

where

$$\begin{aligned} \boldsymbol{S}_k &= \boldsymbol{H}_k\boldsymbol{P}_{k|k-1}\boldsymbol{H}_k^{\mathrm{T}} + \boldsymbol{R}_k \\ \boldsymbol{K}_k &= \boldsymbol{P}_{k|k-1}\boldsymbol{H}_k^{\mathrm{T}}\boldsymbol{S}_k^{-1} \end{aligned} \tag{2.109}$$

Before to derive Eq. (2.107), two properties of jointly Gaussian PDF are given (see Appendix C5.2 for detail derivations):

Property A (From jointly PDF to conditional PDF) *if* $\boldsymbol{x}$ *and* $\boldsymbol{z}$ *are jointly Gaussian distribution, then* $p(\boldsymbol{x}|\boldsymbol{z})$ *is also a Gaussian PDF given by*

$$p(\boldsymbol{x}|\boldsymbol{z}) = \mathcal{N}(\boldsymbol{x}; \bar{\boldsymbol{x}}_{x|z}, \boldsymbol{P}_{xx|z}) \tag{2.110}$$

where

$$\begin{aligned}\bar{\boldsymbol{x}}_{x|z} &\triangleq \mathrm{E}\{\boldsymbol{x}|\boldsymbol{z}\} = \bar{\boldsymbol{x}} + \boldsymbol{P}_{xz}\boldsymbol{P}_{zz}^{-1}(\boldsymbol{z} - \bar{\boldsymbol{z}}) \\ \boldsymbol{P}_{xx|z} &\triangleq \mathrm{E}\{(\boldsymbol{x} - \mathrm{E}\{\boldsymbol{x}|\boldsymbol{z}\})(\boldsymbol{x} - \mathrm{E}\{\boldsymbol{x}|\boldsymbol{z}\})^{\mathrm{T}}|\boldsymbol{z}\} \\ &= \boldsymbol{P}_{xx} - \boldsymbol{P}_{xz}\boldsymbol{P}_{zz}^{-1}\boldsymbol{P}_{zx}\end{aligned} \tag{2.111}$$

Property B (From conditional PDF to jointly PDF) *Consider the linear equation*

$$\boldsymbol{\zeta} = \boldsymbol{M}\boldsymbol{\xi} + \boldsymbol{v} \tag{2.112}$$

Denote $\boldsymbol{y} \triangleq [\boldsymbol{\xi}^{\mathrm{T}}, \boldsymbol{\zeta}^{\mathrm{T}}]^{\mathrm{T}}$. If

$$p(\boldsymbol{\xi}) = \mathcal{N}(\boldsymbol{\xi}; \bar{\boldsymbol{\xi}}, \boldsymbol{P}_{\xi\xi}) \tag{2.113}$$

and $p(\boldsymbol{v}) = \mathcal{N}(\boldsymbol{v}; \boldsymbol{0}, \boldsymbol{P}_{vv})$, $\mathrm{E}\{\boldsymbol{\xi}\boldsymbol{v}^{\mathrm{T}}\} = \boldsymbol{0}$, then

$$\begin{aligned}p(\boldsymbol{\zeta}|\boldsymbol{\xi}) &= \mathcal{N}(\boldsymbol{\zeta}; \boldsymbol{M}\boldsymbol{\xi}, \boldsymbol{P}_{vv}) \\ p(\boldsymbol{y}) &= \mathcal{N}\left(\begin{bmatrix}\boldsymbol{\xi} \\ \boldsymbol{\zeta}\end{bmatrix}; \begin{bmatrix}\bar{\boldsymbol{\xi}} \\ \boldsymbol{M}\bar{\boldsymbol{\xi}}\end{bmatrix}, \begin{bmatrix}\boldsymbol{P}_{\xi\xi} & \boldsymbol{P}_{\xi\xi}\boldsymbol{M}^{\mathrm{T}} \\ \boldsymbol{M}\boldsymbol{P}_{\xi\xi} & \boldsymbol{M}\boldsymbol{P}_{\xi\xi}\boldsymbol{M}^{\mathrm{T}} + \boldsymbol{P}_{vv}\end{bmatrix}\right)\end{aligned} \tag{2.114}$$

Given the above two properties, Eq. (2.107) can be proved:
Assuming

$$p(\boldsymbol{x}_{k-1}|\boldsymbol{z}_{1:k-1}) = \mathcal{N}(\boldsymbol{x}_{k-1}; \hat{\boldsymbol{x}}_{k-1}, \boldsymbol{P}_{k-1}) \tag{2.115}$$

Note that for a first-order HMM, we have $p(\boldsymbol{x}_k|\boldsymbol{x}_{1:k-1}, \boldsymbol{z}_{1:k-1}) = p(\boldsymbol{x}_k|\boldsymbol{x}_{k-1})$ and $p(\boldsymbol{z}_k|\boldsymbol{x}_{1:k}, \boldsymbol{z}_{1:k-1}) = p(\boldsymbol{z}_k|\boldsymbol{x}_k)$. The following equation can be derived using the first equation of Eq. (2.106)

$$\begin{aligned}p(\boldsymbol{x}_k|\boldsymbol{x}_{k-1}, \boldsymbol{z}_{1:k-1}) &= p(\boldsymbol{x}_k|\boldsymbol{x}_{k-1}) \\ &= \mathcal{N}(\boldsymbol{x}_k; \boldsymbol{F}_k\boldsymbol{x}_{k-1}, \boldsymbol{Q}_k)\end{aligned} \tag{2.116}$$

Equation (2.115) corresponds to Eqs. (2.113) and (2.116) corresponds to the first equation of Eq. (2.114). Then from the second equation of Eq. (2.114), we obtain

$$p(\boldsymbol{x}_{k-1}, \boldsymbol{x}_k|\boldsymbol{z}_{1:k-1}) = \mathcal{N}\left(\begin{bmatrix} \boldsymbol{x}_{k-1} \\ \boldsymbol{x}_k \end{bmatrix}; \begin{bmatrix} \hat{\boldsymbol{x}}_{k-1} \\ \boldsymbol{F}_k\hat{\boldsymbol{x}}_{k-1} \end{bmatrix}, \begin{bmatrix} \boldsymbol{P}_{k-1} & \boldsymbol{P}_{k-1}\boldsymbol{F}_k^{\mathrm{T}} \\ \boldsymbol{F}_k\boldsymbol{P}_{k-1} & \boldsymbol{F}_k\boldsymbol{P}_{k-1}\boldsymbol{F}_k^{\mathrm{T}} + \boldsymbol{Q}_{k-1} \end{bmatrix}\right) \tag{2.117}$$

and then

$$p(\boldsymbol{x}_k|\boldsymbol{z}_{1:k-1}) = \mathcal{N}(\boldsymbol{x}_k; \boldsymbol{F}_k\hat{\boldsymbol{x}}_{k-1}, \boldsymbol{F}_k\boldsymbol{P}_{k-1}\boldsymbol{F}_k^{\mathrm{T}} + \boldsymbol{Q}_{k-1}) \tag{2.118}$$

Denote

$$\begin{aligned} \hat{\boldsymbol{x}}_{k|k-1} &\triangleq \mathrm{E}\{\boldsymbol{x}_k|\boldsymbol{z}_{1:k-1}\} \\ \boldsymbol{P}_{k|k-1} &\triangleq \mathrm{var}(\boldsymbol{x}_k|\boldsymbol{z}_{1:k-1}) \end{aligned} \tag{2.119}$$

Then Eq. (2.118) can be written in the form of the first equation of Eq. (2.107) with

$$\begin{aligned} \hat{\boldsymbol{x}}_{k|k-1} &= \boldsymbol{F}_k\hat{\boldsymbol{x}}_{k-1} \\ \boldsymbol{P}_{k|k-1} &= \boldsymbol{F}_k\boldsymbol{P}_{k-1}\boldsymbol{F}_k^{\mathrm{T}} + \boldsymbol{Q}_{k-1} \end{aligned} \tag{2.120}$$

From the second equation of Eqs. (2.114) and (2.106), we obtain

$$\begin{aligned} p(\boldsymbol{x}_k, \boldsymbol{z}_k|\boldsymbol{z}_{1:k-1}) &= p(\boldsymbol{z}_k|\boldsymbol{x}_k)p(\boldsymbol{x}_k|\boldsymbol{z}_{1:k-1}) \\ &= \mathcal{N}\left(\begin{bmatrix} \boldsymbol{x}_k \\ \boldsymbol{z}_k \end{bmatrix}; \begin{bmatrix} \hat{\boldsymbol{x}}_{k|k-1} \\ \boldsymbol{H}_k\hat{\boldsymbol{x}}_{k|k-1} \end{bmatrix}, \begin{bmatrix} \boldsymbol{P}_{k|k-1} & \boldsymbol{P}_{k|k-1}\boldsymbol{H}_k^{\mathrm{T}} \\ \boldsymbol{H}_{kk}\boldsymbol{P}_{k|k-1} & \boldsymbol{H}_{kk}\boldsymbol{P}_{k|k-1}\boldsymbol{H}_k^{\mathrm{T}} + \boldsymbol{R}_k \end{bmatrix}\right) \end{aligned} \tag{2.121}$$

and then from Eq. (2.110), we obtain

$$\begin{aligned} p(\boldsymbol{x}_k, |\boldsymbol{z}_k, \boldsymbol{z}_{1:k-1}) &= p(\boldsymbol{x}_k, |\boldsymbol{z}_{1:k}) \\ &= \mathcal{N}\left(\boldsymbol{x}_k; \hat{\boldsymbol{x}}_{k|k-1} + \boldsymbol{K}_k(\boldsymbol{z}_k - \boldsymbol{H}_k\hat{\boldsymbol{x}}_{k|k-1}), \boldsymbol{P}_{k|k-1} - \boldsymbol{K}_k\boldsymbol{S}_k\boldsymbol{K}_k^{\mathrm{T}}\right) \end{aligned} \tag{2.122}$$

where $\boldsymbol{S}_k$ and $\boldsymbol{K}_k$ are given by Eq. (2.109).

Denote

$$\begin{aligned} \hat{\boldsymbol{x}}_k &\triangleq \mathrm{E}\{\boldsymbol{x}_k|\boldsymbol{z}_{1:k}\} \\ \boldsymbol{P}_k &\triangleq \mathrm{var}(\boldsymbol{x}_k|\boldsymbol{z}_{1:k}) \end{aligned} \tag{2.123}$$

Then

$$\begin{aligned} \hat{\boldsymbol{x}}_k &= \hat{\boldsymbol{x}}_{k|k-1} + \boldsymbol{K}_k(\boldsymbol{z}_k - \boldsymbol{H}_k\hat{\boldsymbol{x}}_{k|k-1}) \\ \boldsymbol{P}_k &= \boldsymbol{P}_{k|k-1} - \boldsymbol{K}_k\boldsymbol{S}_k\boldsymbol{K}_k^{\mathrm{T}} \end{aligned} \tag{2.124}$$

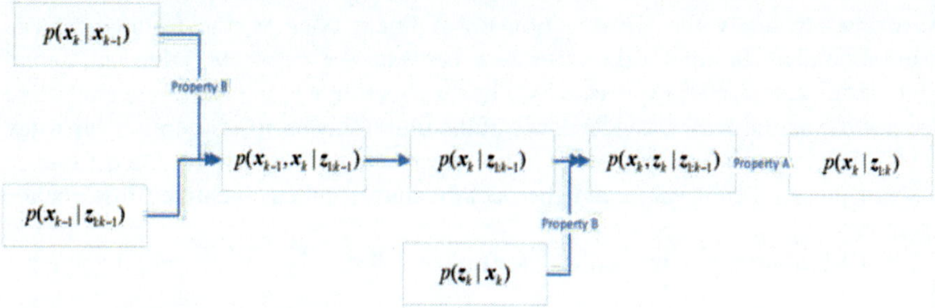

Fig. 2.2 Flowchart of KF deviation

which completes the derivation. The flowchart of the KF deviation is shown in Fig. 2.2. It should be pointed out that the proof replies on the assumption of Eq. (2.115) and this assumption can be validated using inductive methods.

2.4.3 Extended Kalman Filtering

The KF only applies to linear dynamic systems and linear measurement equations. However, the mathematical models encountered in engineering practice are often nonlinear. For example, the orbital dynamics in spacecraft autonomous navigation systems are nonlinear, and optical measurement equations are also nonlinear. In order to extend the KF algorithm to nonlinear systems, extended Kalman filter (EKF) was proposed.

The EKF was originally proposed by Stanley Schmidt [14]. The basic idea is to linearize the nonlinear system and measurement equations and then apply the KF for the linearized model to obtain state estimation. It is the most widely used algorithm for state estimation of nonlinear systems.

The nonlinear system model for the EKF algorithm is

$$\begin{aligned} \boldsymbol{x}_k &= \boldsymbol{f}(\boldsymbol{x}_{k-1}) + \boldsymbol{w}_k \\ \boldsymbol{z}_k &= \boldsymbol{h}(\boldsymbol{x}_k) + \boldsymbol{v}_k \end{aligned} \tag{2.125}$$

where $\boldsymbol{x}_k \in \mathbb{R}^n$ is the state vector, $\boldsymbol{z}_k \in \mathbb{R}^m$ is the measurement, $\boldsymbol{f} : \mathbb{R}^n \to \mathbb{R}^n$ is the state transition function, $\boldsymbol{h} : \mathbb{R}^n \to \mathbb{R}^m$ is the measurement function; $\boldsymbol{w}_k$ and $\boldsymbol{v}_k$ are both zero-mean white noise that satisfy

$$\begin{aligned} \mathrm{E}(\boldsymbol{w}_k \boldsymbol{w}_j^{\mathrm{T}}) &= \begin{cases} \boldsymbol{Q}_k, & k = j \\ \boldsymbol{0}, & k \neq j \end{cases} \\ \mathrm{E}(\boldsymbol{v}_k \boldsymbol{v}_j^{\mathrm{T}}) &= \begin{cases} \boldsymbol{R}_k, & k = j \\ \boldsymbol{0}, & k \neq j \end{cases} \end{aligned} \tag{2.126}$$

In order to apply the KF equation to nonlinear systems, the following basic assumption shall be made: the difference between the predicted value and actual value of the state can be expressed by a linear equation which is accurate enough to describe the actual error characteristics of the filter. This basic assumption can often be satisfied in engineering practice. The linear equation describing the difference between the predicted value and the actual value is called linearized disturbance equation.

The EKF algorithm equation is described as follows. Define the prediction error as

$$\Delta \boldsymbol{x}_k = \boldsymbol{x}_k - \hat{\boldsymbol{x}}_k^n \tag{2.127}$$

where $\hat{\boldsymbol{x}}_k^n$ is the predicted value of state. The linearized disturbance equation can be written as

$$\begin{aligned} \Delta \boldsymbol{x}_k &= \boldsymbol{F}_k \Delta \boldsymbol{x}_{k-1} + \boldsymbol{w}_k \\ \Delta \boldsymbol{z}_k &= \boldsymbol{H}_k \Delta \boldsymbol{x}_k + \boldsymbol{v}_k \end{aligned} \tag{2.128}$$

where $\boldsymbol{F}_k = \left.\frac{\partial \boldsymbol{f}}{\partial \boldsymbol{x}}\right|_{\boldsymbol{x}=\hat{\boldsymbol{x}}_{k-1}}$ and $\boldsymbol{H}_k = \left.\frac{\partial \boldsymbol{h}}{\partial \boldsymbol{x}}\right|_{\boldsymbol{x}=\hat{\boldsymbol{x}}_{k|k-1}}$ are called Jacobian matrix, and $\Delta \boldsymbol{z}_k = \boldsymbol{z}_k - \boldsymbol{h}(\hat{\boldsymbol{x}}_k^n)$. Applying the KF to Eq. (2.128) gives

$$\begin{aligned} \Delta \hat{\boldsymbol{x}}_{k|k-1} &= \boldsymbol{F}_k \Delta \hat{\boldsymbol{x}}_{k-1} \\ \boldsymbol{P}_{k|k-1} &= \boldsymbol{F}_k \boldsymbol{P}_{k-1} \boldsymbol{F}_k^{\mathrm{T}} + \boldsymbol{Q}_k \\ \boldsymbol{K}_k &= \boldsymbol{P}_{k|k-1} \boldsymbol{H}_k^{\mathrm{T}} (\boldsymbol{H}_k \boldsymbol{P}_{k|k-1} \boldsymbol{H}_k^{\mathrm{T}} + \boldsymbol{R}_k)^{-1} \\ \Delta \hat{\boldsymbol{x}}_k &= \Delta \hat{\boldsymbol{x}}_{k|k-1} + \boldsymbol{K}_k (\Delta \boldsymbol{z}_k - \boldsymbol{H}_k \Delta \hat{\boldsymbol{x}}_{k|k-1}) \\ \boldsymbol{P}_k &= (\boldsymbol{I} - \boldsymbol{K}_k \boldsymbol{H}_k) \boldsymbol{P}_{k|k-1} (\boldsymbol{I} - \boldsymbol{K}_k \boldsymbol{H}_k)^{\mathrm{T}} + \boldsymbol{K}_k \boldsymbol{R}_k \boldsymbol{K}_k^{\mathrm{T}} \end{aligned} \tag{2.129}$$

It is worth noting that the deviation $\Delta \boldsymbol{x}_k$ is often not predicted in the application. Instead, $\Delta \hat{\boldsymbol{x}}_{k|k-1}$ is set to be zero. Therefore,

$$\Delta \hat{\boldsymbol{x}}_k = \boldsymbol{K}_k \Delta \boldsymbol{z}_k = \boldsymbol{K}_k [\boldsymbol{z}_k - \boldsymbol{h}(\hat{\boldsymbol{x}}_k^n)] \tag{2.130}$$

Usually $\hat{\boldsymbol{x}}_k^n$ is obtained from the first equation of Eq. (2.125) as

$$\hat{\boldsymbol{x}}_k^n = \hat{\boldsymbol{x}}_{k|k-1} = \boldsymbol{f}(\hat{\boldsymbol{x}}_{k-1}) \tag{2.131}$$

and then $\Delta \hat{\boldsymbol{x}}_k$ is used to correct $\hat{\boldsymbol{x}}_k^n$ as

$$\hat{\boldsymbol{x}}_k = \hat{\boldsymbol{x}}_{k|k-1} + \Delta \hat{\boldsymbol{x}}_k = \hat{\boldsymbol{x}}_{k|k-1} + \boldsymbol{K}_k [\boldsymbol{z}_k - \boldsymbol{h}(\hat{\boldsymbol{x}}_{k|k-1})] \tag{2.132}$$

In application, the factors such as the degree of system nonlinearity, the weight and accuracy requirements of system noise and measurement noise, as well as computation complexity should be comprehensively considered to design or select an appropriate method for state prediction and measurement update. When the dynamic change of the system is not obvious or is periodic, such as satellite attitude determination systems based on gyro and star sensors, a fixed gain or periodic gain can be used, avoiding the recursive updating of the filter gain matrix and reducing the complexity of the filtering algorithm.

2.4.4 Unscented Kalman Filtering

The EKF has been used in many fields such as spacecraft orbit determination and aircraft navigation. Because the second-order terms and other high-order terms of the Taylor series expansion of the nonlinear function are neglected during the linearization process, the accuracy and stability of the EKF may be influenced by large linearized error if the system has a large initial state uncertainties or large external disturbances. In addition, the EKF requires the computation of Jacobian matrices, which are difficult to compute for some applications or even do not exist [15].

Several researchers have studied others nonlinear filtering to cope with problems mentioned above, such as second-order filtering [16], particle filter (PF) [17], and unscented Kalman filter (UKF) [18–20]. The second-order filtering reduces linearized errors by introducing higher-order terms of Taylor series expansion of nonlinear functions. Although second-order filtering can improve the filtering accuracy, the calculation of the higher derivative of a nonlinear function is often very difficult for a complex nonlinear system. The PF is a Bayesian filtering based on Monte Carlo methods. The key idea is to represent the required posterior density function by a set of random samples with associated weights and to compute estimates based on these samples and weights. The PF is very suitable for high-dimensional high-nonlinear systems. However, the PF requires massive particles for state prediction and correction, bringing a heavy computation burden for onboard computers with limited performance. The UKF applies unscented transform (UT) to nonlinear system and measurement equations to propagate mean and covariance. The UKF can achieve the accuracy of the second-order filter but without the calculation of derivation. In addition, the computation cost of the UKF is the same order of magnitude of the EKF and thus without increasing computer burden. The UKF has been successfully applied in attitude determination system of Naval Postgraduate School Satellite 1 (NPSAT 1) [21].

In this section, the UKF algorithm will be briefly descripted. Different from EKF, UKF uses unscented transformation (UT), instead of linearization, to describe the mean and variance of state variables after passing through nonlinear system equation or measurement equation. UT is a method for calculating the statistics of a random variable that undergoes a nonlinear transformation [22]. Consider a nonlinear function $\boldsymbol{y} = \boldsymbol{g}(\boldsymbol{x})$ where $\boldsymbol{x} \in \mathbb{R}^n$ is a random vector with mean $\bar{\boldsymbol{x}}_k$ and covariance $\boldsymbol{P}_k$.

The UT uses deterministic sampling method to choose several sample points (sigma points) in state space and then applies nonlinear transformation $\boldsymbol{g}(\boldsymbol{x})$ to these sigma points. The ensemble mean and covariance of the transformed points provide a good accuracy to approximate true mean and covariance of $\boldsymbol{y}$.

Steps of UT are as follows:

1. Construct $2n+1$ sigma points χ_i as follows:

$$\begin{aligned}\boldsymbol{\chi}_0 &= \bar{\boldsymbol{x}}_k \\ \boldsymbol{\chi}_i &= \bar{\boldsymbol{x}}_k + \left(\sqrt{(n+\lambda)\boldsymbol{P}_k}\right)_i \quad i = 1, 2, \ldots, n \\ \boldsymbol{\chi}_i &= \bar{\boldsymbol{x}}_k - \left(\sqrt{(n+\lambda)\boldsymbol{P}_k}\right)_{i-n} \quad i = n+1, n+2, \ldots, 2n\end{aligned} \tag{2.133}$$

where

$$\lambda = \alpha^2(n+\kappa) - n \tag{2.134}$$

Usually, α is a small positive number (e.g., $10^{-4} \leq \alpha \leq 1$) to control the spread of sigma points. Values of κ should guarantee that $(n+\lambda)\boldsymbol{P}_k$ is a semi-positive matrix. $\left(\sqrt{(n+\lambda)\boldsymbol{P}_k}\right)_i$ is the ith column of $\sqrt{(n+\lambda)\boldsymbol{P}_k}$ and $\sqrt{(n+\lambda)\boldsymbol{P}_k}$ is the matrix square root of $(n+\lambda)\boldsymbol{P}_k$ defined by $\left(\sqrt{(n+\lambda)\boldsymbol{P}_k}\right)\left(\sqrt{(n+\lambda)\boldsymbol{P}_k}\right)^T = (n+\lambda)\boldsymbol{P}_k$

2. Compute the transformed sigma points: $2n+1$ sigma points χ_i as follows:

$$\mathcal{Y}_i = \boldsymbol{g}(\boldsymbol{\chi}_i) \quad i = 0, \ldots, 2n \tag{2.135}$$

3. Approximate the mean and covariance of $\boldsymbol{y}$ using sample mean and covariance of $\mathcal{Y}_i$

$$\begin{aligned}\bar{\boldsymbol{y}} &\triangleq \mathrm{E}\{\boldsymbol{y}\} \approx \sum_{i=0}^{2n} W_i^{(m)} \mathcal{Y}_i \\ \boldsymbol{P}_y &\triangleq \mathrm{E}\{(\boldsymbol{y}-\bar{\boldsymbol{y}})(\boldsymbol{y}-\bar{\boldsymbol{y}})^{\mathrm{T}}\} \approx \sum_{i=0}^{2n} W_i^{(c)} (\mathcal{Y}_i - \bar{\boldsymbol{y}})(\mathcal{Y}_i - \bar{\boldsymbol{y}})^{\mathrm{T}}\end{aligned} \tag{2.136}$$

where W_i are weights given by [23]

$$\begin{aligned}W_0^{(m)} &= \frac{\lambda}{n+\lambda} \\ W_0^{(c)} &= \frac{\lambda}{n+\lambda} + 1 - \alpha^2 + \beta \\ W_i^{(m)} = W_i^{(c)} &= \frac{1}{2(n+\lambda)} \quad i = 1, \ldots, 2n\end{aligned} \tag{2.137}$$

Here β is used to incorporate prior knowledge of the distribution of $\boldsymbol{x}$.

Now, we consider a nonlinear system described by Eq. (2.125). Since both of state and measurement equations are of the form of $\boldsymbol{y} = \boldsymbol{g}(\boldsymbol{x})$, the UT can be applied to them to derive the UKF. The UKF algorithm is described as follows:

1. Initialization. Assume the state has a prior distribution of mean $\hat{\boldsymbol{x}}_0$ and covariance $\boldsymbol{P}_0$
2. Calculate sigma points. Denote $\hat{\boldsymbol{x}}_{k-1}$ and $\boldsymbol{P}_{k-1}$ as the state estimate and state error covariance at t_{k-1}. Then the sigma points are

$$\begin{cases} \boldsymbol{\chi}_{i,k-1} = \hat{\boldsymbol{x}}_{k-1} & i = 0 \\ \boldsymbol{\chi}_{i,k-1} = \hat{\boldsymbol{x}}_{k-1} + (\sqrt{(n+\lambda)\boldsymbol{P}_{k-1}})_i & i = 1, 2, \ldots, n \\ \boldsymbol{\chi}_{i,k-1} = \hat{\boldsymbol{x}}_{k-1} - (\sqrt{(n+\lambda)\boldsymbol{P}_{k-1}})_{i-l} & i = n+1, n+2, \ldots, 2n \end{cases}$$

   where λ is given by Eq. (2.134).
3. Time update. Substituting sigma points into the state transition function gives

$$\boldsymbol{\chi}_{i,k|k-1} = \boldsymbol{f}(\boldsymbol{\chi}_{i,k-1}), \quad i = 0, 1, \ldots, 2n \tag{2.138}$$

   and predicted state and covariance are given by

$$\begin{aligned} \hat{\boldsymbol{x}}_{k|k-1} &= \sum_{i=0}^{2n} W_i^{(m)} \boldsymbol{\chi}_{i,k|k-1} \\ \boldsymbol{P}_{k|k-1} &= \sum_{i=0}^{2n} W_i^{(c)} (\boldsymbol{\chi}_{i,k|k-1} - \hat{\boldsymbol{x}}_{k|k-1})(\boldsymbol{\chi}_{i,k|k-1} - \hat{\boldsymbol{x}}_{k|k-1})^{\mathrm{T}} + \boldsymbol{Q}_k \end{aligned} \tag{2.139}$$

   where W_i is given by (2.137)
4. Measurement update. According to $\hat{\boldsymbol{x}}_{k|k-1}$ and $\boldsymbol{P}_{k|k-1}$, new sigma points can be selected

$$\begin{cases} \boldsymbol{\chi}'_{i,k|k-1} = \hat{\boldsymbol{x}}_{k|k-1} & i = 0 \\ \boldsymbol{\chi}'_{i,k|k-1} = \hat{\boldsymbol{x}}_{k|k-1} + (\sqrt{(n+\lambda)\boldsymbol{P}_{k|k-1}})_i & i = 1, 2, \ldots, n \\ \boldsymbol{\chi}'_{i,k|k-1} = \hat{\boldsymbol{x}}_{k|k-1} - (\sqrt{(n+\lambda)\boldsymbol{P}_{k|k-1}})_{i-l} & i = n+1, n+2, \ldots, 2n \end{cases} \tag{2.140}$$

   Substituting the new sigma points into the measurement equation gives

$$\boldsymbol{\gamma}_{i,k} = \boldsymbol{h}(\boldsymbol{\chi}'_{i,k|k-1}) \quad i = 0, 1, \ldots, 2l \tag{2.141}$$

and the predicted measurement and its covariance are given by

$$\hat{z}_k = \sum_{i=0}^{2n} W_i^{(m)} \boldsymbol{\gamma}_{i,k}$$

$$\boldsymbol{P}_{zz} = \sum_{i=0}^{2n} W_i^{(c)} (\boldsymbol{\gamma}_{i,k} - \hat{\boldsymbol{z}}_k)(\boldsymbol{\gamma}_{i,k} - \hat{\boldsymbol{z}}_k)^{\mathrm{T}} + \boldsymbol{R}_k \tag{2.142}$$

The cross-covariance of the state and the measurement, denoted by $\boldsymbol{P}_{xz}$, are given by

$$\boldsymbol{P}_{xz} = \sum_{i=0}^{2n} W_i^{(c)} (\boldsymbol{\chi}_{i,k|k-1} - \hat{\boldsymbol{x}}_{k|k-1})(\boldsymbol{\gamma}_{i,k} - \hat{\boldsymbol{z}}_k)^{\mathrm{T}} \tag{2.143}$$

Finally, according to the LMMSE estimator derivation, the state estimate and its covariance is obtained

$$\hat{\boldsymbol{x}}_k = \hat{\boldsymbol{x}}_{k|k-1} + \boldsymbol{K}_k (\boldsymbol{z}_k - \hat{\boldsymbol{z}}_k) \tag{2.144}$$

$$\boldsymbol{P}_k = \boldsymbol{P}_{k|k-1} - \boldsymbol{K}_k \boldsymbol{P}_{zz} \boldsymbol{K}_k^{\mathrm{T}} \tag{2.145}$$

$$\boldsymbol{K}_k = \boldsymbol{P}_{xz} \boldsymbol{P}_{zz}^{-1} \tag{2.146}$$

2.4.5 *Constrained Kalman Filtering*

2.4.5.1 Problem Description

In a number of practical situations, the state of the system is subject to some constraints. Examples of such systems include camera tracking, target tracking, vision-based systems, attitude determination, and orbit determination. If the state constraints are either ignored or dealt with heuristically, the use of the Kalman filter may result in a non-optimal estimation [24]. Therefore, it is necessary to incorporate the state constraints into the Kalman filter. Many researchers have studied Kalman filters for linear and nonlinear state constraints, and for equality and inequality constraints, and proposed many methods to deal with constraints, such as model reduction [25, 26], perfect measurements [27–29], estimate projection [30, 31], gain projection [32], and Lagrangian multiplier technique [33–35]. In this section, we only consider equality constraints and apply the Lagrangian multiplier technique to derive constrained filters. Specifically, only two typical equality constraints are considered: linear equality constraint, (LEC) and quadratic equality constraint, (QEC). For a complete review about the constrained Kalman filtering, the readers may refer to [36].

Consider an unknown parameter $\boldsymbol{x} \in \mathbb{R}^n$. A linear equality constraint can be written as

$$Dx = d \tag{2.147}$$

where $D \in \mathbb{R}^{s \times n}$; $d \in \mathbb{R}^s$, and $s \leq n$ is the number of dimensions of the constraint. Usually, it is assumed that D is a full row rank matrix. Otherwise, there are redundant constraints and we can remove them to produce a constraint with a full row rank D.

A quadratic equality constraint can be written as

$$x^{\mathrm{T}} A x = l \tag{2.148}$$

where $A = A^{\mathrm{T}} \in \mathbb{R}^{n \times n}$.

Let $\hat{x}^-$ be a prior state estimation of x before employing the measurement $z(z \in \mathbb{R}^m)$ and let P_x^- the corresponding error covariance matrix. The problem of linear or quadratic equality constrained estimation is to find a posteriori estimation $\hat{x}^+$ that satisfies the constraints (2.147) or (2.148) and minimizes the following performance index

$$J = \frac{1}{2}\mathrm{Tr}[P_x^+] = \frac{1}{2}\mathrm{Tr}\left[\mathrm{E}\left\{(\hat{x}^+ - x)(\hat{x}^+ - x)^{\mathrm{T}}\right\}\right] \tag{2.149}$$

Here, we assume that the measurement is a linear combination of state with the form

$$z = Hx + \eta \tag{2.150}$$

where η is zero-mean measurement noise with the covariance R.

Let $\hat{x}^+$ be a posteriori estimation of x given by

$$\hat{x}^+ = Kz + N\hat{x}^- + n \tag{2.151}$$

where $K \in \mathbb{R}^{n \times m}$, $N \in \mathbb{R}^{n \times n}$ and $n \in \mathbb{R}^n$ are matrices to be determined. If $\hat{x}^- = x$ and $\eta = 0$, then $\hat{x}^+ = x$, which means that

$$x = (KH + N)x + n \tag{2.152}$$

Thus K, N and n satisfy constraints.

$$\begin{aligned} KH + N &= I \\ n &= 0 \end{aligned} \tag{2.153}$$

Substituting Eq. (2.153) into Eq. (2.151) gives

$$\hat{x}^+ = \hat{x}^- + K\epsilon \tag{2.154}$$

where $\epsilon = z - H\hat{x}^-$.

2.4.5.2 State LEC Filtering

For a linear equality constrained problem, the performance index can be written as:

$$J = \frac{1}{2}\mathrm{Tr}\left[\mathrm{E}\left\{(\hat{\boldsymbol{x}}^{+} - \boldsymbol{x})(\hat{\boldsymbol{x}}^{+} - \boldsymbol{x})^{\mathrm{T}}\right\}\right] + \boldsymbol{\lambda}^{\mathrm{T}}(\boldsymbol{D}\hat{\boldsymbol{x}}^{+} - \boldsymbol{d}) \tag{2.155}$$

Assuming that the measurement noise is uncorrelated with $\hat{\boldsymbol{x}}^{-}$ and substituting Eq. (2.154) into Eq. (2.155) yields

$$J = \frac{1}{2}\mathrm{Tr}\left[\boldsymbol{K}\boldsymbol{R}\boldsymbol{K}^{\mathrm{T}} + (\boldsymbol{I} - \boldsymbol{K}\boldsymbol{H})\boldsymbol{P}^{-}(\boldsymbol{I} - \boldsymbol{K}\boldsymbol{H})^{\mathrm{T}}\right] + \boldsymbol{\lambda}^{\mathrm{T}}(\boldsymbol{D}(\hat{\boldsymbol{x}}^{-} + \boldsymbol{K}\boldsymbol{\epsilon}) - \boldsymbol{d}) \tag{2.156}$$

Using the optimal necessary condition in Appendix D, we obtain

$$\begin{aligned} \frac{\partial J}{\partial \boldsymbol{K}} &= \boldsymbol{0} \Rightarrow \boldsymbol{K}\boldsymbol{R} - (\boldsymbol{I} - \boldsymbol{K}\boldsymbol{H})\boldsymbol{P}^{-}\boldsymbol{H}^{\mathrm{T}} + \boldsymbol{D}^{\mathrm{T}}\boldsymbol{\lambda}\boldsymbol{\epsilon}^{\mathrm{T}} = \boldsymbol{0} \\ \frac{\partial J}{\partial \boldsymbol{\lambda}} &= \boldsymbol{0} \Rightarrow \boldsymbol{D}(\hat{\boldsymbol{x}}^{-} + \boldsymbol{K}\boldsymbol{\epsilon}) - \boldsymbol{d} \end{aligned} \tag{2.157}$$

and then $\boldsymbol{K}$ can be solved from the first equation of Eq. (2.157) as

$$\boldsymbol{K} = (\boldsymbol{P}^{-}\boldsymbol{H}^{\mathrm{T}} - \boldsymbol{D}^{\mathrm{T}}\boldsymbol{\lambda}\boldsymbol{\epsilon}^{\mathrm{T}})\boldsymbol{W}^{-1} \tag{2.158}$$

where

$$\boldsymbol{W} = \boldsymbol{R} + \boldsymbol{H}\boldsymbol{P}^{-}\boldsymbol{H}^{\mathrm{T}} \tag{2.159}$$

Substituting Eq. (2.158) into the second equation of Eq. (2.157) gives

$$\boldsymbol{D}(\boldsymbol{P}^{-}\boldsymbol{H}^{\mathrm{T}} - \boldsymbol{D}^{\mathrm{T}}\boldsymbol{\lambda}\boldsymbol{\epsilon}^{\mathrm{T}})\boldsymbol{W}^{-1}\boldsymbol{\epsilon} = \boldsymbol{d} - \boldsymbol{D}\hat{\boldsymbol{x}}^{-} \tag{2.160}$$

After proper simplification, we obtain

$$\boldsymbol{\lambda} = \varepsilon^{-1}(\boldsymbol{D}\boldsymbol{D}^{-1})(\boldsymbol{D}\boldsymbol{P}\boldsymbol{H}^{\mathrm{T}} + \boldsymbol{D}\hat{\boldsymbol{x}}^{-} - \boldsymbol{d}) \tag{2.161}$$

where

$$\varepsilon = \boldsymbol{\epsilon}^{T}\boldsymbol{W}^{-1}\boldsymbol{\epsilon} \tag{2.162}$$

Substituting Eq. (2.161) into Eq. (2.158) gives the Kalman filter gain matrix for the constrained estimate, and the posterior error covariance is given by

$$\boldsymbol{P}^{+} = (\boldsymbol{I} - \boldsymbol{K}\boldsymbol{H})\boldsymbol{P}^{-}(\boldsymbol{I} - \boldsymbol{K}\boldsymbol{H})^{\mathrm{T}} + \boldsymbol{K}\boldsymbol{R}\boldsymbol{K}^{\mathrm{T}} \tag{2.163}$$

2.4.5.3 State QEC Filtering

The performance index for a QEC estimate can be defined as

$$J = \frac{1}{2}\mathrm{Tr}\big[\mathrm{E}\{(\hat{\boldsymbol{x}}^+ - \boldsymbol{x})(\hat{\boldsymbol{x}}^+ - \boldsymbol{x})^{\mathrm{T}}\}\big] + \frac{1}{2}\lambda((\hat{\boldsymbol{x}}^+)^{\mathrm{T}}\boldsymbol{A}\hat{\boldsymbol{x}}^+ - l) \tag{2.164}$$

where λ is a Lagrange multiplier with the constraint in Eq. (2.148). Assuming that the measurement noise is not correlated with $\hat{\boldsymbol{x}}^-$ and substituting Eqs. (2.150) and (2.154) into Eq. (2.164) gives

$$\begin{aligned} J = &\frac{1}{2}\mathrm{Tr}\big[\boldsymbol{K}\boldsymbol{R}\boldsymbol{K}^{\mathrm{T}} + (\boldsymbol{I} - \boldsymbol{K}\boldsymbol{H})\boldsymbol{P}^-(\boldsymbol{I} - \boldsymbol{K}\boldsymbol{H})^{\mathrm{T}}\big] \\ &+ \frac{1}{2}\lambda[(\hat{\boldsymbol{x}}^- + \boldsymbol{K}\boldsymbol{\epsilon})^{\mathrm{T}}\boldsymbol{A}(\hat{\boldsymbol{x}}^- + \boldsymbol{K}\boldsymbol{\epsilon}) - l] \end{aligned} \tag{2.165}$$

The necessary conditions for minimizing J in Eq. (2.165) are

$$\begin{aligned} \frac{\partial J}{\partial \boldsymbol{K}} &= \boldsymbol{K}\boldsymbol{R} - (\boldsymbol{I} - \boldsymbol{K}\boldsymbol{H})\boldsymbol{P}^-\boldsymbol{H}^{\mathrm{T}} + \lambda\boldsymbol{A}(\hat{\boldsymbol{x}}^- + \boldsymbol{K}\boldsymbol{\epsilon})\boldsymbol{\epsilon}^{\mathrm{T}} = \boldsymbol{0} \\ \frac{\partial J}{\partial \lambda} &= (\hat{\boldsymbol{x}}^+)^{\mathrm{T}}\boldsymbol{A}\hat{\boldsymbol{x}}^+ - l = 0 \end{aligned} \tag{2.166}$$

Equation (2.166) can be rearranged as

$$\boldsymbol{K} + \lambda\boldsymbol{A}\boldsymbol{K}\boldsymbol{\epsilon}\boldsymbol{\epsilon}^{\mathrm{T}}\boldsymbol{W}^{-1} = (\boldsymbol{P}^-\boldsymbol{H}^{\mathrm{T}} - \lambda\boldsymbol{A}\hat{\boldsymbol{x}}^-\boldsymbol{\epsilon}^{\mathrm{T}})\boldsymbol{W}^{-1} \tag{2.167}$$

$$\boldsymbol{\epsilon}^{\mathrm{T}}\boldsymbol{K}^{\mathrm{T}}\boldsymbol{A}\boldsymbol{K}\boldsymbol{\epsilon} + 2\boldsymbol{\epsilon}^{\mathrm{T}}\boldsymbol{K}^{\mathrm{T}}\boldsymbol{A}\hat{\boldsymbol{x}}^- + (\hat{\boldsymbol{x}}^-)^{\mathrm{T}}\boldsymbol{A}\hat{\boldsymbol{x}}^- - l = 0 \tag{2.168}$$

where $\boldsymbol{W}$ is derived from Eq. (2.159). Equation (2.167) has two unknown parameters to be solved: $\boldsymbol{K}$ and λ. A possible way is to solve $\boldsymbol{K}$ in terms of λ and other variables from Eq. (2.167) and then solve λ from Eq. (2.168) by substituting the solution of $\boldsymbol{K}$ into Eq. (2.168). $\boldsymbol{K}$ clearly depends on the prior estimate of the state and measurement of residuals, which indicates that $\boldsymbol{K}$ and λ fluctuate with the noise.

Equation (2.167) is known as the discrete-time Sylvester equation for $\boldsymbol{K}$. If $\boldsymbol{A} = \boldsymbol{I}$, $\boldsymbol{K}$ can be solved directly from Eq. (2.167), as presented in [33]. However, for a general $\boldsymbol{A}$, the solution of $\boldsymbol{K}$ cannot be obtained intuitively. For a m dimensional space, there exist $m - 1$ vectors $\boldsymbol{\beta}_i (i = 1, 2, \ldots, m - 1)$ such that $\boldsymbol{\beta}_i \perp \boldsymbol{\beta}_j (i \neq j)$ and $\boldsymbol{\epsilon} \perp \boldsymbol{\beta}_i$. Therefore, the m eigenvalues of $\lambda\boldsymbol{\epsilon}\boldsymbol{\epsilon}^{\mathrm{T}}\boldsymbol{W}^{-1}$ are $0, 0, \ldots, 0, \lambda\varepsilon$ with ε given by Eq. (2.162), and the eigenvectors corresponding to these m eigenvalues are given by $\boldsymbol{W}\boldsymbol{\beta}_1, \boldsymbol{W}\boldsymbol{\beta}_2, \ldots, \boldsymbol{W}\boldsymbol{\beta}_{m-1}, \boldsymbol{\epsilon}$.

Suppose there exist m $\alpha_i \in \boldsymbol{R}$, $(i = 1, 2, \ldots, m)$ such that

$$\alpha_1\boldsymbol{\epsilon} + \alpha_2\boldsymbol{W}\boldsymbol{\beta}_1 + \cdots + \alpha_m\boldsymbol{W}\boldsymbol{\beta}_{m-1} = \boldsymbol{0} \tag{2.169}$$

Because $W > 0$, a left-multiplication of. Equation (2.169) by $\epsilon^{\mathrm{T}} W^{-1}$ gives $\alpha_1 = 0$ and then a left-multiplication of Eq. (2.169) by $\beta_j^{\mathrm{T}} W^{-1} (j = 1, 2, \ldots, m-1)$ gives

$$\alpha_i = 0 \ (i = 2, 3, \ldots, m) \tag{2.170}$$

which indicates that the m eigenvectors of $\lambda\epsilon\epsilon^{\mathrm{T}} W^{-1}$ are linearly independent. In other words, $\lambda\epsilon\epsilon^{\mathrm{T}} W^{-1}$ can be diagonalized as

$$\lambda\epsilon\epsilon^{\mathrm{T}} W^{-1} = V \Gamma V^{-1} \tag{2.171}$$

where $\Gamma \triangleq \mathrm{diag}([\gamma_1, \gamma_2, \ldots, \gamma_m]) = \mathrm{diag}([0, 0, \ldots, 0, \lambda\varepsilon])$, $V = [W\beta_1, W\beta_2, \ldots, W\beta_{m-1}, \epsilon]$. Noting that a symmetrical matrix can be diagonalized by an orthogonal matrix, A can be written as,

$$A = U \Xi U^{\mathrm{T}} \tag{2.172}$$

where U is an orthogonal matrix, and $\Xi = \mathrm{diag}([\xi_1, \xi_2, \ldots, \xi_n])$. Substituting Eq. (2.172) into Eq. (2.167), and multiplying it on the left by U^{T} and on the right by V, we obtain

$$\tilde{K} + \Xi \tilde{K} \Gamma = \tilde{C} \tag{2.173}$$

where $\tilde{K} = U^{\mathrm{T}} K V$; $\tilde{C} = U^{\mathrm{T}} C V$; $C = (P^{-} H^{\mathrm{T}} - \lambda A \hat{x}^{-} \epsilon^{\mathrm{T}}) W^{-1}$. Denoting the (i, j)th elements of $\tilde{K}$ and $\tilde{C}$ as $\tilde{k}_{ij}$ and $\tilde{c}_{ij}$, Then Eq. (2.173) reads as

$$\tilde{k}_{ij} + \xi_i \tilde{k}_{ij} \gamma_j = \tilde{c}_{ij} \tag{2.174}$$

which means that

$$\tilde{k}_{ij} = \frac{\tilde{c}_{ij}}{1 + \xi_i \gamma_j} \tag{2.175}$$

Once $\tilde{K}$ is calculated, $\tilde{K}$ is given by

$$K = U \tilde{K} V^{-1} \tag{2.176}$$

Following the extended Kalman filter (EKF) derivation assumption, the posteriori error covariance is given by

$$P^{+} = (I - KH) P^{-} (I - KH)^{\mathrm{T}} + K R K^{\mathrm{T}} \tag{2.177}$$

Noting that $\tilde{c}_{ij}$ and γ_j depend on λ, λ should be calculated before computing K. Here, an analytical solution of λ will be derived.

A right-multiplication of Eq. (2.167) by ϵ gives

$$\boldsymbol{K\epsilon} + \lambda\varepsilon\boldsymbol{AK\epsilon} = \boldsymbol{d} - \lambda\varepsilon\boldsymbol{A}\hat{\boldsymbol{x}}^{-} \tag{2.178}$$

where $\boldsymbol{d} = \boldsymbol{P}^{-}\boldsymbol{H}^{T}\boldsymbol{W}^{-1}\boldsymbol{\epsilon}$. Then

$$\boldsymbol{K\epsilon} = [\boldsymbol{I} + \lambda\varepsilon\boldsymbol{A}]^{-1}[\boldsymbol{d} - \lambda\varepsilon\boldsymbol{A}\hat{\boldsymbol{x}}^{-}] \tag{2.179}$$

Substituting Eq. (2.179) into Eq. (2.168) gives

$$\begin{aligned}&[\boldsymbol{d} - \lambda\varepsilon\boldsymbol{A}\hat{\boldsymbol{x}}^{-}]^{\mathrm{T}}[\boldsymbol{I} + \lambda\varepsilon\boldsymbol{A}]^{-1}\boldsymbol{A}[\boldsymbol{I} + \lambda\varepsilon\boldsymbol{A}]^{-1}[\boldsymbol{d} - \lambda\varepsilon\boldsymbol{A}\hat{\boldsymbol{x}}^{-}]\\ &\quad + 2[\boldsymbol{d} - \lambda\varepsilon\boldsymbol{A}\hat{\boldsymbol{x}}^{-}]^{\mathrm{T}}[\boldsymbol{I} + \lambda\varepsilon\boldsymbol{A}]^{-1}\boldsymbol{A}\hat{\boldsymbol{x}}^{-} = l - (\hat{\boldsymbol{x}}^{-})^{\mathrm{T}}\boldsymbol{A}\hat{\boldsymbol{x}}^{-}\end{aligned} \tag{2.180}$$

A further simplification of Eq. (2.180) gives

$$\boldsymbol{h}^{T}(\boldsymbol{I} + \lambda\varepsilon\boldsymbol{\Xi})^{-2}\boldsymbol{\Xi}\boldsymbol{h} = l \tag{2.181}$$

where $\boldsymbol{h} = \widetilde{\boldsymbol{d}} + \widetilde{\boldsymbol{x}}$; $\widetilde{\boldsymbol{d}} = \boldsymbol{U}^{\mathrm{T}}\boldsymbol{d}$; $\widetilde{\boldsymbol{x}}^{-} = \boldsymbol{U}^{\mathrm{T}}\hat{\boldsymbol{x}}^{-}$.

Equation (2.181) can be written in a scalar form

$$s_1(\tilde{\lambda}) = \sum_{j=1}^{n}\frac{h_j^2\xi_j}{(1+\tilde{\lambda}\xi_j)^2} - l = 0 \tag{2.182}$$

where $\tilde{\lambda} = \lambda\varepsilon$. Equation (2.182) is similar to the constraint equation expressed in terms of a Lagrangian multiplier in [34] for a constrained least-square problem and Newton's iteration method is proposed for solving λ by starting with $\lambda_0 = 0$ in [34]. However, this method can only find one root of the constraint equation and this root may not be the one corresponding to the minimum value of the loss function because the number of the roots of Eq. (2.182) is $2p$, where

$$p = \begin{cases} q & l \neq 0 \\ q-1 & l = 0 \end{cases} \tag{2.183}$$

$p = q - 1$ and q is the number of distinct nonzero eigenvalues of $\boldsymbol{A}$. Therefore, a more effective method should be developed to find $2p$ roots of Eq. (2.182) and determine the root that minimizes the loss function.

Let $\tilde{\xi}_1, \tilde{\xi}_1, \ldots, \tilde{\xi}_q$ be q different eigenvalues of $\boldsymbol{A}$, and define the polynomial function as

$$s_2(\tilde{\lambda}) = \begin{cases} \frac{1}{\prod_{j=1}^{q}\tilde{\xi}_j^2}\prod_{j=1}^{q}(1+\tilde{\lambda}\tilde{\xi}_j)^2\left(1 - \frac{1}{l}\sum_{i=1}^{n}\frac{h_i^2\xi_i}{(1+\tilde{\lambda}\xi_i)^2}\right) & l \neq 0 \\ \frac{1}{\prod_{j=1}^{q}\tilde{\xi}_j^2}\prod_{j=1}^{q}(1+\tilde{\lambda}\tilde{\xi}_j)^2\sum_{i=1}^{n}\frac{h_i^2\xi_i}{(1+\tilde{\lambda}\xi_i)^2} & l = 0 \end{cases} \tag{2.184}$$

Equation (2.184) can be rewritten as

$$s_2(\tilde{\lambda}) = \beta_0 + \beta_1\tilde{\lambda} + \beta_2\tilde{\lambda}^2 + \cdots + \tilde{\lambda}^{2p} \tag{2.185}$$

where $\beta_i = \frac{1}{i!}\frac{\mathrm{d}^i}{\mathrm{d}\tilde{\lambda}^i}s_2(0)$ can be calculated by expanding Eq. (2.184) or by using difference methods. It is noticed that the roots of function (2.185) are the same as the solutions of Eq. (2.182) and the Frobenius companion matrix corresponding to (2.185) is defined by

$$\boldsymbol{G} = \begin{bmatrix} 0 \; 0 \cdots 0 & -\beta_0 \\ 1 \; 0 \cdots 0 & -\beta_1 \\ 0 \; 1 \cdots 0 & -\beta_2 \\ \vdots \; \vdots \; \ddots \; \vdots & \vdots \\ 0 \; 0 \cdots 1 & -\beta_{2p-1} \end{bmatrix} \tag{2.186}$$

Because the eigenvalues of the matrix $\boldsymbol{G}$ are the roots of function (2.184) the $2p$ roots of (2.184) can be calculated by performing an eigenvalue decomposition of $\boldsymbol{G}$. Compared with the Newton method, the proposed method does not require an iteration process and can find all roots of (2.184).

Because $\boldsymbol{G}$ has $2p$ eigenvalues, one has to determine the value that minimizes the performance index.

The following lemma, derived from the second-order sufficient condition for a constrained optimization problem, can be used to select the Lagrangian multiplier corresponding to the local minimum point.

Lemma 2.2 [37] *If $\tilde{\lambda}^*$ is an eigenvalue of* (2.186)*, and* $\prod_{j=1}^{r}(1+\tilde{\lambda}^*\xi_j) > 0 \quad j \neq i$ *for* $\forall i \in [1, r]$ *with* $r = 2, 3, \ldots, n$*, then $\tilde{\lambda}^*$ corresponds to a minimal performance index.*

Proof The loss function (2.165) is rewritten in terms of $\boldsymbol{K}$ as

$$J = f(\boldsymbol{K}) + \lambda g(\boldsymbol{K}) \tag{2.187}$$

where

$$\begin{aligned} f(\boldsymbol{K}) &= \frac{1}{2}\mathrm{Tr}\left[\boldsymbol{K}\boldsymbol{R}\boldsymbol{K}^{\mathrm{T}} + (\boldsymbol{I} - \boldsymbol{K}\boldsymbol{H})\boldsymbol{P}^{-}(\boldsymbol{I} - \boldsymbol{K}\boldsymbol{H})^{\mathrm{T}}\right] \\ g(\boldsymbol{K}) &= \frac{1}{2}\left[\boldsymbol{\epsilon}^{\mathrm{T}}\boldsymbol{K}^{\mathrm{T}}\boldsymbol{A}\boldsymbol{K}\boldsymbol{\epsilon} + 2\boldsymbol{\epsilon}^{\mathrm{T}}\boldsymbol{K}^{\mathrm{T}}\boldsymbol{A}\hat{\boldsymbol{x}}^{-} + (\hat{\boldsymbol{x}}^{-})^{\mathrm{T}}\boldsymbol{A}\hat{\boldsymbol{x}}^{-} - l\right] \end{aligned} \tag{2.188}$$

Denote $L(\boldsymbol{K}^*, \lambda^*) = \nabla^2_{KK}J(\boldsymbol{K}^*, \lambda^*) = \boldsymbol{W} \otimes \boldsymbol{I}_n + \lambda\boldsymbol{\epsilon}\boldsymbol{\epsilon}^{\mathrm{T}} \otimes \boldsymbol{A}$, where $\otimes$ is the Kronecker product (see Appendix B.2 for the definition). Suppose $\boldsymbol{K}^*$ is a solution of Eq. (2.178) and λ^* is given by Eq. (2.184). Then, according to Appendix D, $\boldsymbol{K}^*$ is a local minimum point of $f(\boldsymbol{K})$ subject to the constraint $g(\boldsymbol{K}^*) = 0$ if

$$Q(z) = \text{vec}(z)^{\mathrm{T}} L(\boldsymbol{K}^*, \boldsymbol{\lambda}^*)\text{vec}(z) > 0 \tag{2.189}$$

for all z satisfying $z \cdot \nabla_K g(\boldsymbol{K}^*) = (\boldsymbol{I}_m \otimes \nabla g^{\mathrm{T}})\text{vec}(z) = 0$, where

$$\nabla_K g(\boldsymbol{K}^*) = \boldsymbol{A}(\boldsymbol{I} + \lambda\varepsilon\boldsymbol{A})^{-1}(\boldsymbol{d} + \hat{\boldsymbol{x}}^-)\boldsymbol{\epsilon}^{\mathrm{T}} \tag{2.190}$$

To determine the definiteness of $Q(z)$ in Eq. (2.189) in the tangent space of $\nabla g(\boldsymbol{K}^*)$, the following matrix is constructed:

$$\boldsymbol{S}_r = \begin{bmatrix} \boldsymbol{0} & (\boldsymbol{I}_m \otimes \nabla g^{\mathrm{T}})_r \\ (\boldsymbol{I}_m \otimes \nabla g)_r & \boldsymbol{L}_{rr}(\boldsymbol{K}^*, \boldsymbol{\lambda}^*) \end{bmatrix} \tag{2.191}$$

where $\boldsymbol{L}_{rr}(\boldsymbol{K}^*, \boldsymbol{\lambda}^*)$ is the $r \times r$ matrix in the top left corner of $\boldsymbol{L}(\boldsymbol{K}^*, \boldsymbol{\lambda}^*)$ and $(\boldsymbol{I}_m \otimes \nabla g^{\mathrm{T}})_r$ is the matrix whose columns are the first r columns of $\boldsymbol{I}_m \otimes \nabla g^{\mathrm{T}}$. Then $Q(z) > 0$ for all $z \cdot \nabla g(\boldsymbol{K}^*) = 0$ if and only if (see pp. 61–62 of [38])

$$(-1)^m \det(\boldsymbol{S}_r) > 0 \quad (r = m(m+1), m(m+2), \ldots, mn) \tag{2.192}$$

Only a scalar measurement situation ($m = 1$) is considered in this note for analyzing $\det(\boldsymbol{S}_r)$, as there is no essential difference between a scalar measurement and vector measurements. Equation (2.191) can be written as

$$\boldsymbol{S}_r = \begin{bmatrix} \boldsymbol{0} & \boldsymbol{h}_r^{\mathrm{T}}\epsilon(\boldsymbol{I}_r + \lambda\varepsilon\boldsymbol{A}_r)^{-1}\boldsymbol{A}_r \\ \boldsymbol{A}_r(\boldsymbol{I}_r + \lambda\varepsilon\boldsymbol{A}_r)^{-1}\boldsymbol{h}_r\epsilon & W(\boldsymbol{I}_r + \lambda\varepsilon\boldsymbol{A}_r) \end{bmatrix} \tag{2.193}$$

$\boldsymbol{S}_r$ is a symmetrical partitioned matrix and its determinant is given by

$$\begin{aligned} \det(\boldsymbol{S}_r) &= -\varepsilon\det(\boldsymbol{W})\det(\boldsymbol{I}_r + \lambda\varepsilon\boldsymbol{A}_r)\boldsymbol{h}_r^{\mathrm{T}}(\boldsymbol{I}_r + \lambda\varepsilon\boldsymbol{A}_r)^{-1} \\ &\quad \boldsymbol{A}_r(\boldsymbol{I}_r + \lambda\varepsilon\boldsymbol{A}_r)^{-1}\boldsymbol{A}_r(\boldsymbol{I}_r + \lambda\varepsilon\boldsymbol{A}_r)^{-1}\boldsymbol{h}_r \\ &= -\varepsilon\boldsymbol{W}\det(\boldsymbol{I}_r + \lambda\varepsilon\boldsymbol{\Xi}_r)\boldsymbol{h}_r^{\mathrm{T}}\boldsymbol{U}_r(\boldsymbol{I}_r + \lambda\varepsilon\boldsymbol{\Xi}_r)^{-3}\boldsymbol{\Xi}_r^2\boldsymbol{U}_r^{\mathrm{T}}\boldsymbol{h}_r \end{aligned} \tag{2.194}$$

Generally, the sign of $\det(\boldsymbol{S}_r)$ should be verified using Eq. (2.194), but it can be seen that a sufficient condition of $\det(\boldsymbol{S}_r) < 0$ is

$$\det(\boldsymbol{I}_r + \lambda\varepsilon\boldsymbol{\Xi}_r)(\boldsymbol{I}_r + \lambda\varepsilon\boldsymbol{\Xi}_r)^{-1} > 0 \tag{2.195}$$

Because the ith eigenvalue of $\det(\boldsymbol{I}_r + \lambda\varepsilon\boldsymbol{\Xi}_r)(\boldsymbol{I}_r + \lambda\varepsilon\boldsymbol{\Xi}_r)^{-1}$ is given by $\prod_{j=1}^{r}(1 + \varepsilon\lambda\xi_j)\ j \neq i$. Therefore, Lemma 2.2 is derived.

Remark

1. The implementation process of state QEC filtering is shown in Fig. 2.3.
2. If $\boldsymbol{A} = \boldsymbol{I}$, the constraint reduces to a norm constraint. Because $\xi = 1$ and $q = 1$, Eq. (2.184) can be simplified as

Fig. 2.3 Simulation flowchart of state QEC filtering [37]

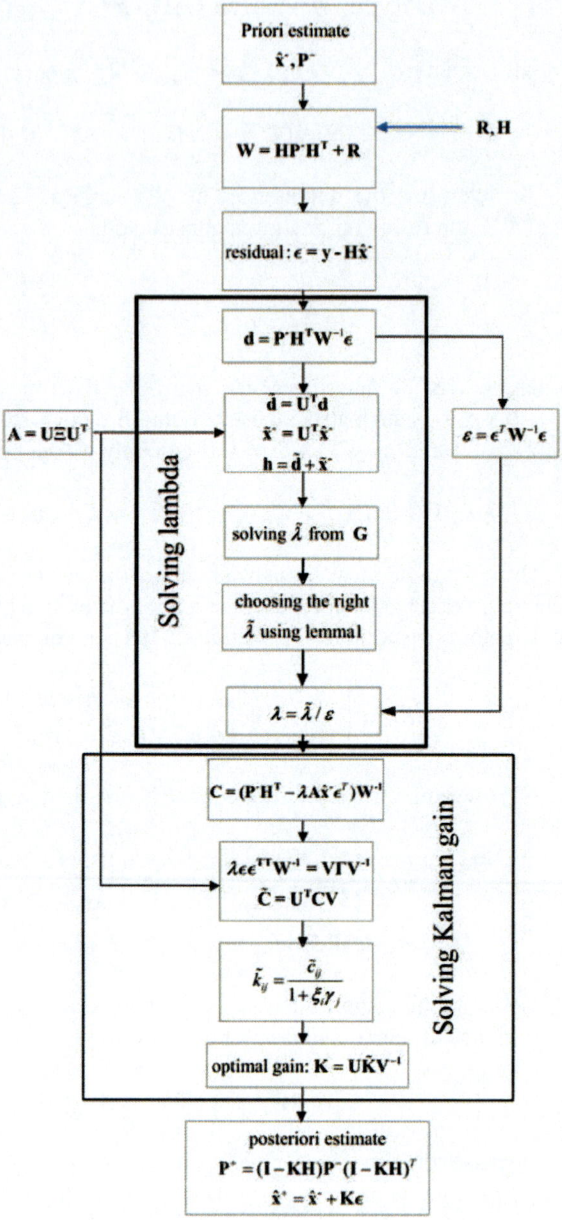

$$s_2(\lambda) = \frac{1}{\varepsilon^2}\left[(1+\lambda\epsilon)^2 - \frac{\boldsymbol{h}^{\mathrm{T}}\boldsymbol{h}}{l}\right] = 0 \tag{2.196}$$

and its solution is given by

$$\lambda = -\frac{1}{\varepsilon} \pm \frac{|\boldsymbol{h}|}{\varepsilon\sqrt{l}} \tag{2.197}$$

It is easy to verify from Lemma 2.2 that the plus sign in Eq. (2.197) corresponds a minimum performance. The Kalman filter gain can be directly computed from Eq. (2.167), which is the same as the gain derived in [33]. Moreover, the proposed constrained Kalman filter (CKF) is mathematically equivalent to the NCKF for a positive-semidefinite $\boldsymbol{A}$.

3. A transformation $\tilde{\boldsymbol{x}}_Y = \boldsymbol{\varUpsilon}\tilde{\boldsymbol{x}}$ with the (i, j) element of $\boldsymbol{\varUpsilon}$ given by

$$\varUpsilon(i, j) = \begin{cases} 1, & \xi_i = 0, i = j \\ \sqrt{|\xi_i|}, & \xi_i \neq 0, i = j \\ 0 & i \neq j \end{cases} \tag{2.198}$$

can be introduced to simplify Eq. (2.148) as

$$\tilde{\boldsymbol{x}}_Y^T \tilde{\boldsymbol{\varXi}} \tilde{\boldsymbol{x}}_Y = l \tag{2.199}$$

where $\tilde{\boldsymbol{\varXi}}$ is a diagonal matrix whose elements include only 0, -1, and 1. For variable $\tilde{\boldsymbol{x}}_Y$,there are at most two distinct nonzero eigenvalues for its quadratic-form matrix, which indicates that there exists at most four roots for the Lagrangian multiplier.

4. For a quadratic constraint, it is difficult to compare the performance of the constrained and unconstrained estimate. But the constrained estimate would provide a more realistic estimate. For example: consider attitude estimation problem of a spinning spacecraft. Because of the need to follow rapidly varying state vector elements and the lack of three-axis rate measurements form gyros, Markley variables [39, 40] may be employed instead of a quaternion-based representation for a spinning spacecraft. The state for the Markley variables includes $\boldsymbol{L}^I$ and $\boldsymbol{L}^B$, where $\boldsymbol{L}^I$ and $\boldsymbol{L}^B$ are the angular momentum components in an inertial frame and spacecraft's body frame, respectively. In this case, we would expect the estimates of $\boldsymbol{L}^I$ and $\boldsymbol{L}^B$, denoted $\hat{\boldsymbol{L}}^I$ and $\hat{\boldsymbol{L}}^B$, respectively, satisfy $\|\boldsymbol{L}^I\| = \|\boldsymbol{L}^B\|$.

2.5 Brief Summary

This chapter mainly focuses on basic point estimation theory. We firstly introduce general parameter estimators and then derive special estimators under certain assumptions or for some special cases. This is helpful for readers to understand the estimation problem. The common parameter estimators presented in Sect. 2.2 are foundations of the estimators proposed in Sects. 2.3 and 2.4. Under certain assumptions, closed form parameter estimators are derived in Sect. 2.3. These estimators are further extended to the state estimators of dynamics systems in Sect. 2.4. The recursive forms of the state estimators enable them to be implemented in real-time systems conveniently.

References

1. Bar-Shalom, Y., X.R. Li, and T. Kirubarajan. 2004. *Estimation with applications to tracking and navigation: Theory algorithms and software.* New York: Wiley.
2. Li, X.R. 1999. *Probability, random signals, and statistics.* New York: CRC Press.
3. Verbeek, M. 2008. *A guide to modern econometrics.* New York: Wiley.
4. Han, C., H. Zhu, and Z. Duan. 2009. *Multi-source information fusion.* Beijing: Tsinghua University Press.
5. Greene, W.H. 2003. *Econometric analysis.* Pearson Education India.
6. Zhou, H.Y., et al. 2013. *Fusion theory and methods for satellite state estimation.* Beijing: Science Press.
7. Crassidis, J.L., and J.L. Junkins. 2011. *Optimal estimation of dynamic systems.* New York: CRC Press.
8. Leon-Garcia, A. 2008. *Probability, statistics, and random processes for electrical engineering.* New Jersey: Pearson Prentice Hall.
9. Kailath, T., A.H. Sayed, and B. Hassibi. 2000. *Linear estimation.* New Jersey: Prentice Hall.
10. Haug, A.J. 2012. *Bayesian estimation and tracking: A practical guide.* New York: Wiley.
11. Kalman, R.E., and R.S. Bucy. 1961. New results in linear filtering and prediction theory. *Journal of Basic Engineering* 83 (1): 95–108.
12. Kalman, R.E. 1960. A new approach to linear filtering and prediction problems. *Journal of Basic Engineering* 82 (1): 35–45.
13. Särkkä, S. 2013. *Bayesian filtering and smoothing*, vol. 3. Cambridge University Press.
14. Smith, G.L., S.F. Schmidt, and L.A. McGee. 1962. *Application of statistical filter theory to the optimal estimation of position and velocity on board a circumlunar vehicle.* NASA Technical Report TR R-135.
15. Julier, S.J., and J.K. Uhlmann. 2004. Unscented filtering and nonlinear estimation. *Proceedings of the IEEE* 92 (3): 401–422.
16. Gelb, A. 1974. *Applied optimal estimation.* Cambridge, MA: MIT Press.
17. Arulampalam, M.S., et al. 2002. A tutorial on particle filters for online nonlinear/non-Gaussian Bayesian tracking. *IEEE Transactions on Signal Processing* 50 (2): 174–188.
18. Julier, S.J., J.K. Uhlmann, and H.F. Durrant-Whyte. 1995. A new approach for filtering nonlinear systems. In *American Control Conference, Proceedings of the 1995.* IEEE.
19. Julier, S., J. Uhlmann, and H.F. Durrant-Whyte. 2000. A new method for the nonlinear transformation of means and covariances in filters and estimators. *IEEE Transactions on Automatic Control* 45 (3): 477–482.

20. Van Der Merwe, R., and E.A. Wan. 2001. The square-root unscented Kalman filter for state and parameter-estimation. In *2001 IEEE International Conference on Acoustics, Speech, and Signal Processing, 2001. Proceedings (ICASSP'01)*. IEEE.
21. Soken, H.E., and C. Hajiyev. 2011. In flight magnetometer calibration via unscented Kalman filter. In *2011 5th International Conference on Recent Advances in Space Technologies (RAST)*. IEEE.
22. Julier, S.J., and J.K. Uhlmann. 1997. New extension of the Kalman filter to nonlinear systems. In *Signal Processing, Sensor Fusion, and Target Recognition VI*. International Society for Optics and Photonics.
23. Haykin, S. 2004. *Kalman filtering and neural networks*, vol. 47. New York: Wiley.
24. Massicotte, D., R.Z. Morawski, and A. Barwicz. 1995. Incorporation of a positivity constraint into a Kalman-filter-based algorithm for correction of spectrometric data. *IEEE Transactions on Instrumentation and Measurement* 44 (1): 2–7.
25. Duan, Z., and X.R. Li. 2015. Analysis, design, and estimation of linear equality-constrained dynamic systems. *IEEE Transactions on Aerospace and Electronic Systems* 51 (4): 2732–2746.
26. Wen, W., and H.F. Durrant-Whyte. 1992. Model-based multi-sensor data fusion. In *1992 IEEE International Conference on Robotics and Automation, 1992. Proceedings*. IEEE.
27. Alouani, A.T., and W.D. Blair. 1993. Use of a kinematic constraint in tracking constant speed, maneuvering targets. *IEEE Transactions on Automatic Control* 38 (7): 1107–1111.
28. Simon, D., and T.L. Chia. 2002. Kalman filtering with state equality constraints. *IEEE Transactions on Aerospace and Electronic Systems* 38 (1): 128–136.
29. Duan, Z., and X.R. Li. 2013. The role of pseudo measurements in equality-constrained state estimation. *IEEE Transactions on Aerospace and Electronic Systems* 49 (3): 1654–1666.
30. De Ruiter, A.H. 2016. Quadratically constrained least squares with aerospace applications. *Journal of Guidance, Control and Dynamics* 39 (1): 487–497.
31. Chia, T.-L. 1985. *Parameter identification and state estimation of constained systems*. Case Western Reserve University.
32. Gupta, N., and R. Hauser. 2007. *Kalman filtering with equality and inequality state constraints*. arXiv preprint arXiv:0709.2791.
33. Zanetti, R., et al. 2009. Norm-constrained Kalman filtering. *Journal of Guidance, Control and Dynamics* 32 (5): 1458–1465.
34. Yang, C., and E. Blasch. 2009. Kalman filtering with nonlinear state constraints. *IEEE Transactions on Aerospace and Electronic Systems* 45 (1): 70–84.
35. Fu, X., and Y. Jia. 2012. H_∞ filtering with combined linear and nonlinear constraints. *IEEE Transactions on Aerospace and Electronic Systems* 48 (4): 3347–3362.
36. Simon, D. 2010. Kalman filtering with state constraints: a survey of linear and nonlinear algorithms. *IET Control Theory and Applications* 4 (8): 1303–1318.
37. Wang, D.Y., et al. 2014. Kalman filtering for a quadratic form state equality constraint. *Journal of Guidance, Control and Dynamics* 37 (3): 951–958.
38. Magnus, J.R., and H. Neudecker. 1988. *Matrix differential calculus with applications in statistics and econometrics*. Wiley series in probability and mathematical statistics. New York: Wiley.
39. Markley, L. 1993. New dynamic variables for momentum-bias spacecraft. *Journal of the Astronautical Sciences* 41 (4): 557–567.
40. Markley, F.L., and J.E. Sedlak. 2008. Kalman filter for spinning spacecraft attitude estimation. *Journal of Guidance, Control and Dynamics* 31 (6): 1750–1760.

Chapter 3
Estimation Fusion Algorithm

Multiple source information-based autonomous navigation is essentially an estimation fusion problem. This chapter will go deep into the estimation fusion algorithm. Section 3.1 presents linear models and algorithms, mainly including linear unified model and its fusion algorithm as well as covariance intersection algorithm in the distributed fusion, which can be used to estimate the parameters of a static system and states of a dynamic system. Sections 3.2 and 3.3 describe the centralized Kalman filter and distributed Kalman filter of a dynamic system, respectively, and discuss several implementation models of centralized filtering and typical models of distributed filtering.

3.1 Linear Fusion Models and Algorithms

This section considers a linear estimation fusion problem in which the measurement equation is linear. It should be mentioned that, in a distributed or hybrid fusion architecture, local estimates can be seen as virtual measurements, and the corresponding virtual measurement equations are linear. In this case, the problem can also be formulated as a linear fusion problem.

3.1.1 Linear Unified Model

In Chap. 1, three architectures for information fusion were introduced. Distributed and centralized fusion can be seen as special forms of hybrid fusion, so this section directly discusses hybrid fusion architecture. Following from [1], this section proposes a unified linear model for hybrid fusion architecture with linear measurement equation.

D. Wang et al., *Spacecraft Autonomous Navigation Technologies Based on Multi-source Information Fusion*, Space Science and Technologies, https://doi.org/10.1007/978-981-15-4879-6_3

Suppose that m distributed-fusion nodes and additional n sensor measurements are fused in the hybrid manner. Denote $\hat{\boldsymbol{x}}^{(i)}$ and $\boldsymbol{P}^{(i)}$ as the ith node's estimate and its error covariance. Then

$$\hat{\boldsymbol{x}}^{(i)} = \boldsymbol{x} + \boldsymbol{v}_x^{(i)} \tag{3.1}$$

where $i = 1, 2, \ldots, m$ and $\mathrm{E}\{\boldsymbol{v}_x^{(i)}(\boldsymbol{v}_x^{(i)})^{\mathrm{T}}\} = \boldsymbol{P}^{(i)}$. Suppose that the measurement equations of n sensors are linear and given by

$$\boldsymbol{z}^{(j)} = \boldsymbol{H}^{(j)}\boldsymbol{x} + \boldsymbol{v}_z^{(j)} \tag{3.2}$$

with $j = 1, 2, \ldots, n$.

Equations (3.1) and (3.2) can be written in a unified form as

$$\boldsymbol{y} = \mathcal{H}\boldsymbol{x} + \boldsymbol{v} \tag{3.3}$$

where

$$\boldsymbol{y} \triangleq \begin{bmatrix} \hat{\boldsymbol{x}}^{(1)} \\ \vdots \\ \hat{\boldsymbol{x}}^{(m)} \\ \boldsymbol{z}^{(1)} \\ \vdots \\ \boldsymbol{z}^{(n)} \end{bmatrix}, \quad \boldsymbol{v} \triangleq \begin{bmatrix} \boldsymbol{v}_x^{(1)} \\ \vdots \\ \boldsymbol{v}_x^{(m)} \\ \boldsymbol{v}_z^{(1)} \\ \vdots \\ \boldsymbol{v}_z^{(n)} \end{bmatrix}, \quad \mathcal{H} \triangleq \begin{bmatrix} \boldsymbol{I} \\ \vdots \\ \boldsymbol{I} \\ \boldsymbol{H}^{(1)} \\ \vdots \\ \boldsymbol{H}^{(n)} \end{bmatrix} \tag{3.4}$$

3.1.2 *Fusion Algorithm from the Linear Unified Model*

According to Theorem 2.2, if the noise of Eq. (3.3) satisfies $\mathrm{E}\{\boldsymbol{v}\} = \boldsymbol{0}$, $\mathrm{E}\{\boldsymbol{v}\boldsymbol{v}^{\mathrm{T}}\} = \mathcal{R}$, then the LMMSE estimate of $\boldsymbol{x}$, denoted by $\hat{\boldsymbol{x}}^{\mathrm{LMMSE}}$, and its error covariance, denoted by $\boldsymbol{P}_{\tilde{\boldsymbol{x}}^{\mathrm{LMMSE}}}$, are given by

$$\begin{aligned} \hat{\boldsymbol{x}}^{\mathrm{LMMSE}} &= (\mathcal{H}^{\mathrm{T}}\mathcal{R}^{-1}\mathcal{H})^{-1}\mathcal{H}^{\mathrm{T}}\mathcal{R}^{-1}\boldsymbol{y} \\ \boldsymbol{P}_{\tilde{\boldsymbol{x}}^{\mathrm{LMMSE}}} &= (\mathcal{H}^{\mathrm{T}}\mathcal{R}^{-1}\mathcal{H})^{-1} \end{aligned} \tag{3.5}$$

Corollary 3.1 *Given a local estimate* $\hat{\boldsymbol{x}}^{-}$ *with a measurement* $\boldsymbol{z}$

$$\begin{aligned} \hat{\boldsymbol{x}}^{-} &= \boldsymbol{x} + \boldsymbol{w} \\ \boldsymbol{z} &= \boldsymbol{H}\boldsymbol{x} + \boldsymbol{v} \end{aligned} \tag{3.6}$$

and

$$\begin{aligned} \mathrm{E}\{\boldsymbol{v}\} &= \mathbf{0}, \mathrm{E}\{\boldsymbol{v}\boldsymbol{v}^{\mathrm{T}}\} = \boldsymbol{R} \\ \mathrm{E}\{\boldsymbol{w}\} &= \mathbf{0}, \mathrm{E}\{\boldsymbol{w}\boldsymbol{w}^{\mathrm{T}}\} = \boldsymbol{P}^{-} \\ \mathrm{E}\{\boldsymbol{w}\boldsymbol{v}^{\mathrm{T}}\} &= \mathbf{0}, \end{aligned} \tag{3.7}$$

then the LMMSE estimate to fuse $\hat{\boldsymbol{x}}^{-}$ *with* $\boldsymbol{z}$ *is given by*

$$\hat{\boldsymbol{x}} = \left[(\boldsymbol{P}^{-})^{-1} + \boldsymbol{H}^{\mathrm{T}}\boldsymbol{R}^{-1}\boldsymbol{H}\right]^{-1}\left[(\boldsymbol{P}^{-})^{-1}\hat{\boldsymbol{x}}^{-} + \boldsymbol{H}^{\mathrm{T}}\boldsymbol{R}^{-1}\boldsymbol{z}\right] \tag{3.8}$$

and

$$\boldsymbol{P} = \left[(\boldsymbol{P}^{-})^{-1} + \boldsymbol{H}^{\mathrm{T}}\boldsymbol{R}^{-1}\boldsymbol{H}\right]^{-1} \tag{3.9}$$

where $\boldsymbol{P}$ *is the error covariance of* $\hat{\boldsymbol{x}}$.

Remark Equations (3.8) and (3.9) correspond to the measurement update process of the Kalman filter.

Corollary 3.2 *For the centralized fusion* ($m = 0$), *if the measurement noises are independent with each other and satisfy*

$$\mathrm{E}\{\boldsymbol{v}_z^{(i)}\} = \mathbf{0}, \mathrm{E}\{\boldsymbol{v}_z^{(i)}(\boldsymbol{v}_z^{(i)})^{\mathrm{T}}\} = \boldsymbol{R}^{(i)}, \mathrm{E}\{\boldsymbol{v}_z^{(i)}(\boldsymbol{v}_z^{(j)})^{\mathrm{T}}\} = \mathbf{0} \quad i \neq j \tag{3.10}$$

Then, according to Eq. (3.5), *the LMMSE estimate to fuse measurements is given by*

$$\begin{aligned} \hat{\boldsymbol{x}}^{\mathrm{LMMSE}} &= \left(\sum_{i=1}^{n}(\boldsymbol{H}^{(i)})^{\mathrm{T}}(\boldsymbol{R}^{(i)})^{-1}\boldsymbol{H}^{(i)}\right)^{-1}\left(\sum_{i=1}^{n}(\boldsymbol{H}^{(i)})^{\mathrm{T}}(\boldsymbol{R}^{(i)})^{-1}\boldsymbol{z}^{(i)}\right) \\ \boldsymbol{P}_{\hat{x}^{\mathrm{LMMSE}}} &= \left(\sum_{i=1}^{n}(\boldsymbol{H}^{(i)})^{\mathrm{T}}(\boldsymbol{R}^{(i)})^{-1}\boldsymbol{H}^{(i)}\right)^{-1} \end{aligned} \tag{3.11}$$

Specially, if $\boldsymbol{R}^{(i)} = \sigma_i^2\boldsymbol{I}$, Eq. (3.11) *can be simplified as*

$$\begin{aligned} \hat{\boldsymbol{x}}^{\mathrm{LMMSE}} &= \left(\sum_{i=1}^{n}\sigma_i^{-2}(\boldsymbol{H}^{(i)})^{\mathrm{T}}\boldsymbol{H}^{(i)}\right)^{-1}\left(\sum_{i=1}^{n}\sigma_i^{-2}(\boldsymbol{H}^{(i)})^{\mathrm{T}}\boldsymbol{z}^{(i)}\right) \\ \boldsymbol{P}_{\hat{x}^{\mathrm{LMMSE}}} &= \left(\sum_{i=1}^{n}\sigma_i^{-2}(\boldsymbol{H}^{(i)})^{\mathrm{T}}\boldsymbol{H}^{(i)}\right)^{-1} \end{aligned} \tag{3.12}$$

Remark

1. In fact, the optimal estimate given by Eq. (3.12) is the weighted sum of the observation data z_i from all the sensors, and the weight is [2]

$$\omega_i = \left[\sum_{i=1}^{n} \sigma_i^{-2}(\boldsymbol{H}^{(i)})^{\mathrm{T}}\boldsymbol{H}^{(i)}\right]^{-1} \sigma_i^{-2}(\boldsymbol{H}^{(i)})^{\mathrm{T}} \quad i = 1, \cdots, n \tag{3.13}$$

   If the state to be estimated is a scalar and can be measured by each sensor, the optimal estimation fusion of the state is the weighted average of measurements from all the sensors, and the weight is given by

$$\omega_i = \frac{\frac{1}{\sigma_i^2}}{\sum_{i=1}^{n} \frac{1}{\sigma_i^2}} \tag{3.14}$$

   Obviously, $\sum_{i=1}^{n} \omega_i = 1$, and the numerator in Eq. (3.14) represents each sensor's measurement accuracy.
2. The accuracy of the LMMSE estimation of multi-sensors is better than that of any single sensor, i.e.,

$$\mathrm{var}(\hat{\boldsymbol{x}}^{\mathrm{LMMSE}}) < \sum_{i=1}^{n} \sigma_i^{-2}(\boldsymbol{H}^{(i)})^{\mathrm{T}}\boldsymbol{H}^{(i)} \tag{3.15}$$

Corollary 3.3 *for distributed fusion* ($n = 0$), *denote the cross-covariance of the* ith *node and the* jth *node as*

$$\boldsymbol{P}^{(ij)} \triangleq \mathrm{E}\{\tilde{\boldsymbol{x}}^{(i)}(\tilde{\boldsymbol{x}}^{(j)})^{\mathrm{T}}\} \quad i \neq j \tag{3.16}$$

where $\tilde{\boldsymbol{x}}^{(i)} = \boldsymbol{x} - \hat{\boldsymbol{x}}^{(i)}$, then $\mathcal{R}$, $\mathcal{H}$, $\boldsymbol{y}$ *in* Eq. (3.5) *can be written as*

$$\mathcal{R} = \begin{bmatrix} \boldsymbol{P}^{(1)} & \boldsymbol{P}^{(1,2)} & \dots & \boldsymbol{P}^{(1,m)} \\ \boldsymbol{P}^{(2,1)} & \boldsymbol{P}^{(2)} & \dots & \boldsymbol{P}^{(2,m)} \\ \vdots & \vdots & \ddots & \vdots \\ \boldsymbol{P}^{(m,1)} & \boldsymbol{P}^{(m,2)} & \dots & \boldsymbol{P}^{(m)} \end{bmatrix}$$

$$\mathcal{H} = \begin{bmatrix} \boldsymbol{I} \\ \vdots \\ \boldsymbol{I} \\ \boldsymbol{I} \end{bmatrix}, \boldsymbol{y} = \begin{bmatrix} \hat{\boldsymbol{x}}^{(1)} \\ \vdots \\ \hat{\boldsymbol{x}}^{(m-1)} \\ \hat{\boldsymbol{x}}^{(m)} \end{bmatrix} \tag{3.17}$$

Specially,

1. *If* $\boldsymbol{P}^{(ij)} = 0$, Eq. (3.5) *can be simplified as*

$$
\begin{aligned}
\hat{\boldsymbol{x}}^{\text{LMMSE}} &= \left(\sum_{i=1}^{m} (\boldsymbol{P}^{(i)})^{-1}\right)^{-1} \sum_{i=1}^{m} \left((\boldsymbol{P}^{(i)})^{-1} \hat{\boldsymbol{x}}^{(i)}\right) \\
\boldsymbol{P}_{\tilde{\boldsymbol{x}}^{\text{LMMSE}}} &= \left(\sum_{i=1}^{m} (\boldsymbol{P}^{(i)})^{-1}\right)^{-1}
\end{aligned}
\tag{3.18}
$$

2. *If* $m = 2$, *then*

$$
\begin{aligned}
\hat{\boldsymbol{x}}^{\text{LMMSE}} =& (\boldsymbol{P}^{(2)} - \boldsymbol{P}^{(2,1)})(\boldsymbol{P}^{(1)} + \boldsymbol{P}^{(2)} - \boldsymbol{P}^{(1,2)} - \boldsymbol{P}^{(2,1)})^{-1} \hat{\boldsymbol{x}}^{(1)} \\
&+ (\boldsymbol{P}^{(1)} - \boldsymbol{P}^{(1,2)})(\boldsymbol{P}^{(1)} + \boldsymbol{P}^{(2)} - \boldsymbol{P}^{(1,2)} - \boldsymbol{P}^{(2,1)})^{-1} \hat{\boldsymbol{x}}^{(2)} \\
\boldsymbol{P}_{\tilde{\boldsymbol{x}}^{\text{LMMSE}}} =& \boldsymbol{P}^{(1)} - (\boldsymbol{P}^{(1)} - \boldsymbol{P}^{(1,2)})(\boldsymbol{P}^{(1)} + \boldsymbol{P}^{(2)} - \boldsymbol{P}^{(1,2)} \\
&- \boldsymbol{P}^{(2,1)})^{-1} (\boldsymbol{P}^{(1)} - \boldsymbol{P}^{(2,1)})
\end{aligned}
\tag{3.19}
$$

Equation (3.19) *is also called Bar-Shalom Campo (BC) fusion algorithm* [3].

3.1.3 Covariance Intersection Algorithm in the Distributed Fusion

The LMMSE estimation for distributed or hybrid fusion requires the calculation of cross-covariance $\boldsymbol{P}^{(ij)}$. However, in practice, $\boldsymbol{P}^{(ij)}$ may be unknown [4], or its calculation process may be very complicated and difficult [5]. To overcome these difficulties, Simon Julier et al. [6–9] proposed the covariance intersection (CI) algorithm, which has been widely used in the field of information fusion.

The CI algorithm is a robust distributed estimation fusion algorithm that can produce a consistent estimate to cope with unknown cross-covariance. Before introducing the CI algorithm, estimated consistency is defined as follows: Denote $\hat{\boldsymbol{x}}$ and $\hat{\boldsymbol{P}}$ as the estimates of $\boldsymbol{x}$ and its error variance. If

$$
\hat{\boldsymbol{P}} \geq \mathrm{E}\{(\boldsymbol{x} - \bar{\boldsymbol{x}})(\boldsymbol{x} - \bar{\boldsymbol{x}})^{\mathrm{T}}\}
\tag{3.20}
$$

, then $\hat{\boldsymbol{x}}$ is consistent [10]. A consistent estimate can prevent the filter to diverge.

For convenience, CI algorithm is illustrated by taking two distributed nodes as an example. Suppose that the two local estimates are $\hat{\boldsymbol{x}}^{(1)}$, $\hat{\boldsymbol{x}}^{(2)}$, and their error covariances are $\boldsymbol{P}^{(1)}$, $\boldsymbol{P}^{(2)}$, respectively. In practice, only the estimate of $\boldsymbol{P}^{(i)}$ is known (denoted as $\hat{\boldsymbol{P}}^{(i)}$). To meet the consistency,

$$
\hat{\boldsymbol{P}}^{(i)} - \boldsymbol{P}^{(i)} \geq 0
\tag{3.21}
$$

The estimation after fusion should meet the consistency, i.e.,

$$\hat{\boldsymbol{P}} - \boldsymbol{P} \geq 0 \tag{3.22}$$

Through a convex combination of the estimates of local state and covariance in CI algorithm, the global state estimate and covariance estimate can be obtained as below [9]

$$\begin{aligned}
(\hat{\boldsymbol{P}})^{-1}\hat{\boldsymbol{x}} &= \omega(\hat{\boldsymbol{P}}^{(1)})^{-1}\hat{\boldsymbol{x}}^{(1)} + (1-\omega)(\hat{\boldsymbol{P}}^{(2)})^{-1}\hat{\boldsymbol{x}}^{(2)} \\
(\hat{\boldsymbol{P}})^{-1} &= \omega(\hat{\boldsymbol{P}}^{(1)})^{-1} + (1-\omega)(\hat{\boldsymbol{P}}^{(2)})^{-1}
\end{aligned} \tag{3.23}$$

where $\omega \in [0, 1]$ is a weighting factor.

The optimal value of ω can be obtained by optimizing the trace or determinant of $\hat{\boldsymbol{P}}$, for example

$$\begin{aligned}
\omega &= \text{argmin Tr}\left[\hat{\boldsymbol{P}}\right] \\
&= \text{argmin Tr}\left[\left(\omega(\hat{\boldsymbol{P}}^{(1)})^{-1} + (1-\omega)(\hat{\boldsymbol{P}}^{(2)})^{-1}\right)^{-1}\right]
\end{aligned} \tag{3.24}$$

Since the optimization index is a convex function of ω and has only one optimal point, ω can be solved by Newton iterative algorithm or other nonlinear optimization algorithms [11].

Theorem 3.1 [9]: *for any* ω, *the estimation after the fusion by CI algorithm meets the consistency, i.e.,*

$$\hat{\boldsymbol{P}} - \boldsymbol{P} \geq 0 \tag{3.25}$$

For a distributed-fusion architecture with multiple local nodes, the CI algorithm is given below

$$\begin{aligned}
(\hat{\boldsymbol{P}})^{-1}\hat{\boldsymbol{x}} &= \omega_1(\hat{\boldsymbol{P}}^{(1)})^{-1}\hat{\boldsymbol{x}}^{(1)} + \cdots + \omega_m(\hat{\boldsymbol{P}}^{(m)})^{-1}\hat{\boldsymbol{x}}^{(m)} \\
(\hat{\boldsymbol{P}})^{-1} &= \omega_1(\hat{\boldsymbol{P}}^{(1)})^{-1} + \cdots + \omega_m(\hat{\boldsymbol{P}}^{(m)})^{-1}
\end{aligned} \tag{3.26}$$

where $\sum_{i=1}^{m} \omega_i = 1$ and $\omega_i \geq 0$. The optimal weight can be solved by minimizing the trace of $\hat{\boldsymbol{P}}$, i.e.,

$$\begin{aligned}
\omega &= \text{argmin Tr}\left[\hat{\boldsymbol{P}}\right] \\
&= \text{argmin Tr}\left[\left(\sum \omega_i(\hat{\boldsymbol{P}}^{(i)})^{-1}\right)^{-1}\right]
\end{aligned} \tag{3.27}$$

3.2 Centralized-Fusion Kalman Filtering for a Dynamic System

Consider the following discrete state equation

$$\boldsymbol{x}_k = \boldsymbol{F}_{k,k-1}\boldsymbol{x}_{k-1} + \boldsymbol{w}_k \tag{3.28}$$

where $\boldsymbol{F}_{k,k-1}$ is the state transition matrix from t_{k-1} to t_k and $\boldsymbol{w}_k$ is the process noise.

Assuming that the system has n sensor nodes. The observation equation of the ith sensor node is

$$\boldsymbol{z}_k^{(i)} = \boldsymbol{H}_k^{(i)}\boldsymbol{x}_k + \boldsymbol{v}_k^{(i)} \tag{3.29}$$

where $\boldsymbol{H}_k^{(i)}$ is the measurement sensitivity matrix, $\boldsymbol{v}_k^{(i)}$ is the measurement noise, and the superscript (i) corresponds to the ith node.

For the dynamic system composed of Eqs. (3.28) and (3.29), the following assumptions are made:

1. Both the process noise and the measurement noise are zero-mean Gaussian noise and are uncorrelated with each other, i.e.,

$$\begin{aligned}
\mathrm{E}\{\boldsymbol{w}_k\} &= \boldsymbol{0}\\
\mathrm{E}\{\boldsymbol{v}_k^{(i)}\} &= \boldsymbol{0}\\
\mathrm{E}\{\boldsymbol{w}_k\boldsymbol{w}_p^{\mathrm{T}}\} &= \begin{cases} \boldsymbol{Q}_k, & k = p\\ \boldsymbol{0}, & k \neq p \end{cases}\\
\mathrm{E}\{\boldsymbol{v}_k^{(i)}(\boldsymbol{v}_p^{(i)})^{\mathrm{T}}\} &= \begin{cases} \boldsymbol{R}_k^{(i)}, & k = p\\ \boldsymbol{0}, & k \neq p \end{cases}\\
\mathrm{E}\{\boldsymbol{w}_k(\boldsymbol{v}_k^{(j)})^{\mathrm{T}}\} &= \boldsymbol{0}
\end{aligned} \tag{3.30}$$

where $\boldsymbol{Q}_k$ and $\boldsymbol{R}_k^{(i)}$ are the process and measurement noises covariances, respectively.

2. The initial state $\boldsymbol{x}_0$ is a Gaussian random variable. Its mean and variance are given by

$$\begin{aligned}
\mathrm{E}\{\boldsymbol{x}_0\} &= \hat{\boldsymbol{x}}_0\\
\mathrm{var}\{\boldsymbol{x}_0\} &= \boldsymbol{P}_0
\end{aligned} \tag{3.31}$$

3. The initial state is independent of process noise and measurement noise, i.e.,

$$\begin{aligned}
\mathrm{cov}\{\boldsymbol{x}_0, \boldsymbol{w}_k\} &= 0\\
\mathrm{cov}\{\boldsymbol{x}_0, \boldsymbol{v}_k^{(i)}\} &= 0
\end{aligned} \tag{3.32}$$

4. The observation noise of the *i*th sensor node is also independent of that of the *j*th sensor node, i.e.,

$$E\{v_k^{(i)}(v_k^{(j)})^T\} = 0 \quad i \neq j \tag{3.33}$$

The centralized-fusion Kalman filtering is usually implemented in three ways: parallel filtering, sequential filtering, and data compression filtering [12]. Among them, parallel filtering integrates all the observations into a large one for filtering. Sequential filtering updates each observation in turn and take the estimates of state and variance after updating as the final estimates. Compression filtering firstly processes the observations and compresses and extracts the data, and then takes the compressed data as observations for filtering. These three methods will be separately discussed below.

3.2.1 *Parallel Filtering*

Figure 3.1 shows the implementation process of parallel filtering. Suppose that all the measurements are obtained at the same time, the generalized measurement equation can be obtained from

$$z = Hx + v \tag{3.34}$$

where

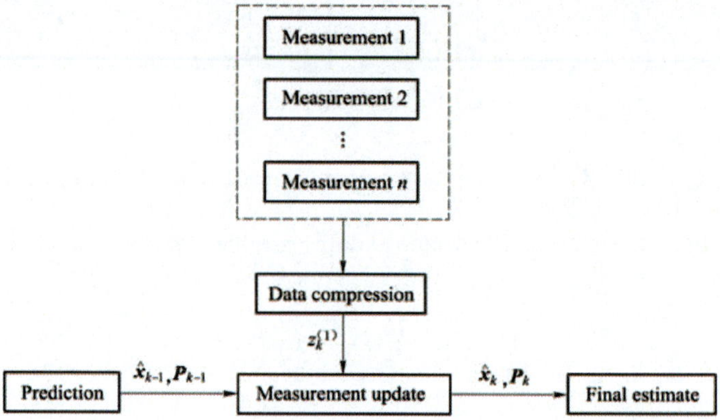

Fig. 3.1 Implementation process of parallel filtering

$$\boldsymbol{z} \triangleq \begin{bmatrix} \boldsymbol{z}^{(1)} \\ \vdots \\ \boldsymbol{z}^{(n)} \end{bmatrix}, \quad \boldsymbol{v} \triangleq \begin{bmatrix} \boldsymbol{v}^{(1)} \\ \vdots \\ \boldsymbol{v}^{(n)} \end{bmatrix}, \quad \boldsymbol{H} \triangleq \begin{bmatrix} \boldsymbol{H}^{(1)} \\ \vdots \\ \boldsymbol{H}^{(n)} \end{bmatrix} \tag{3.35}$$

It can be known from Eq. (3.30) that

$$\begin{aligned} \mathrm{E}\{\boldsymbol{w}_k\} &= \boldsymbol{0} \\ \mathrm{E}\{\boldsymbol{v}_k\} &= \boldsymbol{0} \\ \boldsymbol{R}_k = \mathrm{var}\{\boldsymbol{v}_k\} &= \mathrm{diag}\left[\boldsymbol{R}_k^{(1)}, \boldsymbol{R}_k^{(2)}, \cdots, \boldsymbol{R}_k^{(n)}\right] \\ \mathrm{cov}\{\boldsymbol{v}_k, \boldsymbol{w}_k\} &= \boldsymbol{0} \\ \mathrm{cov}\{\boldsymbol{x}_0, \boldsymbol{v}_k\} &= \boldsymbol{0} \end{aligned} \tag{3.36}$$

The dynamic system composed of Eqs. (3.28) and (3.34) can be fused directly by using the Kalman filter equation, as follows:

1. State prediction:

$$\hat{\boldsymbol{x}}_{k|k-1} = \boldsymbol{F}_{k,k-1}\hat{\boldsymbol{x}}_{k-1} \tag{3.37}$$

where $\hat{\boldsymbol{x}}_{k-1} \triangleq \hat{\boldsymbol{x}}_{k-1|k-1}$, $\boldsymbol{P}_{k-1} \triangleq \boldsymbol{P}_{k-1|k-1}$.

2. Error covariance prediction:

$$\boldsymbol{P}_{k|k-1} = \boldsymbol{F}_{k,k-1}\boldsymbol{P}_{k-1}\boldsymbol{F}_{k,k-1}^{\mathrm{T}} + \boldsymbol{Q}_{k-1} \tag{3.38}$$

3. Error covariance update:

$$\begin{aligned} \boldsymbol{P}_k^{-1} &= \boldsymbol{P}_{k|k-1}^{-1} + \boldsymbol{H}_k^{\mathrm{T}}\boldsymbol{R}_k^{-1}\boldsymbol{H}_k \\ &= \boldsymbol{P}_{k|k-1}^{-1} + \sum_{i=1}^{n}\left(\boldsymbol{H}_k^{(i)}\right)^{\mathrm{T}}\left(\boldsymbol{R}_k^{(i)}\right)^{-1}\boldsymbol{H}_k^{(i)} \end{aligned} \tag{3.39}$$

4. Measurement update

$$\hat{\boldsymbol{x}}_k = \hat{\boldsymbol{x}}_{k|k-1} + \boldsymbol{K}_k(\boldsymbol{z}_k - \boldsymbol{H}_k\hat{\boldsymbol{x}}_{k|k-1}) \tag{3.40}$$

where

$$\boldsymbol{K}_k = \boldsymbol{P}_k\boldsymbol{H}_k^{\mathrm{T}}\boldsymbol{R}_k^{-1} \tag{3.41}$$

By substituting $\boldsymbol{R}_k$ of Eq. (3.36) into Eq. (3.41), and then substituting Eq. (3.41) into Eq. (3.40), the following equation can be obtained

$$\hat{\boldsymbol{x}}_k = \hat{\boldsymbol{x}}_{k|k-1} + \boldsymbol{P}_k \sum_{i=1}^{n} \left(\boldsymbol{H}_k^{(i)}\right)^{\mathrm{T}} \left(\boldsymbol{R}_k^{(i)}\right)^{-1} (\boldsymbol{z}_k^{(i)} - \boldsymbol{H}_k^{(i)} \hat{\boldsymbol{x}}_{k|k-1}) \tag{3.42}$$

Remark

1. The centralized fusion is optimal and usually be used a reference to compare with the performance of other fusion algorithms [2]. However, as long as one subsystem fails and is not timely detected and isolated, this fault will spread rapidly to the whole system, contaminate all the state parameters, and even disable the fusion system. So the fault-tolerance of this system is poor [13].
2. The derivation of parallel filtering is based on the assumption that each measurement is independent of the other.

3.2.2 Sequential Filtering

Sequential filtering updates each measurement in turn, as implemented in Fig. 3.2. Its algorithm is as follows:

1. State prediction:

$$\hat{\boldsymbol{x}}_{k|k-1} = \boldsymbol{F}_{k,k-1}\hat{\boldsymbol{x}}_{k-1|} \tag{3.43}$$

2. Error covariance prediction

$$\boldsymbol{P}_{k|k-1} = \boldsymbol{F}_{k,k-1}\boldsymbol{P}_{k-1}\boldsymbol{F}_{k,k-1}^{\mathrm{T}} + \boldsymbol{Q}_{k-1} \tag{3.44}$$

3. Update n measurements in turn

① Error covariance update

$$(\boldsymbol{P}_k^{(i)})^{\mathrm{T}} = (\boldsymbol{P}_k^{(i-1)})^{\mathrm{T}} + \left(\boldsymbol{H}_k^{(i)}\right)^{\mathrm{T}} \left(\boldsymbol{R}_k^{(i)}\right)^{-1} \boldsymbol{H}_k^{(i)} \tag{3.45}$$

where $i = 1, 2, \ldots, n$; $\boldsymbol{P}_k^{(0)} = \boldsymbol{P}_{k|k-1}$.

② Measurement update

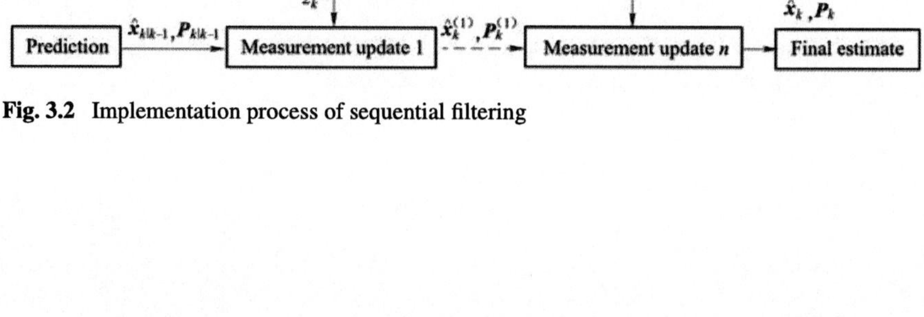

Fig. 3.2 Implementation process of sequential filtering

$$\hat{x}_k^{(i)} = \hat{x}_k^{(i-1)} + K_k^{(i)}(z_k^{(i)} - H_k^{(i)}\hat{x}_k^{(i-1)}) \tag{3.46}$$

where $i = 1, \ldots, n$, $\hat{x}_k^{(0)} = \hat{x}_{k|k-1}$, and

$$K_k^{(i)} = P_k^{(i)}(H_k^{(i)})^{\mathrm{T}}(R_k^{(i)})^{-1} \tag{3.47}$$

4. The final estimates given by the fusion center are

$$\begin{aligned} \hat{x}_k &= \hat{x}_k^{(n)} \\ P_k &= P_k^{(n)} \end{aligned} \tag{3.48}$$

3.2.3 Data Compression Filtering

Data compression filtering firstly compresses and processes a number of measured data and then treats the processed data as virtual observations for filter measurement update. There are commonly three data compression methods, one of which will be introduced in this chapter. For other data compression algorithms, refer to the literature [14]. Figure 3.3 shows the implementation process of data compression filtering. During the measurement update, a virtual observation is first obtained by the filter through data compression and then updated. Here, the LMMSE estimate of state, which can be obtained through Eq. (3.11), is taken as virtual observation. The specific algorithm is as follows:

1. State prediction:

$$\hat{x}_{k|k-1} = F_{k,k-1}\hat{x}_{k-1} \tag{3.49}$$

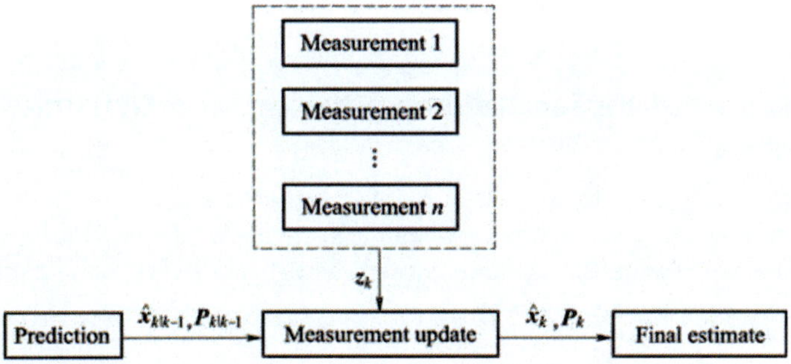

Fig. 3.3 Implementation process of data compression filtering

2. Error covariance prediction

$$P_{k|k-1} = F_{k,k-1}P_{k-1}F_{k,k-1}^{\mathrm{T}} + Q_{k-1} \tag{3.50}$$

3. The constructed virtual observation is

$$\begin{aligned} z_k^a &\triangleq \hat{x}^{\mathrm{LMMSE}} = \left(\sum_{i=1}^{n}(H_k^{(i)})^{\mathrm{T}}(R_k^{(i)})^{-1}H_k^{(i)}\right)^{-1}\left(\sum_{i=1}^{n}(H_k^{(i)})^{\mathrm{T}}(R_k^{(i)})^{-1}z_k^{(i)}\right) \\ R_k^a &\triangleq P_{\tilde{x}^{\mathrm{LMMSE}}} = \left(\sum_{i=1}^{n}(H_k^{(i)})^{\mathrm{T}}(R_k^{(i)})^{-1}H_k^{(i)}\right)^{-1} \\ H_k^a &= I \end{aligned} \tag{3.51}$$

4. Error covariance update:

$$P_k^{-1} = P_{k|k-1}^{-1} + \left(H_k^a\right)^{\mathrm{T}}\left(R_k^a\right)^{-1}H_k^a \tag{3.52}$$

5. Measurement update:

$$\hat{x}_k = \hat{x}_{k|k-1} + K_k^a(z_k^a - H_k^a\hat{x}_{k|k-1}) \tag{3.53}$$

where

$$K_k^a = P_k\left(H_k^a\right)^{\mathrm{T}}\left(R_k^a\right)^{-1} \tag{3.54}$$

Note

1. Data compression filtering first obtains the LMMSE estimate of state and then fuse with the predicted state. From this point of view, data compression filtering can also be considered as a distributed filtering.
2. Data compression filtering can be further expanded to the case of correlated measurement noises or the cases where only parts of states are compressed.

3.3 Distributed-Fusion Kalman Filtering for a Dynamic System

In this section, we consider the same dynamic system with the previous section. For convenience, we rewrite the dynamic system here. The discrete state equation and measurement equation are given by

$$\begin{aligned} x_k &= F_{k,k-1}x_{k-1} + w_k \\ z_k^{(i)} &= H_k^{(i)}x_k + v_k^{(i)} \quad i = 1, 2, \cdots, n \end{aligned} \tag{3.55}$$

The process noise $\boldsymbol{w}_k$ and measurement noise $\boldsymbol{v}_k^{(i)}$ are zero-mean Gaussian noises with covariance $\boldsymbol{Q}_k$ and $\boldsymbol{R}_k^{(i)}$, respectively. In addition, the following assumption are also made:

$$\begin{aligned} \mathrm{E}\{\boldsymbol{w}_k(\boldsymbol{v}_k^{(j)})^{\mathrm{T}}\} &= \boldsymbol{0} \\ \mathrm{E}\{\boldsymbol{x}_0\} &= \hat{\boldsymbol{x}}_0 \\ \mathrm{var}\{\boldsymbol{x}_0\} &= \boldsymbol{P}_0 \\ \mathrm{cov}\{\boldsymbol{x}_0, \boldsymbol{w}_k\} &= \boldsymbol{0} \\ \mathrm{cov}\{\boldsymbol{x}_0, \boldsymbol{v}_k^{(i)}\} &= \boldsymbol{0} \\ \mathrm{E}\{\boldsymbol{v}_k^{(i)}(\boldsymbol{v}_k^{(j)})^{\mathrm{T}}\} &= \boldsymbol{0} \quad i \neq j \end{aligned} \tag{3.56}$$

For the distributed fusion of a dynamic system, local estimates and their error covariances are obtained first. The Kalman filtering algorithm of the ith local node is as follows:

1. State prediction

$$\hat{\boldsymbol{x}}_{k|k-1}^{(i)} = \boldsymbol{F}_{k,k-1}\hat{\boldsymbol{x}}_{k-1|}^{(i)} \tag{3.57}$$

2. Covariance prediction

$$\boldsymbol{P}_{k|k-1}^{(i)} = \boldsymbol{F}_{k,k-1}\boldsymbol{P}_{k-1}^{(i)}\boldsymbol{F}_{k,k-1}^{\mathrm{T}} + \boldsymbol{Q}_{k-1} \tag{3.58}$$

3. Covariance update

$$\left(\boldsymbol{P}_k^{(i)}\right)^{-1} = \left(\boldsymbol{P}_{k|k-1}^{(i)}\right)_k^{-1} + \left(\boldsymbol{H}_k^{(i)}\right)^{\mathrm{T}}\left(\boldsymbol{R}_k^{(i)}\right)^{-1}\boldsymbol{H}_k^{(i)} \tag{3.59}$$

4. Measurement update

$$\hat{\boldsymbol{x}}_k^{(i)} = \hat{\boldsymbol{x}}_{k|k-1}^{(i)} + \boldsymbol{K}_k^{(i)}(\boldsymbol{z}_k^{(i)} - \boldsymbol{H}_k^{(i)}\hat{\boldsymbol{x}}_{k|k-1}^{i}) \tag{3.60}$$

where

$$\boldsymbol{K}_k^{(i)} = \boldsymbol{P}_k^{(i)}(\boldsymbol{H}_k^{(i)})^{\mathrm{T}}(\boldsymbol{R}_k^{(i)})^{-1} \tag{3.61}$$

3.3.1 Standard Distributed Kalman Filtering

For the distributed Kalman filtering without feedback (global information is not fed back to local estimators), the fusion center can perform the optimal fusion for local estimates without calculating the cross-covariance between local nodes, provided that each sensor can provide its own updated and predicted estimates:

$$\begin{aligned}&\hat{\boldsymbol{x}}_{k|k-1}^{(i)}, \boldsymbol{P}_{k|k-1}^{(i)}\\&\hat{\boldsymbol{x}}_{k}^{(i)}, \boldsymbol{P}_{k}^{(i)}\end{aligned} \tag{3.62}$$

The measurement update process of local estimates can be written as (refer to Eqs. 3.8 and 3.9)

$$\begin{aligned}\left(\boldsymbol{P}_k^{(i)}\right)^{-1}\hat{\boldsymbol{x}}_k^{(i)} &= \left(\boldsymbol{P}_{k|k-1}^{(i)}\right)^{-1}\hat{\boldsymbol{x}}_{k|k-1}^{(i)} + (\boldsymbol{H}_k^{(i)})^{\mathrm{T}}(\boldsymbol{R}_k^{(i)})^{-1}\boldsymbol{z}_k^{(i)}\\ \left(\boldsymbol{P}_k^{(i)}\right)^{-1} &= \left(\boldsymbol{P}_{k|k-1}^{(i)}\right)^{-1} + (\boldsymbol{H}_k^{(i)})^{\mathrm{T}}(\boldsymbol{R}_k^{(i)})^{-1}(\boldsymbol{H}_k^{(i)})\end{aligned} \tag{3.63}$$

From the first equation of Eq. (3.63), the following equation can be derived as

$$(\boldsymbol{H}_k^{(i)})^{\mathrm{T}}(\boldsymbol{R}_k^{(i)})^{-1}\boldsymbol{z}_k^{(i)} = \left(\boldsymbol{P}_k^{(i)}\right)^{-1}\hat{\boldsymbol{x}}_k^{(i)} - \left(\boldsymbol{P}_{k|k-1}^{(i)}\right)^{-1}\hat{\boldsymbol{x}}_{k|k-1}^{(i)} \tag{3.64}$$

In addition, from Eqs. (3.39) and (3.42) of centralized Kalman filtering, the following equation can be derived as

$$\begin{aligned}\boldsymbol{P}_k^{-1}\hat{\boldsymbol{x}}_k &= \boldsymbol{P}_{k|k-1}^{-1}\hat{\boldsymbol{x}}_{k|k-1} + \sum_{i=1}^{n}(\boldsymbol{H}_k^{(i)})^{\mathrm{T}}(\boldsymbol{R}_k^{(i)})^{-1}\boldsymbol{z}_k^{(i)}\\ \boldsymbol{P}_k^{-1} &= \boldsymbol{P}_{k|k-1}^{-1} + \sum_{i=1}^{n}(\boldsymbol{H}_k^{(i)})^{\mathrm{T}}(\boldsymbol{R}_k^{(i)})^{-1}(\boldsymbol{H}_k^{(i)})\end{aligned} \tag{3.65}$$

Substituting Eq. (3.63) into Eq. (3.65) gives

$$\begin{aligned}\boldsymbol{P}_k^{-1}\hat{\boldsymbol{x}}_k &= \boldsymbol{P}_{k|k-1}^{-1}\hat{\boldsymbol{x}}_{k|k-1} + \sum_{i=1}^{n}\left[\left(\boldsymbol{P}_k^{(i)}\right)^{-1}\hat{\boldsymbol{x}}_k^{(i)} - \left(\boldsymbol{P}_{k|k-1}^{(i)}\right)^{-1}\hat{\boldsymbol{x}}_{k|k-1}^{(i)}\right]\\ \boldsymbol{P}_k^{-1} &= \boldsymbol{P}_{k|k-1}^{-1} + \sum_{i=1}^{n}\left[\left(\boldsymbol{P}_k^{(i)}\right)^{-1} - \left(\boldsymbol{P}_{k|k-1}^{(i)}\right)^{-1}\right]\end{aligned} \tag{3.66}$$

Thus, the Kalman filtering algorithm based on distributed fusion is obtained. For convenience, Table 3.1 summarizes the standard distributed Kalman filtering, whose implementation process is shown in Fig. 3.4.

Remark

1. The derivation of the standard distributed Kalman filtering algorithm without feedback is completely based on centralized filtering algorithm and therefore is also the global optimal.
2. By comparing Eq. (3.63) with (3.65), it is easy to find that the optimality of the standard distributed Kalman filter is based on the assumption of uncorrelated measurement noises between local filters so that local estimates can be decoupled.

Table 3.1 Standard distributed Kalman filtering algorithm without feedback

Estimation by local nodes	
State prediction	$\hat{\boldsymbol{x}}_{k\|k-1}^{(i)} = \boldsymbol{F}_{k,k-1}\hat{\boldsymbol{x}}_{k-1}^{(i)}$
Covariance prediction	$\boldsymbol{P}_{k\|k-1}^{(i)} = \boldsymbol{F}_{k,k-1}\boldsymbol{P}_{k-1}^{(i)}\boldsymbol{F}_{k,k-1}^{\mathrm{T}} + \boldsymbol{Q}_{k-1}$
Covariance update	$\left(\boldsymbol{P}_k^{(i)}\right)^{-1} = \left(\boldsymbol{P}_{k\|k-1}^{(i)}\right)_k^{-1} + \left(\boldsymbol{H}_k^{(i)}\right)^{\mathrm{T}}\left(\boldsymbol{R}_k^{(i)}\right)^{-1}\boldsymbol{H}_k^{(i)}$
Measurement update	$\hat{\boldsymbol{x}}_k^{(i)} = \hat{\boldsymbol{x}}_{k\|k-1}^{(i)} + \boldsymbol{K}_k^{(i)}(\boldsymbol{z}_k^{(i)} - \boldsymbol{H}_k^{(i)}\hat{\boldsymbol{x}}_{k\|k-1}^{i})$ $\boldsymbol{K}_k^{(i)} = \boldsymbol{P}_k^{(i)}(\boldsymbol{H}_k^{(i)})^{\mathrm{T}}(\boldsymbol{R}_k^{(i)})^{-1}$
Fusion by the fusion center	
Covariance fusion	$\boldsymbol{P}_k^{-1} = \boldsymbol{P}_{k\|k-1}^{-1} + \sum_{i=1}^{n}\left[\left(\boldsymbol{P}_k^{(i)}\right)^{-1} - \left(\boldsymbol{P}_{k\|k-1}^{(i)}\right)^{-1}\right]$
State fusion	$\boldsymbol{P}_k^{-1}\hat{\boldsymbol{x}}_k = \boldsymbol{P}_{k\|k-1}^{-1}\hat{\boldsymbol{x}}_{k\|k-1} + \sum_{i=1}^{n}\left[\left(\boldsymbol{P}_k^{(i)}\right)^{-1}\hat{\boldsymbol{x}}_k^{(i)} - \left(\boldsymbol{P}_{k\|k-1}^{(i)}\right)^{-1}\hat{\boldsymbol{x}}_{k\|k-1}^{(i)}\right]$

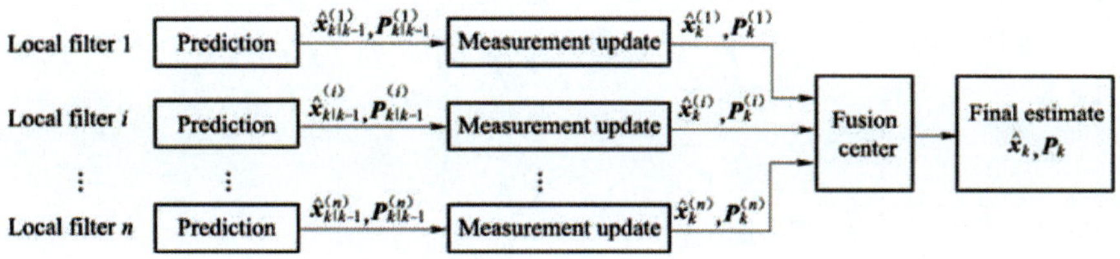

Fig. 3.4 Implementation process of the standard distributed Kalman filtering without feedback

In the case with feedback, the fusion center will feed the final estimate back to local filters, so

$$
\begin{aligned}
\breve{\boldsymbol{x}}_{k|k-1}^{(i)} &= \boldsymbol{F}_{k,k-1}\breve{\boldsymbol{x}}_{k-1|} = \breve{\boldsymbol{x}}_{k|k-1} \\
\breve{\boldsymbol{P}}_{k|k-1}^{(i)} &= \breve{\boldsymbol{P}}_{k|k-1}
\end{aligned} \tag{3.67}
$$

Here, $\breve{\bullet}$ represents a quantity estimate with feedback. Substituting Eq. (3.67) into Eq. (3.66) gives [15]

$$
\begin{aligned}
\breve{\boldsymbol{P}}_k^{-1}\breve{\boldsymbol{x}}_k &= \breve{\boldsymbol{P}}_{k|k-1}^{-1}\breve{\boldsymbol{x}}_{k|k-1} + \sum_{i=1}^{n}\left[\left(\breve{\boldsymbol{P}}_k^{(i)}\right)^{-1}\breve{\boldsymbol{x}}_k^{(i)} - \left(\breve{\boldsymbol{P}}_{k|k-1}\right)^{-1}\breve{\boldsymbol{x}}_{k|k-1}\right] \\
&= \sum_{i=1}^{n}\left(\breve{\boldsymbol{P}}_k^{(i)}\right)^{-1}\breve{\boldsymbol{x}}_k^{(i)} - (n-1)\breve{\boldsymbol{P}}_{k|k-1}^{-1}\breve{\boldsymbol{x}}_{k|k-1}
\end{aligned}
$$

Table 3.2 Standard distributed Kalman filtering algorithm with feedback

Prediction by the fusion center	
State prediction	$\breve{\boldsymbol{x}}_{k\|k-1} = \boldsymbol{F}_{k,k-1}\breve{\boldsymbol{x}}_{k-1}$
Covariance prediction	$\breve{\boldsymbol{P}}_{k\|k-1} = \boldsymbol{F}_{k,k-1}\breve{\boldsymbol{P}}_{k-1}\boldsymbol{F}^{\mathrm{T}}_{k,k-1} + \boldsymbol{Q}_{k-1}$
The fusion center feeds the state and variance back to local nodes	
	$\breve{\boldsymbol{x}}^{(i)}_{k\|k-1} = \breve{\boldsymbol{x}}_{k\|k-1}, \breve{\boldsymbol{P}}^{(i)}_{k\|k-1} = \breve{\boldsymbol{P}}_{k\|k-1}$
Measurement update by local nodes	
Covariance update	$\left(\breve{\boldsymbol{P}}^{(i)}_{k}\right)^{-1} = \left(\breve{\boldsymbol{P}}^{(i)}_{k\|k-1}\right)^{-1}_{k} + \left(\boldsymbol{H}^{(i)}_{k}\right)^{T}\left(\boldsymbol{R}^{(i)}_{k}\right)^{-1}\boldsymbol{H}^{(i)}_{k}$
Measurement update	$\breve{\boldsymbol{x}}^{(i)}_{k} = \breve{\boldsymbol{x}}^{(i)}_{k\|k-1} + \boldsymbol{K}^{(i)}_{k}(\boldsymbol{z}^{(i)}_{k} - \boldsymbol{H}^{(i)}_{k}\breve{\boldsymbol{x}}^{i}_{k\|k-1})$ $\breve{\boldsymbol{K}}^{(i)}_{k} = \breve{\boldsymbol{P}}^{(i)}_{k}(\boldsymbol{H}^{(i)}_{k})^{\mathrm{T}}(\boldsymbol{R}^{(i)}_{k})^{-1}$
Fusion by the fusion center	
Covariance fusion	$\breve{\boldsymbol{P}}^{-1}_{k} = \sum_{i=1}^{n}\left(\breve{\boldsymbol{P}}^{(i)}_{k}\right)^{-1} - (n-1)\breve{\boldsymbol{P}}^{-1}_{k\|k-1}$
State fusion	$\breve{\boldsymbol{P}}^{-1}_{k}\breve{\boldsymbol{x}}_{k} = \sum_{i=1}^{n}\left(\breve{\boldsymbol{P}}^{(i)}_{k}\right)^{-1}\breve{\boldsymbol{x}}^{(i)}_{k} - (n-1)\breve{\boldsymbol{P}}^{-1}_{k\|k-1}\breve{\boldsymbol{x}}_{k\|k-1}$

$$\begin{aligned}\breve{\boldsymbol{P}}^{-1}_{k} &= \breve{\boldsymbol{P}}^{-1}_{k|k-1} + \sum_{i=1}^{n}\left[\left(\breve{\boldsymbol{P}}^{(i)}_{k}\right)^{-1} - \left(\breve{\boldsymbol{P}}_{k|k-1}\right)^{-1}\right] \\ &= \sum_{i=1}^{n}\left(\breve{\boldsymbol{P}}^{(i)}_{k}\right)^{-1} - (n-1)\breve{\boldsymbol{P}}^{-1}_{k|k-1}\end{aligned} \quad (3.68)$$

The standard distributed Kalman filtering algorithm with feedback is summarized in Table 3.2, and its implementation process is shown in Fig. 3.5.

Remark It has been proved in [16] that the distributed-fusion filtering with feedback described by Eqs. (3.67) and (3.68) has the same performance as centralized-fusion filtering. The main advantage of feedback is the reduce of the error covariances of local estimates.

3.3.2 Covariance Intersection Algorithm

Compared with standard covariance intersection (CI) algorithm, the CI algorithm for a dynamic system should consider the prediction and measurement update of local filters. In addition, since the CI is a distributed-fusion algorithm, it is necessary to

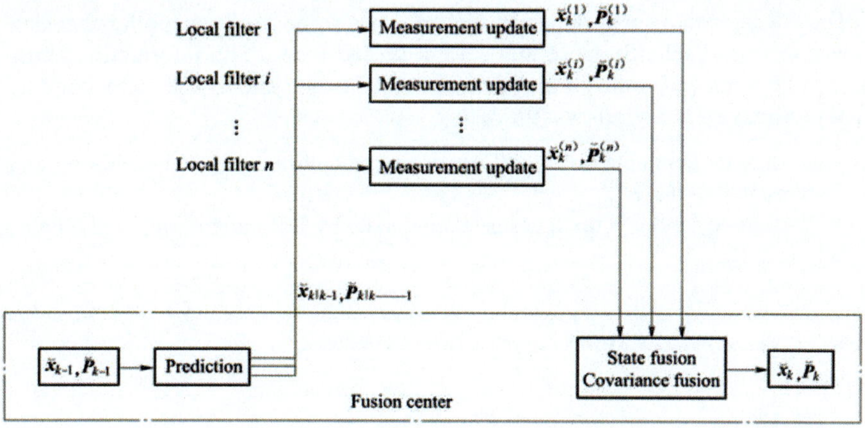

Fig. 3.5 Implementation process of the standard distributed Kalman filtering with feedback

consider whether to provide feedback. Figures 3.6 and 3.7, respectively, show the implementation processes of CI algorithms without feedback and with feedback for a dynamic system, whose specific algorithms are given in Tables 3.3 and 3.4.

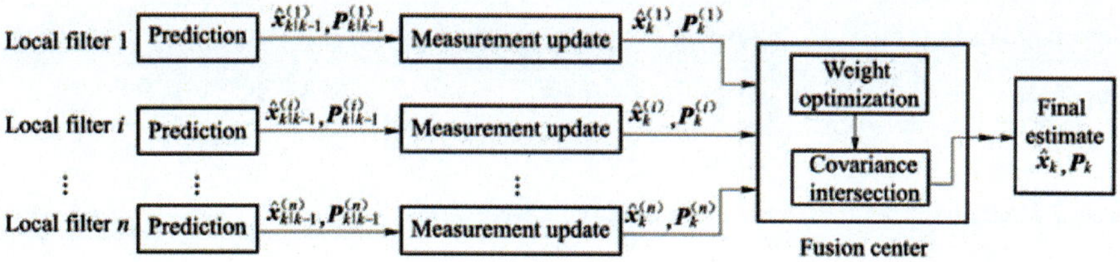

Fig. 3.6 Implementation processes of CI algorithm without feedback for a dynamic system

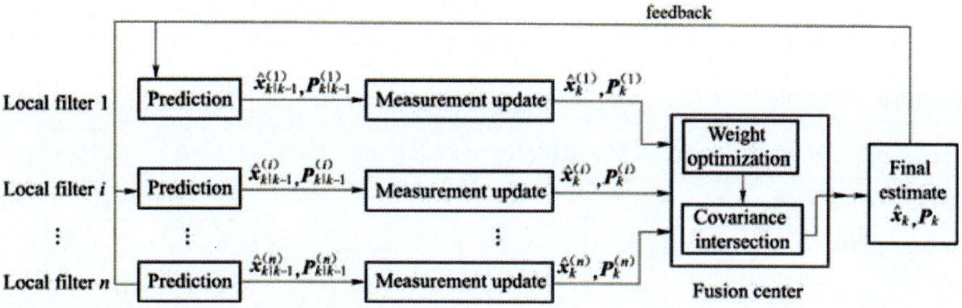

Fig. 3.7 Implementation processes of CI algorithm with feedback for a dynamic system

The CI algorithm for a dynamic system can also be used when global fusion nodes are unavailable. Suppose that several sensor nodes, with information transmission between each other, are contained in the fusion system. The corresponding CI algorithm will be as follows [9]:

1. The ith node conducts the state/Covariance prediction by Kalman filtering and obtains $\hat{\boldsymbol{x}}_{k|k-1}^{(i)}$ and $\boldsymbol{P}_{k|k-1}^{(i)}$.
2. The ith node conducts the measurement update by Kalman filtering and obtains $\hat{\boldsymbol{x}}_{k}^{(i)}$ and $\boldsymbol{P}_{k}^{(i)}$.

Table 3.3 CI algorithm for a dynamic system (without feedback)

Estimation by local nodes	
State prediction	$\hat{\boldsymbol{x}}_{k\vert k-1}^{(i)} = \boldsymbol{F}_{k,k-1}\hat{\boldsymbol{x}}_{k-1}^{(i)}$
Covariance prediction	$\boldsymbol{P}_{k\vert k-1}^{(i)} = \boldsymbol{F}_{k,k-1}\boldsymbol{P}_{k-1}^{(i)}\boldsymbol{F}_{k,k-1}^{\mathrm{T}} + \boldsymbol{Q}_{k-1}$
Covariance update	$\left(\boldsymbol{P}_{k}^{(i)}\right)^{-1} = \left(\boldsymbol{P}_{k\vert k-1}^{(i)}\right)_{k}^{-1} + \left(\boldsymbol{H}_{k}^{(i)}\right)^{\mathrm{T}}\left(\boldsymbol{R}_{k}^{(i)}\right)^{-1}\boldsymbol{H}_{k}^{(i)}$
Measurement update	$\hat{\boldsymbol{x}}_{k}^{(i)} = \hat{\boldsymbol{x}}_{k\vert k-1}^{(i)} + \boldsymbol{K}_{k}^{(i)}(\boldsymbol{z}_{k}^{(i)} - \boldsymbol{H}_{k}^{(i)}\hat{\boldsymbol{x}}_{k\vert k-1}^{i})$ $\boldsymbol{K}_{k}^{(i)} = \boldsymbol{P}_{k}^{(i)}(\boldsymbol{H}_{k}^{(i)})^{\mathrm{T}}(\boldsymbol{R}_{k}^{(i)})^{-1}$
Fusion by the fusion center	
Covariance fusion	$(\hat{\boldsymbol{P}}_{k})^{-1} = \omega_1(\hat{\boldsymbol{P}}_{k}^{(1)})^{-1} + \cdots + \omega_m(\hat{\boldsymbol{P}}_{k}^{(m)})^{-1}$
State fusion	$(\hat{\boldsymbol{P}}_{k})^{-1}\hat{\boldsymbol{x}}_{k} = \omega_1(\hat{\boldsymbol{P}}_{k}^{(1)})^{-1}\hat{\boldsymbol{x}}_{k}^{(1)} + \cdots + \omega_m(\hat{\boldsymbol{P}}_{k}^{(m)})^{-1}\hat{\boldsymbol{x}}_{k}^{(m)}$

Table 3.4 CI algorithm for a dynamic system (with feedback)

Estimation by local nodes	
State prediction	$\hat{\boldsymbol{x}}_{k\vert k-1}^{(i)} = \boldsymbol{F}_{k,k-1}\hat{\boldsymbol{x}}_{k-1}^{(i)}$
Covariance prediction	$\boldsymbol{P}_{k\vert k-1}^{(i)} = \boldsymbol{F}_{k,k-1}\boldsymbol{P}_{k-1}^{(i)}\boldsymbol{F}_{k,k-1}^{\mathrm{T}} + \boldsymbol{Q}_{k-1}$
Covariance update	$\left(\boldsymbol{P}_{k}^{(i)}\right)^{-1} = \left(\boldsymbol{P}_{k\vert k-1}^{(i)}\right)_{k}^{-1} + \left(\boldsymbol{H}_{k}^{(i)}\right)^{\mathrm{T}}\left(\boldsymbol{R}_{k}^{(i)}\right)^{-1}\boldsymbol{H}_{k}^{(i)}$
Measurement update	$\hat{\boldsymbol{x}}_{k}^{(i)} = \hat{\boldsymbol{x}}_{k\vert k-1}^{(i)} + \boldsymbol{K}_{k}^{(i)}(\boldsymbol{z}_{k}^{(i)} - \boldsymbol{H}_{k}^{(i)}\hat{\boldsymbol{x}}_{k\vert k-1}^{i})$ $\boldsymbol{K}_{k}^{(i)} = \boldsymbol{P}_{k}^{(i)}(\boldsymbol{H}_{k}^{(i)})^{\mathrm{T}}(\boldsymbol{R}_{k}^{(i)})^{-1}$
Fusion by the fusion center	
Covariance fusion	$(\hat{\boldsymbol{P}}_{k})^{-1} = \omega_1(\hat{\boldsymbol{P}}_{k}^{(1)})^{-1} + \cdots + \omega_m(\hat{\boldsymbol{P}}_{k}^{(m)})^{-1}$
State fusion	$(\hat{\boldsymbol{P}}_{k})^{-1}\hat{\boldsymbol{x}}_{k} = \omega_1(\hat{\boldsymbol{P}}_{k}^{(1)})^{-1}\hat{\boldsymbol{x}}_{k}^{(1)} + \cdots + \omega_m(\hat{\boldsymbol{P}}_{k}^{(m)})^{-1}\hat{\boldsymbol{x}}_{k}^{(m)}$
Feedback	
State feedback	$\hat{\boldsymbol{x}}_{k}^{(i)} = \hat{\boldsymbol{x}}_{k}$
Covariance feedback	$\hat{\boldsymbol{P}}_{k}^{(i)} = \hat{\boldsymbol{P}}_{k}$

3. The ith node broadcasts $\hat{\boldsymbol{x}}_k^{(i)}$ and $\boldsymbol{P}_k^{(i)}$ to other sensor nodes and receives the state/covariance information from them.
4. The ith node performs the CI fusion, so do other nodes.

3.3.3 Federated Filtering Algorithm

A federated filter is a two-stage filter composed of a main filter and several local filters. The output of the reference system is, on the one hand, directly channeled to the main filter, and on the other hand, to various local filters as measured values. The output of each subsystem is only channeled to the corresponding local filter. The estimate $\hat{\boldsymbol{x}}^{(i)}$ of each local filter (in common state, $i = 1, 2, \ldots N$) and its covariance matrix $\boldsymbol{P}^{(i)} \triangleq \boldsymbol{P}^{(ii)}$ are sent to the main filter and then fused with the state of main filter to obtain the global estimate. The federated filtering algorithm makes use of the technique of covariance upper-bound and the principle of information allocation.

3.3.3.1 Derivation of the Key Formula of Federated Filtering

The derivation follows from [17]. For the dynamic system described by Eqs. (3.28) and (3.29), suppose that there are n local filters, and the state of the ith local filter is $\boldsymbol{x}^{(i)}$, and

$$\boldsymbol{x}^{(i)} = \begin{bmatrix} \boldsymbol{x}^{(ci)} \\ \boldsymbol{x}^{(bi)} \end{bmatrix} \tag{3.71}$$

where $\boldsymbol{x}^{(ci)}$ is the estimate of common system state of local filters; $\boldsymbol{x}^{(bi)}$ is the unique estimate of the state to the ith local filter (such as sensor biases).

If the state of main filter is $\boldsymbol{x}^m$, the states of local filters and the state of main filter can be combined into

$$\boldsymbol{X} = \begin{bmatrix} \boldsymbol{x}^{(1)} \\ \vdots \\ \boldsymbol{x}^{(N)} \\ \boldsymbol{x}^m \end{bmatrix} \tag{3.72}$$

where $\hat{\boldsymbol{x}}^{(i)}$ is the estimate of $\boldsymbol{x}^{(i)}$, $\hat{\boldsymbol{x}}^m$ is the estimate of $\boldsymbol{x}^m$, $\hat{\boldsymbol{X}}$ is the estimate of $\boldsymbol{X}$, and

$$\mathcal{P} = \begin{bmatrix} \boldsymbol{P}^{(11)} & \ldots & \boldsymbol{P}^{(1N)} & \boldsymbol{P}^{(1)m} \\ \vdots & \ddots & \vdots & \vdots \\ \boldsymbol{P}^{(N1)} & \ldots & \boldsymbol{P}^{(NN)} & \boldsymbol{P}^{(N)m} \\ \boldsymbol{P}^{m(1)} & \ldots & \boldsymbol{P}^{m(N)} & \boldsymbol{P}^{mm} \end{bmatrix} \tag{3.73}$$

The fusion framework based on Kalman filtering mainly includes three steps: prediction, measurement update, and fusion.

1. Prediction

The state prediction equation of $\boldsymbol{X}$ can be written as

$$\begin{bmatrix} \boldsymbol{x}_{k|k-1}^{(1)} \\ \vdots \\ \boldsymbol{x}_{k|k-1}^{(N)} \\ \boldsymbol{x}_{k|k-1}^{m} \end{bmatrix} = \begin{bmatrix} \boldsymbol{F}_{k,k-1} & & & \\ & \ddots & & \\ & & \ddots & \\ & & & \boldsymbol{F}_{k,k-1} \end{bmatrix} \begin{bmatrix} \boldsymbol{x}_{k-1}^{(1)} \\ \vdots \\ \boldsymbol{x}_{k-1}^{(N)} \\ \boldsymbol{x}_{k-1}^{m} \end{bmatrix} + \begin{bmatrix} \boldsymbol{I} \\ \vdots \\ \vdots \\ \boldsymbol{I} \end{bmatrix} \boldsymbol{w}_k \tag{3.74}$$

The corresponding covariance prediction can be written as

$$\begin{aligned} \begin{bmatrix} \boldsymbol{P}_{k|k-1}^{(11)} & \cdots & \boldsymbol{P}_{k|k-1}^{(1N)} & \boldsymbol{P}_{k|k-1}^{(1)m} \\ \vdots & \ddots & \vdots & \vdots \\ \boldsymbol{P}_{k|k-1}^{(N1)} & \cdots & \boldsymbol{P}_{k|k-1}^{(NN)} & \boldsymbol{P}_{k|k-1}^{(N)m} \\ \boldsymbol{P}_{k|k-1}^{m(1)} & \cdots & \boldsymbol{P}_{k|k-1}^{m(N)} & \boldsymbol{P}_{k|k-1}^{mm} \end{bmatrix} &= \begin{bmatrix} \boldsymbol{F}_{k,k-1} & & & \\ & \ddots & & \\ & & \ddots & \\ & & & \boldsymbol{F}_{k,k-1} \end{bmatrix} \\ &\quad \begin{bmatrix} \boldsymbol{P}_{k-1}^{(11)} & \cdots & \boldsymbol{P}_{k-1}^{(1N)} & \boldsymbol{P}_{k-1}^{(1)m} \\ \vdots & \ddots & \vdots & \vdots \\ \boldsymbol{P}_{k-1}^{(N1)} & \cdots & \boldsymbol{P}_{k-1}^{(NN)} & \boldsymbol{P}_{k-1}^{(N)m} \\ \boldsymbol{P}_{k-1}^{m(1)} & \cdots & \boldsymbol{P}_{k-1}^{m(N)} & \boldsymbol{P}_{k-1}^{mm} \end{bmatrix} \begin{bmatrix} \boldsymbol{F}_{k,k-1}^{\mathrm{T}} & & & \\ & \ddots & & \\ & & \ddots & \\ & & & \boldsymbol{F}_{k,k-1}^{\mathrm{T}} \end{bmatrix} \\ &\quad + \begin{bmatrix} \boldsymbol{Q}_{k-1} & \cdots & & \boldsymbol{Q}_{k-1} \\ \vdots & \ddots & & \vdots \\ \vdots & & \ddots & \vdots \\ \boldsymbol{Q}_{k-1} & \cdots & \cdots & \boldsymbol{Q}_{k-1} \end{bmatrix} \end{aligned} \tag{3.75}$$

It can be seen from Eq. (3.75) that $\boldsymbol{P}_{k-1}^{(ij)} = 0, (i \neq j)$ does not implies that $\boldsymbol{P}_{k|k-1}^{(ij)} \neq 0$. From B-39 of Appendix B.7, the following inequality can be derived

$$\begin{bmatrix} \boldsymbol{Q}_{k-1} & \cdots & & \boldsymbol{Q}_{k-1} \\ \vdots & \ddots & & \vdots \\ \vdots & & \ddots & \vdots \\ \boldsymbol{Q}_{k-1} & \cdots & \cdots & \boldsymbol{Q}_{k-1} \end{bmatrix} \leq \begin{bmatrix} \gamma_1 \boldsymbol{Q}_{k-1} & & & \\ & \ddots & & \\ & & \gamma_N \boldsymbol{Q}_{k-1} & \\ & & & \gamma_m \boldsymbol{Q}_{k-1} \end{bmatrix} \tag{3.76}$$

where

$$\gamma_i = \beta_i^{-1}, i = 1, 2, \cdots, N, m$$

$$\begin{aligned} \beta_1 + \cdots + \beta_N + \beta_m = 1 \\ 0 \le \beta_i \le 1 \end{aligned} \tag{3.77}$$

Substituting Eq. (3.76) into Eq. (3.75), the following inequality can be obtained

$$\begin{bmatrix} \boldsymbol{P}_{k|k-1}^{(11)} & \cdots & \boldsymbol{P}_{k|k-1}^{(1N)} & \boldsymbol{P}_{k|k-1}^{(1)m} \\ \vdots & \ddots & \vdots & \vdots \\ \boldsymbol{P}_{k|k-1}^{(N1)} & \cdots & \boldsymbol{P}_{k|k-1}^{(NN)} & \boldsymbol{P}_{k|k-1}^{(N)m} \\ \boldsymbol{P}_{k|k-1}^{m(1)} & \cdots & \boldsymbol{P}_{k|k-1}^{m(N)} & \boldsymbol{P}_{k|k-1}^{mm} \end{bmatrix} \le \begin{bmatrix} \boldsymbol{F}_{k,k-1} & & \\ & \ddots & \\ & & \ddots & \\ & & & \boldsymbol{F}_{k,k-1} \end{bmatrix}$$

$$\begin{bmatrix} \boldsymbol{P}_{k-1}^{(11)} & \cdots & \boldsymbol{P}_{k-1}^{(1N)} & \boldsymbol{P}_{k-1}^{(1)m} \\ \vdots & \ddots & \vdots & \vdots \\ \boldsymbol{P}_{k-1}^{(N1)} & \cdots & \boldsymbol{P}_{k-1}^{(NN)} & \boldsymbol{P}_{k-1}^{(N)m} \\ \boldsymbol{P}_{k-1}^{m(1)} & \cdots & \boldsymbol{P}_{k-1}^{m(N)} & \boldsymbol{P}_{k-1}^{mm} \end{bmatrix} \begin{bmatrix} \boldsymbol{F}_{k,k-1}^{\mathrm{T}} & & \\ & \ddots & \\ & & \ddots & \\ & & & \boldsymbol{F}_{k,k-1}^{\mathrm{T}} \end{bmatrix}$$

$$+ \begin{bmatrix} \gamma_1 \boldsymbol{Q}_{k-1} & & & \\ & \ddots & & \\ & & \gamma_N \boldsymbol{Q}_{k-1} & \\ & & & \gamma_m \boldsymbol{Q}_{k-1} \end{bmatrix} \tag{3.78}$$

Consider the equality case for (3.78) and

$$\boldsymbol{P}_{k-1}^{(ij)} = 0, \ (i \ne j) \tag{3.79}$$

then

$$\begin{aligned} \boldsymbol{P}_{k|k-1}^{(ii)} &= \boldsymbol{F}_{k,k-1} \boldsymbol{P}_{k-1}^{(ii)} \boldsymbol{F}_{k,k-1}^{\mathrm{T}} + \gamma_i \boldsymbol{Q}_{k-1}, \quad i = 1, 2, \ldots, N \\ \boldsymbol{P}_{k|k-1}^{mm} &= \boldsymbol{F}_{k,k-1} \boldsymbol{P}_{k-1}^{mm} \boldsymbol{F}_{k,k-1}^{\mathrm{T}} + \gamma_m \boldsymbol{Q}_{k-1} \\ \boldsymbol{P}_{k|k-1}^{(ij)} &= \boldsymbol{F}_{k,k-1} \boldsymbol{P}_{k-1}^{(ij)} \boldsymbol{F}_{k,k-1}^{\mathrm{T}} = 0 \end{aligned} \tag{3.80}$$

Therefore, if the subfiltering estimate of the previous time is uncorrelated, the predicted value can also be made uncorrelated by using the matrix upper-bound inequality. This technique is called covariance upper-bound technique.

2. Measurement update:

The measurement equation Eq. (3.29) can be rewritten as

$$\boldsymbol{z} = \boldsymbol{H}_X \boldsymbol{X} + \boldsymbol{v} \tag{3.81}$$

where

$$\boldsymbol{z} \triangleq \begin{bmatrix} \boldsymbol{z}^{(1)} \\ \vdots \\ \boldsymbol{z}^{(n)} \end{bmatrix}, \quad \boldsymbol{v} \triangleq \begin{bmatrix} \boldsymbol{v}^{(1)} \\ \vdots \\ \boldsymbol{v}^{(n)} \end{bmatrix},$$

$$\boldsymbol{H}_X \triangleq \begin{bmatrix} \boldsymbol{H}^{(1)} & \cdots & 0 \\ \vdots & \ddots & \vdots \\ 0 & \cdots & \boldsymbol{H}^{(n)} \end{bmatrix} \tag{3.82}$$

Give the priori state estimate $\hat{\boldsymbol{X}}_{k|k-1}$ of $\boldsymbol{X}$ and its error covariance $\mathcal{P}_{k|k-1}$. Then, the measurement update of Kalman filtering can be written as

$$\begin{aligned} \hat{\boldsymbol{X}}_{k|k} &= \hat{\boldsymbol{X}}_{k|k-1} + \boldsymbol{K}_k(z_k - \boldsymbol{H}_{X,k}\hat{\boldsymbol{X}}_{k|k-1}) \\ \mathcal{P}_{k|k} &= (\boldsymbol{I} - \boldsymbol{K}_k\boldsymbol{H}_{X,k})\mathcal{P}_{k|k-1} \end{aligned} \tag{3.83}$$

where

$$\boldsymbol{K}_k = \mathcal{P}_{k|k-1}\boldsymbol{H}_{X,k}(\boldsymbol{H}_{X,k}\mathcal{P}_{k|k-1}\boldsymbol{H}_{X,k}^{\mathrm{T}} + \boldsymbol{R}_k)^{-1} \tag{3.84}$$

Note that the measurement noises are uncorrelated, i.e.,

$$\boldsymbol{R}_k = \mathrm{var}\{\boldsymbol{v}_k\} = \mathrm{diag}\Big[\boldsymbol{R}_k^{(1)}, \boldsymbol{R}_k^{(2)}, \cdots, \boldsymbol{R}_k^{(n)}\Big] \tag{3.85}$$

Then, Eq. (3.84) can be rewritten as

$$\begin{aligned} \boldsymbol{K} = &\begin{bmatrix} \boldsymbol{P}_{k|k-1}^{(11)} & \cdots & \boldsymbol{P}_{k|k-1}^{(1N)} \\ \vdots & \ddots & \vdots \\ \boldsymbol{P}_{k|k-1}^{(N1)} & \cdots & \boldsymbol{P}_{k|k-1}^{(NN)} \end{bmatrix} \begin{bmatrix} \boldsymbol{H}_k^{(1)} & \cdots & 0 \\ \vdots & \ddots & \vdots \\ 0 & \cdots & \boldsymbol{H}_k^{(N)} \end{bmatrix} \\ &\left(\begin{bmatrix} \boldsymbol{H}_k^{(1)}\boldsymbol{P}_{k|k-1}^{(11)}(\boldsymbol{H}_k^{(1)})^{\mathrm{T}} & \cdots & \boldsymbol{H}_k^{(1)}\boldsymbol{P}_{k|k-1}^{(1N)}(\boldsymbol{H}_k^{(N)})^{\mathrm{T}} \\ \vdots & \ddots & \vdots \\ \boldsymbol{H}_k^{(N)}\boldsymbol{P}_{k|k-1}^{(N1)}(\boldsymbol{H}_k^{(1)})^{\mathrm{T}} & \cdots & \boldsymbol{H}_k^{(N)}\boldsymbol{P}_{k|k-1}^{(NN)}(\boldsymbol{H}_k^{(N)})^{\mathrm{T}} \end{bmatrix} + \begin{bmatrix} \boldsymbol{R}_k^{(1)} & \cdots & 0 \\ \vdots & \ddots & \vdots \\ 0 & \cdots & \boldsymbol{R}_k^{(N)} \end{bmatrix} \right)^{-1} \end{aligned} \tag{3.86}$$

If

$$\boldsymbol{P}_{k|k-1}^{(ij)} = 0 \tag{3.87}$$

, then

$$
\boldsymbol{K} = \begin{bmatrix} \boldsymbol{P}_{k|k-1}^{(11)} & \cdots & 0 \\ \vdots & \ddots & \vdots \\ 0 & \cdots & \boldsymbol{P}_{k|k-1}^{(NN)} \end{bmatrix} \begin{bmatrix} \boldsymbol{H}_k^{(1)} & \cdots & 0 \\ \vdots & \ddots & \vdots \\ 0 & \cdots & \boldsymbol{H}_k^{(N)} \end{bmatrix}
$$

$$
+ \left(\begin{bmatrix} \boldsymbol{H}_k^{(1)} \boldsymbol{P}_{k|k-1}^{(11)} (\boldsymbol{H}_k^{(1)})^{\mathrm{T}} & \cdots & 0 \\ \vdots & \ddots & \vdots \\ 0 & \cdots & \boldsymbol{H}_k^{(N)} \boldsymbol{P}_{k|k-1}^{(NN)} (\boldsymbol{H}_k^{(N)})^{\mathrm{T}} \end{bmatrix} + \begin{bmatrix} \boldsymbol{R}_k^{(1)} & \cdots & 0 \\ \vdots & \ddots & \vdots \\ 0 & \cdots & \boldsymbol{R}_k^{(N)} \end{bmatrix} \right)^{-1} \tag{3.88}
$$

Substituting Eq. (3.88) into Eq. (3.83), the following equation can be obtained

$$
\begin{aligned}
\hat{\boldsymbol{x}}_{k|k}^{(i)} &= \hat{\boldsymbol{x}}_{k|k-1}^{(i)} + \boldsymbol{K}_k^{(i)} (\boldsymbol{z}_k^{(i)} - \boldsymbol{H}_k^{(i)} \hat{\boldsymbol{x}}_{k|k-1}^{(i)}) \\
\boldsymbol{P}_{k|k}^{(ii)} &= (\boldsymbol{I} - \boldsymbol{K}_k^{(i)} \boldsymbol{H}_k^{(i)}) \boldsymbol{P}_{k|k-1}^{(ii)} \\
\boldsymbol{P}_k^{(ij)} &= 0 \quad i \neq j
\end{aligned} \tag{3.89}
$$

where

$$
\boldsymbol{K}_k^{(i)} = \boldsymbol{P}_{k|k-1}^{(ii)} \boldsymbol{H}_k^{(i)} (\boldsymbol{H}_k^{(i)} \boldsymbol{P}_{k|k-1}^{(ii)} (\boldsymbol{H}_k^{(i)})^{\mathrm{T}} + \boldsymbol{R}_k^{(i)})^{-1} \tag{3.90}
$$

is derived from Eq. (3.88). It can be seen from the above derivation that if local estimates are uncorrelated before the measurement update, the measurement update of Kalman filtering can also be carried out independently.

3. Fusion

Through the measurement update by local filters and the prediction by main filter, we have

$$
\hat{\boldsymbol{X}}_k = \begin{bmatrix} \hat{\boldsymbol{x}}_k^{(1)} \\ \vdots \\ \hat{\boldsymbol{x}}_k^{(N)} \\ \hat{\boldsymbol{x}}_{k|k-1}^{m} \end{bmatrix} \tag{3.91}
$$

and

$$
\mathcal{P} = \begin{bmatrix} \boldsymbol{P}_k^{(11)} & \cdots & \cdots & 0 \\ \vdots & \ddots & \vdots & \vdots \\ \vdots & \cdots & \boldsymbol{P}_k^{(NN)} & \vdots \\ 0 & \cdots & \cdots & \boldsymbol{P}_{k|k-1}^{mm} \end{bmatrix} \tag{3.92}
$$

Therefore, from the LMMSE estimation algorithm, the global estimate, denoted with a subscript g, can be derived as

$$\begin{aligned} \boldsymbol{P}_{g,k} &= \left[\sum_{i=1}^{N} (\boldsymbol{P}_k^{(ii)})^{-1} + (\boldsymbol{P}_{k|k-1}^{mm})^{-1} \right]^{-1} \\ \hat{\boldsymbol{x}}_{g,k} &= \boldsymbol{P}_{k|k} \left[\sum_{i=1}^{N} (\boldsymbol{P}_k^{(ii)})^{-1} \hat{\boldsymbol{x}}_k^{(i)} + (\boldsymbol{P}_{k|k-1}^{mm})^{-1} \hat{\boldsymbol{x}}_{k|k-1}^{m} \right] \end{aligned} \tag{3.93}$$

3.3.3.2 Information Allocation Principle

A dynamic system contains two kinds of information, namely, state equations and measurement equations. Since measurements are independent from each other, no information allocation is needed in the measurement equations. But information allocation in the state equations needs to be considered, including the distribution of process noise covariance and initial state values. The information amount of state equations is inversely proportional to that of process noise variance and thus can be represented by the inverse of $\boldsymbol{Q}_g$. The information amount of initial state values can be represented by the inverse of the error covariance of the initial estimate.

By distributing the total information amount of process noise to local filters and main filter, the following equation can be obtained

$$\boldsymbol{Q}_g^{-1} = \sum_{i=1}^{N} (\boldsymbol{Q}^{(i)})^{-1} + (\boldsymbol{Q}^{\mathrm{m}})^{-1} \tag{3.94}$$

If $\boldsymbol{Q}^{(\mathrm{i})} = \beta_i^{-1} \boldsymbol{Q}_g \quad i = 1, \ldots, N, m$, then $\boldsymbol{Q}_g^{-1} = \sum_{i=1}^{N} \beta_i \boldsymbol{Q}_g^{-1} + \beta_m \boldsymbol{Q}_g^{-1}$, and therefore, $\sum_{i=1}^{N} \beta_i + \beta_m = 1$. Similarly, state estimate error covariance can also be allocated to realize the information conservation:

$$(\boldsymbol{P}^{(ii)})^{-1} = \beta_i \boldsymbol{P}_g^{-1} \tag{3.95}$$

3.3.3.3 Summary of the Federated Filtering

The key steps of federated filtering algorithm are deduced above. By considering the information allocation principle, the federated filtering can be described as follows:

1. Information allocation process.

Information allocation is to allocate the system information between local filters and main filter. The process noise covariance and error covariance of the system are allocated in the following principle

$$
\begin{aligned}
(\boldsymbol{P}_k^{(ii)})^{-1} &= \gamma_i \boldsymbol{P}_{g,k}^{-1} \\
\boldsymbol{Q}_k^{(i)} &= \gamma_i \boldsymbol{Q}_{g,k} \\
\hat{\boldsymbol{x}}_k^{(i)} &= \hat{\boldsymbol{x}}_{g,k}
\end{aligned} \tag{3.96}
$$

where $i = 1, 2, \ldots, N, m$, $\gamma_i = \beta_i^{-1}$, $0 \leq \beta_i \leq 1$, $\beta_1 + \beta_2 + \cdots + \beta_N + \beta_m = 1$

2. Prediction by local filters and main filter

$$
\begin{aligned}
\hat{\boldsymbol{x}}_{k|k-1}^{(i)} &= \boldsymbol{F}_{k,k-1} \hat{\boldsymbol{x}}_{k-1}^{(i)} \\
\boldsymbol{P}_{k|k-1}^{(ii)} &= \boldsymbol{F}_{k,k-1} \boldsymbol{P}_{k-1}^{(ii)} \boldsymbol{F}_{k,k-1}^{\mathrm{T}} + \gamma_i \boldsymbol{Q}_{g,k-1}, \quad i = 1, 2, \ldots, N \\
\hat{\boldsymbol{x}}_{k|k-1}^{m} &= \boldsymbol{F}_{k,k-1} \hat{\boldsymbol{x}}_{k-1}^{m} \\
\boldsymbol{P}_{k|k-1}^{mm} &= \boldsymbol{F}_{k,k-1} \boldsymbol{P}_{k-1}^{mm} \boldsymbol{F}_{k,k-1}^{\mathrm{T}} + \gamma_m \boldsymbol{Q}_{g,k-1}
\end{aligned} \tag{3.97}
$$

3. Measurement update by local filters

$$
\begin{aligned}
\hat{\boldsymbol{x}}_{k|k}^{(i)} &= \hat{\boldsymbol{x}}_{k|k-1}^{(i)} + \boldsymbol{K}_k^{(i)} (\boldsymbol{z}_k^{(i)} - \boldsymbol{H}_k^{(i)} \hat{\boldsymbol{x}}_{k|k-1}^{(i)}) \\
\boldsymbol{P}_{k|k}^{(ii)} &= (\boldsymbol{I} - \boldsymbol{K}_k^{(i)} \boldsymbol{H}_k^{(i)}) \boldsymbol{P}_{k|k-1}^{(ii)}
\end{aligned} \tag{3.98}
$$

where

$$
\boldsymbol{K}_k^{(i)} = \boldsymbol{P}_{k|k-1}^{(ii)} \boldsymbol{H}_k^{(i)} (\boldsymbol{H}_k^{(i)} \boldsymbol{P}_{k|k-1}^{(ii)} (\boldsymbol{H}_k^{(i)})^{\mathrm{T}} + \boldsymbol{R}_k^{(i)})^{-1} \tag{3.99}
$$

4. Information fusion

$$
\begin{aligned}
\boldsymbol{P}_{g,k} &= \left[\sum_{i=1}^{N} (\boldsymbol{P}_k^{(ii)})^{-1} + (\boldsymbol{P}_{k|k-1}^{mm})^{-1} \right]^{-1} \\
\hat{\boldsymbol{x}}_{g,k} &= \boldsymbol{P}_{g,k} \left[\sum_{i=1}^{N} (\boldsymbol{P}_k^{(ii)})^{-1} \hat{\boldsymbol{x}}_k^{(i)} + (\boldsymbol{P}_{k|k-1}^{mm})^{-1} \hat{\boldsymbol{x}}_{k|k-1}^{m} \right]
\end{aligned} \tag{3.100}
$$

3.4 Brief Summary

The design of fusion algorithm should be carried out under the fusion framework. This chapter first discusses linear fusion models and algorithms and presents a linear unified model. Under this model, the centralized, distributed, and hybrid architectures can be written in a unified form, and the fusion algorithm can be derived from LMMSE estimation algorithm. For unknown correlation between local estimates, the CI algorithm is presented. By applying the fusion algorithm to a dynamic system, the Kalman filtering based on centralized and distributed fusion is obtained for the system. For centralized filtering, several specific implementation methods are given. For the Kalman filtering based on distributed-fusion, standard distributed Kalman filtering is first presented to establish the connection between distributed and centralized filtering. Then, by considering the unknown correlation between local estimates, CI algorithm and federated filtering algorithm are presented. For the autonomous navigation system based on multi-source information fusion, the centralized and distributed filtering algorithms constitute its core filtering algorithms.

References

1. Li, X.R., et al. 2003. Optimal linear estimation fusion. I. Unified fusion rules. *IEEE Transactions on Information Theory* 49 (9): 2192–2208.
2. Zhou, H.Y., et al. 2013. *Fusion theory and methods for satellite state estimation*. Beijing: Science Press.
3. Bar-Shalom, Y., and L. Campo. 1986. The effect of the common process noise on the two-sensor fused-track covariance. *IEEE Transactions on Aerospace and Electronic Systems* 6: 803–805.
4. Deng, Z.L., et al. 2008. Self-tuning decoupled information fusion Wiener state component filters and their convergence. *Automatica* 44 (3): 685–695.
5. Sun, X.-J., et al. 2010. Multi-model information fusion Kalman filtering and white noise deconvolution. *Information Fusion* 11 (2): 163–173.
6. Julier, S.J. and J.K. Uhlmann. 1997. A non-divergent estimation algorithm in the presence of unknown correlations. In *American control conference, 1997. Proceedings of the 1997*. IEEE.
7. Uhlmann, J.K. 2003. Covariance consistency methods for fault-tolerant distributed data fusion. *Information Fusion* 4 (3): 201–215.
8. Uhlmann, J.K. 1996. General data fusion for estimates with unknown cross covariances. In *Proceedings of the SPIE aerosense conference*.
9. Julier, S. and J.K. 2009 Uhlmann, General decentralized data fusion with covariance intersection. Handbook of multisensor data fusion: theory and practice, 319–344.
10. Jazwinski, A.H. 2007. *Stochastic processes and filtering theory*. New York: Courier Corporation.
11. Yuan, Y., and W. Sun. 1997. *Optimization theory and methods*. Beijing: Science Press.
12. Willner, D., C. Chang, and K. Dunn. 1976. *Kalman filter algorithms for a multi-sensor system*. In *1976 IEEE conference on decision and control including the 15th symposium on adaptive processes*. IEEE.
13. Gan, Q., and C.J. Harris. 2001. Comparison of two measurement fusion methods for Kalman-filter-based multisensor data fusion. *IEEE Transactions on Aerospace and Electronic Systems* 37 (1): 273–279.
14. Han, C., H. Zhu, and Z. Duan. 2009. *Multi-source information fusion*. Beijing: Tsinghua University Press.
15. Chong, C.-Y. 1990. Distributed multitarget multisensor tracking, in Multitarget-multisensor tracking: advanced applications, 247–296. Artech House: Boston.

16. Zhu, Y., et al. 2001. The optimality for the distributed Kalman filtering fusion with feedback. *Automatica* 37 (9): 1489–1493.
17. Carlson, N.A. 1990. Federated square root filter for decentralized parallel processors. *IEEE Transactions on Aerospace and Electronic Systems* 26 (3): 517–525.

Chapter 4
Performance Analysis

The main analysis methods of system performance include observability analysis, Monte Carlo method, and linear covariance analysis. This chapter starts with the basic concept of observability to present the definition and observability criteria of linear and nonlinear systems, mainly introducing observability analysis methods of nonlinear systems. In the analysis of system observability, methods based on singular value decomposition, observability Gramian, and estimation error covariance are taken as examples for discussion. Then, the definition, implementation method, and general structure of Monte Carlo method are briefly introduced. Finally, the linear covariance technique is introduced, and the general method of establishing the linear propagation formula of error covariance is described for onboard autonomous navigation system.

4.1 Observability of Linear System

This section introduces basic concept of the observability of linear systems, defines the observability of linear time-invariant (LTI) and time-varying systems (LTV), and gives their observability criteria.

4.1.1 Observability Analysis of LTI Systems

4.1.1.1 Observability Definition of LTI Systems

The observability reflects the system's ability to determine its state according to the observed quantity over a finite time. It can be tested by rank condition of observability matrix. The observability matrix of LTI systems is simple and easy to obtain. For LTV systems, their observability can be analyzed by solving the observability matrix

D. Wang et al., *Spacecraft Autonomous Navigation Technologies Based on Multi-source Information Fusion*, Space Science and Technologies, https://doi.org/10.1007/978-981-15-4879-6_4

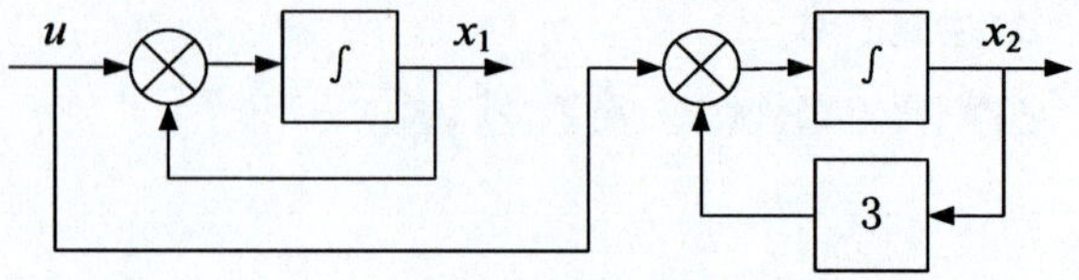

Fig. 4.1 State structure of LTI system

at different times. For nonlinear systems, their observability can be analyzed through singular value decomposition.

Consider an LTI system described by

$$\begin{aligned} \dot{\boldsymbol{x}}(t) &= \boldsymbol{A}\boldsymbol{x}(t) + \boldsymbol{B}\boldsymbol{u}(t) \\ z(t) &= \boldsymbol{C}\boldsymbol{x}(t) \end{aligned} \tag{4.1}$$

where $\boldsymbol{x}(t) \in \mathbb{R}^n$ is state vector, $\boldsymbol{u}(t) \in \mathbb{R}^r$ is input vector, $z(t) \in \mathbb{R}^m$ is output vector, and $\boldsymbol{A}$, $\boldsymbol{B}$, and $\boldsymbol{C}$ are state matrix, input matrix, and output matrix, respectively.

Definition 4.1 For the system (4.1), if the initial state vector $\boldsymbol{x}(t_0)$ can be determined uniquely given the output vector $z(t)$ over a finite time interval $[t_0, t_f]$ for any $\boldsymbol{u}(t)$, then the system state is completely observable (or observable for short).

For the LTI system, as shown in Fig. 4.1, since the state $\boldsymbol{x}_1$ is completely separated from the output, this system cannot be observed.

It should be pointed out that an LTI system is observable only if every state in the state space is observable. For an observable LTI system, once the initial state is determined according to the output vector $z(t)$, the instantaneous state at any time can be determined by using the state transition matrix according to the given control input vector $\boldsymbol{u}(t)$. If an LTI system is unobservable, some of its states may be observable and some may not. In this case, these states can be decomposed into completely observable and unobservable subspaces.

4.1.1.2 Criteria of the Observability of LTI Systems

The system observability can be tested by many criteria. This chapter mainly introduces the rank criterion.

For the LTI system (4.1), its observability matrix is defined as follows.

Definition 4.2 The observability matrix $\boldsymbol{N}$ for the system described by Eq. (4.1) is defined as

$$\boldsymbol{N} \triangleq \begin{bmatrix} \boldsymbol{C} \\ \boldsymbol{C}\boldsymbol{A} \\ \vdots \\ \boldsymbol{C}\boldsymbol{A}^{n-1} \end{bmatrix} \in \mathbb{R}^{mn \times n} \tag{4.2}$$

Given the definition of the observability matrix $\boldsymbol{N}$, the following theorem can be used to test the observability of the LTI system (4.1).

Theorem 4.1 *The LTI system* (4.1) *is observable if and only if* $\boldsymbol{N}$ *is full rank, i.e.,*

$$\text{rank}(\boldsymbol{N}) = n \tag{4.3}$$

In other words, n column vectors of $\boldsymbol{N}$ are linearly independent. If $\text{rank}(\boldsymbol{N}) < n$, the system is unobservable.

Proof If $\boldsymbol{u}(t) = 0$, then the solution for system (4.1) is

$$z(t) = \boldsymbol{C}\exp(\boldsymbol{A}t)\boldsymbol{x}(t_0) \tag{4.4}$$

According to the Cayley–Hamilton theorem [1],

$$\exp(\boldsymbol{A}t) = \sum_{k=0}^{n-1} a_k(t)\boldsymbol{A}^k \tag{4.5}$$

where $a_k(t)$ is a scalar function of time.

Substituting Eq. (4.5) into Eq. (4.4) gives

$$\begin{aligned} z(t) &= \boldsymbol{C}\sum_{k=0}^{n-1} a_k(t)\boldsymbol{A}^k\boldsymbol{x}(t_0) \\ &= [a_0(t)\boldsymbol{I}_m \; a_1(t)\boldsymbol{I}_m \; \cdots \; a_{n-1}(t)\boldsymbol{I}_m] \begin{bmatrix} \boldsymbol{C} \\ \boldsymbol{C}\boldsymbol{A} \\ \vdots \\ \boldsymbol{C}\boldsymbol{A}^{n-1} \end{bmatrix} \boldsymbol{x}(t_0) \end{aligned} \tag{4.6}$$

where $\boldsymbol{I}_m \in \mathbb{R}^m$ is an identity matrix.

Equation (4.6) is a group of m linear equations with n unknowns. If $m < n$, it does not have a unique solution. To determine the initial state vector $\boldsymbol{x}(t_0)$ uniquely, a system of n independent linear equations should be established at different times

$$\begin{bmatrix} z(t_1) \\ z(t_2) \\ \vdots \\ z(t_f) \end{bmatrix} = \begin{bmatrix} a_0(t_1)\boldsymbol{I}_m & a_1(t_1)\boldsymbol{I}_m & \cdots & a_{n-1}(t_1)\boldsymbol{I}_m \\ a_0(t_2)\boldsymbol{I}_m & a_1(t_2)\boldsymbol{I}_m & \cdots & a_{n-1}(t_2)\boldsymbol{I}_m \\ \vdots & \vdots & & \vdots \\ a_0(t_f)\boldsymbol{I}_m & a_1(t_f)\boldsymbol{I}_m & \cdots & a_{n-1}(t_f)\boldsymbol{I}_m \end{bmatrix} \begin{bmatrix} \boldsymbol{C} \\ \boldsymbol{C}\boldsymbol{A} \\ \vdots \\ \boldsymbol{C}\boldsymbol{A}^{n-1} \end{bmatrix} \boldsymbol{x}(t_0) \tag{4.7}$$

The above equation can be rewritten as

$$\bar{z} = \boldsymbol{Q}\boldsymbol{x}(t_0) \tag{4.8}$$

where

$$\bar{z} = \begin{bmatrix} z(t_1) \\ z(t_2) \\ \vdots \\ z(t_f) \end{bmatrix}$$

$$Q = \begin{bmatrix} a_0(t_1)\boldsymbol{I}_m & a_1(t_1)\boldsymbol{I}_m & \cdots & a_{n-1}(t_1)\boldsymbol{I}_m \\ a_0(t_2)\boldsymbol{I}_m & a_1(t_2)\boldsymbol{I}_m & \cdots & a_{n-1}(t_2)\boldsymbol{I}_m \\ \vdots & \vdots & & \vdots \\ a_0(t_f)\boldsymbol{I}_m & a_1(t_f)\boldsymbol{I}_m & \cdots & a_{n-1}(t_f)\boldsymbol{I}_m \end{bmatrix} \begin{bmatrix} \boldsymbol{C} \\ \boldsymbol{CA} \\ \vdots \\ \boldsymbol{CA}^{n-1} \end{bmatrix}$$

To make Eq. (4.8) solvable with a unique solution, the coefficient matrix $\boldsymbol{Q}$ and the augmented matrix $[\ \boldsymbol{Q}\ \bar{z}\ ]$ should satisfy

$$\text{rank}(\boldsymbol{Q}) = \text{rank}([\ \boldsymbol{Q}\ \bar{z}\ ]) = n \tag{4.9}$$

The rank of matrix $\boldsymbol{Q}$ is n if and only if the rank of the observability matrix $\boldsymbol{N}$ is n. Here, over the time interval of $[t_0, t_f]$, the unique solution $\boldsymbol{x}(t_0)$ of Eq. (4.8) can be determined according to the measured output vector $z(t)$, which means the sufficient and necessary condition for the observability of an LTI system is that the rank of the observability matrix $\boldsymbol{N}$ is n. This completes the proof.

4.1.2 Observability Analysis of LTV Systems

4.1.2.1 Definition of the Observability of LTV Systems

Consider the following LTV system

$$\begin{aligned} \dot{\boldsymbol{x}}(t) &= \boldsymbol{A}(t)\boldsymbol{x}(t) + \boldsymbol{B}(t)\boldsymbol{u}(t) \\ z(t) &= \boldsymbol{C}(t)\boldsymbol{x}(t) \end{aligned} \tag{4.10}$$

where $\boldsymbol{x}(t) \in \mathbb{R}^n$ is state vector, $\boldsymbol{u}(t) \in \mathbb{R}^r$ is input vector, $z(t) \in \mathbb{R}^m$ is output vector; $\boldsymbol{A}(t)$, $\boldsymbol{B}(t)$, and $\boldsymbol{C}(t)$ are time-varying state matrix, input matrix, and output matrix, respectively. For the definitions of other symbols, see 4.1.1.

Definition 4.3 The LTV system (4.10) is observable at t_0, if $\boldsymbol{x}(t_0)$ can be determined uniquely by input vector $\boldsymbol{u}(t)$ and output vector $z(t)$ over a finite time interval $[t_0, t_f]$.

Remark For the system (4.10), if $\boldsymbol{C}(t) = \boldsymbol{0}$, then $z(t) \equiv 0$ $(t > t_0)$ which implies that $\boldsymbol{x}(t_0)$ is unobservable at t_0.

4.1.2.2 Observability Criteria of LTV Systems

The observability criterion theorem for LTV systems [1] is as follows.

Theorem 4.2 *The system* (4.10) ***at*** t_0 *is observable if and only if the following observability Gramian is non-singular*:

$$\boldsymbol{W}_o(t_0, t_f) \triangleq \int_{t_0}^{t_f} \boldsymbol{F}^T(\tau, t_0)\boldsymbol{C}^T(\tau)\boldsymbol{C}(\tau)\boldsymbol{F}(\tau, t_0)\mathrm{d}\tau \tag{4.11}$$

where $\boldsymbol{F}(\tau, t_0)$ is the state transition matrix from t_0 to τ, which can be derived from the system state matrix $\boldsymbol{A}(t)$. Note that $\boldsymbol{W}_o(t_0, t_f)$ depends on $\boldsymbol{A}(t)$ and $\boldsymbol{C}(t)$. Therefore, observability is the inherent characteristic of an LTV system determined by the system structure and parameters, independent of the specific transient system behavior. To evaluate the system's observability by this criterion, $\boldsymbol{F}(\tau, t_0)$ must be firstly obtained with a large amount of calculation. If the transition matrix cannot be obtained in a closed-form, it would be difficult to use Theorem 4.2 to test the observability. In this case, the following corollary, derived from Theorem 4.2, is very useful to analyze the observability of LTV systems.

Corollary 4.1 *Assume that the state matrix $\boldsymbol{A}(t)$ and output matrix $\boldsymbol{C}(t)$ are $(n-1)$ times continuously differentiable. The LTV system* (4.10) *is observable at t_0 if a finite t_f exists ($t_f > t_0$) such that*

$$\mathrm{rank}\left(\begin{bmatrix} \boldsymbol{N}_0(t_f) \\ \boldsymbol{N}_1(t_f) \\ \vdots \\ \boldsymbol{N}_{n-1}(t_f) \end{bmatrix}\right) = n \tag{4.12}$$

where

$$\boldsymbol{N}_0(t) = \boldsymbol{C}(t)$$

$$\boldsymbol{N}_{k+1}(t) = \boldsymbol{N}_k(t)\boldsymbol{A}(t) + \frac{\mathrm{d}}{\mathrm{d}t}\boldsymbol{N}_k(t), \quad k = 0, 1, 2, \ldots, n-1$$

However, the LTV systems that do not meet this condition are not necessarily unobservable.

4.2 Observability of Nonlinear Systems

On the basis of the definition and criteria of the linear system observability explained in Sect. 4.1, this section will introduce the definition and criteria of the nonlinear system observability and describe observability analysis method based on singular value decomposition (SVD).

4.2.1 *Definition and Criteria of the Observability of Nonlinear Systems*

4.2.1.1 Definition of the Observability of Nonlinear Systems

To analyze the observability of a nonlinear system, the concept of local weak observability is introduced firstly. Suppose the nonlinear system can be expressed as

$$\begin{aligned} \dot{\boldsymbol{x}} &= \boldsymbol{f}(\boldsymbol{x}) \\ z &= \boldsymbol{h}(\boldsymbol{x}) \end{aligned} \tag{4.13}$$

where $\boldsymbol{x} \in \mathbb{R}^n$ is system state vector; $z \in \mathbb{R}^m$ is observation vector; $\boldsymbol{f}(\boldsymbol{x}) : \mathbb{R}^n \mapsto \mathbb{R}^n$ and $\boldsymbol{h}(\boldsymbol{x}) : \mathbb{R}^n \mapsto \mathbb{R}^m$ are the state function and observation function of the nonlinear system, respectively.

Definition 4.4 The system (4.13) is local weak observable at $\boldsymbol{x}_0$ if an open neighborhood $\boldsymbol{U}$ of $\boldsymbol{x}_0$ exists such that the nonlinear system is observable for any open neighborhood within U. Furthermore, the system is local weak observable if it is local weak observable for every $\boldsymbol{x}$.

4.2.1.2 Criteria of the Observability of Nonlinear Systems

The observability matrix for nonlinear system (4.13) is defined as

$$\boldsymbol{Q}(\boldsymbol{x}) = \begin{bmatrix} \mathrm{d}L_f^0 \boldsymbol{h}(\boldsymbol{x}) \\ \mathrm{d}L_f^1 \boldsymbol{h}(\boldsymbol{x}) \\ \vdots \\ \mathrm{d}L_f^{n-1} \boldsymbol{h}(\boldsymbol{x}) \end{bmatrix} \tag{4.14}$$

where n is the dimension number of the state vector $\boldsymbol{x}$, and $L_f^k \boldsymbol{h}(\boldsymbol{x})$ is the kth-order Lie derivative as defined below

$$L_f^0 \boldsymbol{h}(\boldsymbol{x}) = \boldsymbol{h}(\boldsymbol{x}) \tag{4.15}$$

$$L_f^k \boldsymbol{h}(\boldsymbol{x}) = \frac{\partial (L_f^{k-1} \boldsymbol{h})}{\partial \boldsymbol{x}} f(\boldsymbol{x}), \quad k = 1, 2, \ldots, n-1 \tag{4.16}$$

with

$$\mathrm{d}L_f^k \boldsymbol{h}(\boldsymbol{x}) = \frac{\partial(L_f^k \boldsymbol{h})}{\partial \boldsymbol{x}}, \quad k = 1, 2, \ldots, n-1 \tag{4.17}$$

If $\mathrm{rank}(\boldsymbol{Q}(\boldsymbol{x}_0)) = n$, then the nonlinear system satisfies the rank condition of observability at the point $\boldsymbol{x}_0$. If the observability rank condition is satisfied at each point $\boldsymbol{x}$ within the defined interval, the whole system is considered as satisfying the observability rank condition.

The criteria of the observability of nonlinear systems can be reduced to the following theorems [2].

Theorem 4.3 *If the observability rank condition is satisfied at the point $\boldsymbol{x}_0$, the nonlinear system is local weak observable at $\boldsymbol{x}_0$.*

Theorem 4.4 *If the observability rank condition is satisfied generically, the nonlinear system is local weak observable.*

Consider the following LTI system

$$\begin{aligned} \dot{\boldsymbol{x}} &= \boldsymbol{A}\boldsymbol{x} \\ \boldsymbol{z} &= \boldsymbol{H}\boldsymbol{x} \end{aligned} \tag{4.18}$$

which can be treated as a special form of nonlinear system. Then, according to Eq. (4.14), the system (4.18) is observable if

$$\mathrm{rank}(\boldsymbol{Q}) = n$$

where

$$\boldsymbol{Q} = \begin{bmatrix} \boldsymbol{H} \\ \boldsymbol{H}\boldsymbol{A} \\ \vdots \\ \boldsymbol{H}\boldsymbol{A}^{n-1} \end{bmatrix}$$

It is easy to verify that the observability criteria described in this section are consistent with those given in Sect. 4.1.1 for an LTI system.

4.2.1.3 Case Study for the Observability Analysis of Nonlinear Systems

Ignoring model and observation errors, the orbit dynamics model and observation model of a spacecraft navigation system without unknown model parameters can be expressed as

$$\begin{cases} \dot{\boldsymbol{r}} = \boldsymbol{v} \\ \dot{\boldsymbol{v}} = \boldsymbol{f}(\boldsymbol{r}, \boldsymbol{v}) \end{cases} \tag{4.19}$$

$$\boldsymbol{z} = \boldsymbol{h}(\boldsymbol{r}, \boldsymbol{v}) \tag{4.20}$$

where $\boldsymbol{r}$ and $\boldsymbol{v}$ are position and velocity vectors, respectively; $\boldsymbol{f}(\boldsymbol{r}, \boldsymbol{v})$ is the acceleration acting on the spacecraft; $\boldsymbol{z}$ is observation (measurement) vector; and $\boldsymbol{h}(\boldsymbol{r}, \boldsymbol{v})$ is observation function.

As navigation observations are generally used to correct the spacecraft orbit prediction containing certain errors in the actual autonomous navigation of a spacecraft, it is only necessary to determine whether the autonomous navigation system is observable within a certain neighborhood of the given orbit. Therefore, the above observability rank condition for nonlinear systems can be used to analyze the observability of autonomous navigation system.

In this section, we analyze the observability for an autonomous navigation system in three scenarios, using the rank condition criteria.

Scenario 1: The observation is the position vector and no unknown model parameters are required to be estimated.

In this case, the state vector is $\boldsymbol{x} \triangleq [\boldsymbol{r}^T, \boldsymbol{v}^T]^T \in \mathbb{R}^6$, and the observation equation is given by

$$\boldsymbol{h}(\boldsymbol{r}) = \boldsymbol{r} \tag{4.21}$$

Then,

$$L_f^0 \boldsymbol{h} = \boldsymbol{r}$$

$$\mathrm{d}L_f^0 \boldsymbol{h} = [\boldsymbol{I}_{3\times3}\ \boldsymbol{0}_{3\times3}]$$

$$L_f^1 \boldsymbol{h} = \boldsymbol{v}$$

$$\mathrm{d}L_f^1 \boldsymbol{h} = [\boldsymbol{0}_{3\times3}\ \boldsymbol{I}_{3\times3}]$$

Define $\boldsymbol{Q}_{01} \triangleq \begin{bmatrix} dL_f^0 \boldsymbol{h} \\ dL_f^1 \boldsymbol{h} \end{bmatrix}$, then $\boldsymbol{Q}_{01} = \begin{bmatrix} \boldsymbol{I}_{3\times3} & \boldsymbol{0}_{3\times3} \\ \boldsymbol{0}_{3\times3} & \boldsymbol{I}_{3\times3} \end{bmatrix}$. Since $\text{rank}(\boldsymbol{Q}_{01}) = 6$, implying that the autonomous navigation is local weak observable.

Scenario 2: The observation is the position vector and some unknown model parameters are required to be estimated.

One method to deal with the unknown perturbation acceleration of a spacecraft and the system error of its navigation sensor is to model them as unknown model parameters, which are then augmented as state variables and estimated together with the spacecraft position and velocity. To solve the problem of obtaining the estimates of unknown model parameters, it is necessary to analyze the observability of the autonomous navigation system for the augmented system, the dynamic equation of which is given by

$$\begin{cases} \dot{\boldsymbol{r}} = \boldsymbol{v} \\ \dot{\boldsymbol{v}} = \boldsymbol{f}(\boldsymbol{r}, \boldsymbol{v}, \boldsymbol{\theta}) \\ \dot{\boldsymbol{\theta}} = 0 \end{cases} \tag{4.22}$$

where $\boldsymbol{\theta} \in \mathbb{R}^m$ is the parameter to be estimated.

Define the state vector as $\boldsymbol{x} \triangleq [\boldsymbol{r}^{\mathrm{T}}, \boldsymbol{v}^{\mathrm{T}}, \boldsymbol{\theta}^{\mathrm{T}}]^{\mathrm{T}} \in \mathbb{R}^{m+6}$. Then,

$$L_f^0 \boldsymbol{h} = \boldsymbol{r}$$

$$\mathrm{d}L_f^0 \boldsymbol{h} = [\, \boldsymbol{I}_{3\times3} \; \boldsymbol{0}_{3\times3} \; \boldsymbol{0}_{3\times3} \,]$$

$$L_f^1 \boldsymbol{h} = \boldsymbol{v}$$

$$\mathrm{d}L_f^1 \boldsymbol{h} = [\, \boldsymbol{0}_{3\times3} \; \boldsymbol{I}_{3\times3} \; \boldsymbol{0}_{3\times3} \,]$$

$$L_f^2 \boldsymbol{h} = [\, \boldsymbol{0}_{1\times3} \; \boldsymbol{f}(\boldsymbol{r}, \boldsymbol{v}, \boldsymbol{\theta})^{\mathrm{T}} \; \boldsymbol{0}_{1\times3} \,]^{\mathrm{T}}$$

$$\mathrm{d}L_f^2 \boldsymbol{h} = \begin{bmatrix} \boldsymbol{0}_{3\times3} & \boldsymbol{0}_{3\times3} & \boldsymbol{0}_{3\times m} \\ \frac{\partial \boldsymbol{f}}{\partial \boldsymbol{r}} & \frac{\partial \boldsymbol{f}}{\partial \boldsymbol{v}} & \frac{\partial \boldsymbol{f}}{\partial \boldsymbol{\theta}} \\ \boldsymbol{0}_{3\times3} & \boldsymbol{0}_{3\times3} & \boldsymbol{0}_{3\times m} \end{bmatrix}$$

Define $\boldsymbol{Q}_{012} \triangleq \begin{bmatrix} \mathrm{d}L_f^0 \boldsymbol{h} \\ \mathrm{d}L_f^1 \boldsymbol{h} \\ \mathrm{d}L_f^2 \boldsymbol{h} \end{bmatrix}$, then $\mathrm{rank}(\boldsymbol{Q}_{012}) = \mathrm{rank}\left(\begin{bmatrix} \boldsymbol{I}_{3\times3} & \boldsymbol{0}_{3\times3} & \boldsymbol{0}_{3\times m} \\ \boldsymbol{0}_{3\times3} & \boldsymbol{I}_{3\times3} & \boldsymbol{0}_{3\times3} \\ \frac{\partial \boldsymbol{f}}{\partial \boldsymbol{r}} & \frac{\partial \boldsymbol{f}}{\partial \boldsymbol{v}} & \frac{\partial \boldsymbol{f}}{\partial \boldsymbol{\theta}} \end{bmatrix}\right)$.

If $m \le 3$ and $\mathrm{rank}\left(\frac{\partial \boldsymbol{f}}{\partial \boldsymbol{\theta}}\right) = m$, then $\mathrm{rank}(\boldsymbol{Q}_{012}) = 6 + m = n$, and $\mathrm{rank}(\boldsymbol{Q}) = n$, implying that the corresponding autonomous navigation system is local weak observable. If $m > 3$, it is difficult to analyze $\mathrm{rank}(\boldsymbol{Q})$ analytically, which is not be discussed here.

Scenario 3: The observations are the unit LOS vector and radial velocity.

In this scenario, the observation equation can be written as

$$\boldsymbol{h}(\boldsymbol{r}, \boldsymbol{v}) = \begin{bmatrix} \boldsymbol{r}_{i0} \\ v_r \end{bmatrix} \tag{4.23}$$

where $\boldsymbol{r}_{i0}$ and v_r are the unit LOS vector of the navigation target and the relative radial velocity of the spacecraft, respectively.

The observation model of unit LOS vector is

$$\boldsymbol{r}_{i0} = \boldsymbol{C}_n^s \frac{1}{\sqrt{(x_i - x)^2 + (y_i - y)^2 + (z_i - z)^2}} \begin{bmatrix} x_i - x \\ y_i - y \\ z_i - z \end{bmatrix} \tag{4.24}$$

where $\boldsymbol{C}_n^s$ is the attitude matrix from spacecraft navigation coordinate system to sensor coordinate system, and (x_i, y_i, z_i) is the position of navigation target (or a control point on its surface) in the navigation coordinate system, as observed at the observation point, $i = 1, 2, \ldots, N$. In general, the position of the targeted celestial body is known.

By using the Doppler drift generated by the relative motion between the light source (generally the sun) and the spacecraft, the radial velocity of the spacecraft relative to the measured celestial body can be calculated, and the corresponding observation model can be expressed as

$$v_r = \frac{\boldsymbol{r} \cdot \boldsymbol{v}}{\|\boldsymbol{r}\|} \tag{4.25}$$

where $\boldsymbol{r}$ and $\boldsymbol{v}$ are the position and velocity vectors of the spacecraft, respectively.

Denote $\boldsymbol{v} = [\,\mathrm{d}x \ \mathrm{d}y \ \mathrm{d}z\,]^T$, then

$$L_f^0 \boldsymbol{h} = \boldsymbol{h}(\boldsymbol{r}, \boldsymbol{v}) \tag{4.26}$$

$$\mathrm{d}L_f^0 \boldsymbol{h} = \begin{bmatrix} \frac{-1}{r_i^3} \begin{bmatrix} (y_i - y)^2 + (z_i - z)^2 & -(x_i - x)(y_i - y) & -(x_i - x)(z_i - z) \\ -(x_i - x)(y_i - y) & (x_i - x)^2 + (z_i - z)^2 & -(y_i - y)(z_i - z) \\ -(x_i - x)(z_i - z) & -(y_i - y)(z_i - z) & (x_i - x)^2 + (y_i - y)^2 \end{bmatrix} & \boldsymbol{0}_{3\times 3} \\ \frac{\mathrm{d}x}{r} - \frac{x(\boldsymbol{r}\cdot\boldsymbol{v})}{r^3} \quad \frac{\mathrm{d}y}{r} - \frac{y(\boldsymbol{r}\cdot\boldsymbol{v})}{r^3} \quad \frac{\mathrm{d}z}{r} - \frac{z(\boldsymbol{r}\cdot\boldsymbol{v})}{r^3} & \frac{\boldsymbol{r}^{\mathrm{T}}}{r} \end{bmatrix} \tag{4.27}$$

It is obvious that the expression of $\mathrm{d}L_f^1 \boldsymbol{h}$ is very complicated. Therefore, the observability matrix is too complex to determine the rank of the matrix analytically. To solve this problem, the observability of onboard autonomous navigation system can be analyzed numerically, which will be presented in the next section.

4.2.2 *Observability Analysis Based on Singular Value Decomposition*

In practice, the observability of onboard autonomous navigation system is often analyzed numerically. The observability of an autonomous navigation system depends on the number of measurements, measurement frequency, measurement geometry, and measurement precision. The observability analysis based on SVD is one of the methods based on numerical analysis.

Section 4.2.1 presents the observability matrix of nonlinear systems, as shown in Eq. (4.14). Performing the SVD of $\boldsymbol{Q}_{m\times n}$ at $\boldsymbol{x}_0$ gives

$$\boldsymbol{Q} = \boldsymbol{U}\boldsymbol{S}\boldsymbol{V}^T \tag{4.28}$$

where $\boldsymbol{S} = \begin{bmatrix} \boldsymbol{D} \\ \boldsymbol{O}_{(m-n)\times n} \end{bmatrix}$, $\boldsymbol{D}$ is the diagonal matrix composed of the singular values σ_i $(i = 1, 2, \ldots, n)$ of $\boldsymbol{Q}$, $\boldsymbol{U} \in \mathbb{R}^m$, and $\boldsymbol{V} \in \mathbb{R}^n$ are orthogonal matrices.

From the rank condition of the observability of a nonlinear system, the following conclusions can be drawn [3]: if $\text{rank}(\boldsymbol{D}) = n$, the system is local weak observable at $\boldsymbol{x}_0$; otherwise, the system is incompletely observable at $\boldsymbol{x}_0$.

4.3 Degree of Observability for Autonomous Navigation System

The analysis of the observability of autonomous navigation system can help test whether the navigation parameters could be determined by the selected observation data, but cannot evaluate the estimation accuracy of filtering algorithm. In fact, the degree of observability can be used to evaluate the estimation accuracy. Generally, the higher the degree of observability, the more accurate estimation result would be achieved. Tools used to measure degree of observability in common use include observability Gramian and estimation error covariance. The method based on observability Gramian is mainly used to analyze the degree of observability for different systems or the same system under different operating conditions, whereas the estimation error covariance-based methods can be used to analyze the estimation accuracy of different state variables for a navigation system.

4.3.1 Observability Gramian Based Method

Autonomous navigation system can usually be expressed as a nonlinear system given by

$$\begin{cases} \dot{\boldsymbol{x}} = \boldsymbol{f}(\boldsymbol{x}) + \boldsymbol{w}(t) \\ \boldsymbol{z} = \boldsymbol{h}(\boldsymbol{x}) + \boldsymbol{v}(t) \end{cases} \tag{4.29}$$

where $\boldsymbol{x}$ is state vector, including the position and velocity vector of the spacecraft; $\boldsymbol{z}$ is observation vector; $\boldsymbol{f}(\boldsymbol{x})$ and $\boldsymbol{h}(\boldsymbol{x})$ are the nonlinear functions of the autonomous navigation parameter; $\boldsymbol{w}(t)$ and $\boldsymbol{v}(t)$ are the system and the measurement noises, respectively.

Through the linearization and discretization near the designed nominal orbit, the system can be written as

$$\begin{cases} \Delta \boldsymbol{x}_k = \boldsymbol{F}_{k,k-1} \Delta \boldsymbol{x}_{k-1} + \boldsymbol{w}_{k-1} \\ \Delta z_k = \boldsymbol{H}_k \Delta \boldsymbol{x}_k + \boldsymbol{v}_k \end{cases} \tag{4.30}$$

where $\boldsymbol{F}_{k,k-1} = \boldsymbol{F}(t_k, t_{k-1})$, $\dot{\boldsymbol{F}}(t, t_{k-1}) = \boldsymbol{A}_k \boldsymbol{F}(t, t_{k-1})$, $\boldsymbol{F}(t_{k-1}, t_{k-1}) = \boldsymbol{I}$, $\boldsymbol{A}_k = \left.\frac{\partial \boldsymbol{f}(\boldsymbol{x})}{\partial \boldsymbol{x}}\right|_{\boldsymbol{x}=\boldsymbol{x}_k}$; $\boldsymbol{H}_k = \left.\frac{\partial \boldsymbol{h}(\boldsymbol{x})}{\partial \boldsymbol{x}}\right|_{\boldsymbol{x}=\boldsymbol{x}_k}$; $\boldsymbol{w}_k$; and $\boldsymbol{v}_k$ are the discrete-time system noise and measurement noise, respectively, that satisfy $E\{\boldsymbol{v}_k \boldsymbol{v}_j^T\} = \boldsymbol{R}_k \delta_{kj}$,$E\{\boldsymbol{w}_k \boldsymbol{w}_j^T\} = \boldsymbol{Q}_k \delta_{kj}$,$\delta_{kj} = \begin{cases} 1, \ k = j \\ 0, \ k \neq j \end{cases}$.

According to Eq. (4.30), the observability Gramian for the autonomous navigation system is defined as

$$\boldsymbol{W}_o(0, k) = \sum_{i=0}^{k} \boldsymbol{F}_i^{\mathrm{T}} \boldsymbol{H}_i^{\mathrm{T}} \boldsymbol{R}_i^{-1} \boldsymbol{H}_i \boldsymbol{F}_i \tag{4.31}$$

Its full rank can be used as a condition to judge whether the system is observable. It should be noted that the observability Gramian can analyze not only the observability of the system, but also its degree of observability [4]. The degree of observability is quantified by the following equation

$$\gamma(k) = \frac{n}{\mathrm{Tr}[\boldsymbol{W}_o^{-1}(0, k)]} \tag{4.32}$$

where n is the number of dimensions of the state vector.

For the LMMSE estimator, when the system is completely controllable, the following inequality is hold

$$\mathrm{Tr}[\boldsymbol{P}_k] \leq \frac{n}{\gamma(k)} + \mathrm{Tr}[\boldsymbol{W}_c(0, k)] \tag{4.33}$$

where $\boldsymbol{P}_k$ is the error covariance matrix of the state; $\boldsymbol{W}_c(0, k) = \sum\limits_{i=1}^{k} \boldsymbol{F}_{i,k} \boldsymbol{Q}_{i-1} \boldsymbol{F}_{i,k}^{\mathrm{T}}$. If the system is completely controllable, then $\boldsymbol{W}_c(0, k) > 0$.

According to Eq. (4.33), it is easy to see that a higher degree of observability, or a larger $\gamma(k)$, results in a smaller error covariance matrix, indicating a higher potential estimation accuracy. The degree of observability $\gamma(k)$ describes the dependence of system accuracy upon dynamical model, measurement model, and measurement accuracy.

It can be proved that the relationship between degree of observability and observation time is as follows:

$$\gamma(k+1) \geq \gamma(k) \tag{4.34}$$

That is, the more observation data the same autonomous navigation system uses, or the longer the observation time is, the higher the degree of observability of the system will be.

By analyzing the degree of observability for an autonomous navigation system, the relation between system accuracy and measurement type/frequency/accuracy can be investigated. In general, the more the observations are, the higher the degree of observability will be. However, considering the constraints of actual observation opportunities, the observation frequency is limited. Through the analysis of degree of observability, observation frequency can be chosen properly.

4.3.2 Error Covariance-Based Method

The degree of observability analysis method based on observability Gramian can describe the ability of the whole autonomous navigation system to determine navigation parameters, but cannot describe which states have a higher degree of observability (high theoretical estimation accuracy) and which states have a low degree of observability (low theoretical estimation accuracy) under a certain observation mode. To describe the estimation accuracy achieved by different states of the same system, this section presents a degree of observability analysis method based on the eigenvalues and eigenvectors of estimation error covariance matrix.

In the Kalman filtering algorithm, the convergence of filtering can be identified by observing each element in the estimation error covariance matrix $\boldsymbol{P}$. Similarly, the degree of observability of system can be evaluated based on the eigenvalues and eigenvectors of estimation error covariance matrix. Define the estimation error vector as

$$\tilde{\boldsymbol{x}} = \hat{\boldsymbol{x}} - \boldsymbol{x} \tag{4.35}$$

where $\boldsymbol{x}$ is system state vector, and $\hat{\boldsymbol{x}}$ is state vector estimate. Define the n components of $\tilde{\boldsymbol{x}}$ as $\tilde{x}_1, \tilde{x}_2, \ldots, \tilde{x}_n$, and consider their linear combination

$$w = v_1\tilde{x}_1 + v_2\tilde{x}_2 + \cdots + v_n\tilde{x}_n = \boldsymbol{v}^T\tilde{\boldsymbol{x}} \tag{4.36}$$

where $\boldsymbol{v} = [v_1, v_2, \ldots, v_n]^T$ and $\boldsymbol{v}^T\boldsymbol{v} = 1$. Because $\tilde{x}_1, \tilde{x}_2, \ldots, \tilde{x}_n$ are random variables, so is w. The variance of w is given by

$$\sigma_w^2 = E\{w^2\} = \boldsymbol{v}^T E\{\tilde{\boldsymbol{x}}\tilde{\boldsymbol{x}}^T\}\boldsymbol{v} = \boldsymbol{v}^T\boldsymbol{P}\boldsymbol{v} \tag{4.37}$$

where estimation error covariance matrix $\boldsymbol{P} = E\{\tilde{\boldsymbol{x}}\tilde{\boldsymbol{x}}^T\}$. The value of σ_w^2 directly reflects the degree of observability of the state.

Now consider the problem that determines $\boldsymbol{v}$ under the constraint condition $\boldsymbol{v}^T\boldsymbol{v} = 1$ to maximize σ_w^2. The necessary condition for the constrained optimization problem is given by

$$\frac{\partial}{\partial \boldsymbol{v}}\left[\sigma_w^2 - \lambda(\boldsymbol{v}^T\boldsymbol{v} - 1)\right] = \frac{\partial}{\partial \boldsymbol{v}}\left[\boldsymbol{v}^T\boldsymbol{P}\boldsymbol{v} - \lambda(\boldsymbol{v}^T\boldsymbol{v} - 1)\right] = 0 \tag{4.38}$$

where λ is the Lagrangian multiplier.

From Eq. (4.38), the following equation can be derived

$$(\boldsymbol{P} - \lambda\boldsymbol{I})\boldsymbol{v} = 0 \tag{4.39}$$

As can be seen from Eq. (4.39), λ is the eigenvalue of estimation error covariance matrix $\boldsymbol{P}$. Multiplying both sides of Eq. (4.39) by $\boldsymbol{v}^T$ gives

$$\boldsymbol{v}^T(\boldsymbol{P} - \lambda\boldsymbol{I})\boldsymbol{v} = \boldsymbol{v}^T\boldsymbol{P}\boldsymbol{v} - \boldsymbol{v}^T\lambda\boldsymbol{v} = 0 \tag{4.40}$$

Since $\boldsymbol{v}^T\boldsymbol{P}\boldsymbol{v} = \sigma_w^2$ and $\boldsymbol{v}^T\boldsymbol{v} = 1$, Eq. (4.40) becomes

$$\sigma_w^2 = \lambda \tag{4.41}$$

That is, the extremum of σ_w^2 corresponds to the eigenvalue λ of estimation error covariance matrix $\boldsymbol{P}$, and the linear combination of the relevant states corresponds to the eigenvector $\boldsymbol{v}$.

In general, a smaller σ_w^2 or λ indicates a better degree of observability of linear state combination. Therefore, in the degree of observability analysis method based on estimation error covariance, the eigenvalues and eigenvectors of estimation error covariance matrix $\boldsymbol{P}$ are used to evaluate the degree of observability of system.

If the eigenvalues of $\boldsymbol{P}$ are calculated directly, with no normalization, the resulting range of values can be almost anything. Not only is there no set bound for the eigenvalues, but dimensional homogeneity may not exist among various linear combinations of states in the system. This can cause confusion when making comparisons among the eigenvalues. To solve this problem, the normalization method of $\boldsymbol{P}$ was proposed [5]. The matrix $\boldsymbol{P}$ is normalized through congruent transformation (let $\boldsymbol{A}$ and $\boldsymbol{B}$ be two n-order matrices. If there is an n-order invertible matrix $\boldsymbol{C}$ that makes Equation $\boldsymbol{B} = \boldsymbol{C}^{\mathrm{T}}\boldsymbol{A}\boldsymbol{C}$ true, then it is called the congruent between matrix $\boldsymbol{B}$ and matrix $\boldsymbol{A}$). The dimensionless estimation error covariance matrix $\widetilde{\boldsymbol{P}}_k^+$ can be calculated as follows

$$\widetilde{\boldsymbol{P}}_k^+ = \left(\sqrt{\boldsymbol{P}_0^-}\right)^{-1}\boldsymbol{P}_k^+\left(\sqrt{\boldsymbol{P}_0^-}\right)^{-1} \tag{4.42}$$

where $\boldsymbol{P}_0^-$ is the estimation error covariance matrix of initial state; the superscript "-" means that before the measurement updating; $\boldsymbol{P}_k^+$ is the estimation error covariance matrix of current state; and the superscript " + " means that after the measurement updating. Next, by multiplying $n \big/ \mathrm{Tr}[\widetilde{\boldsymbol{P}}_k^+]$ by the matrix $\widetilde{\boldsymbol{P}}_k^+$, the final normalized matrix can be obtained

$$\boldsymbol{P}_k^{N+} = \frac{n}{\mathrm{Tr}[\widetilde{\boldsymbol{P}}_k^+]} \widetilde{\boldsymbol{P}}_k^+ \tag{4.43}$$

Since the sum of the eigenvalues of a matrix is equal to the trace of the matrix, the eigenvalues of $\boldsymbol{P}_k^{N+}$ are between 0 and n, and n is the number of dimensions of the state vector.

By analyzing the eigenvalues and eigenvectors of the estimation error covariance matrix $\boldsymbol{P}_k^{N+}$, the degree of observability of system state can be obtained. The smaller the eigenvalue of $\boldsymbol{P}_k^{N+}$ is, the smaller the estimation error of the linear state combination corresponding to the eigenvector will be, that is, the higher the degree of observability of the state combination will be.

The degree of observability analysis method based on estimation error covariance can be used to investigate the system observability corresponding to the given observation and the degree of observability of state vectors (such as position and velocity), and then to analyze how to improve the degree of observability of the system through the selection of state variables and combined measurement information, so as to improve the estimation accuracy of navigation filter and provide a reference for the design of integrated navigation system. In addition, in the research of onboard autonomous navigation systems, the autonomous navigation accuracy of spacecraft can be evaluated by using the estimation error covariance matrix output by the navigation filter.

4.4 Monte Carlo Method

Monte Carlo method is a general numerical technique widely used in engineering to analyze complex system behaviors, that is, to solve the problems through random sampling and random variable simulation [6]. The most basic form of Monte Carlo simulation is a heuristic solution of the mathematical model containing at least one random variable. The simulation consists of N tests, in which the random variable value of each test is different and the results of each test are saved. The result set is then statistically analyzed. The accuracy of statistical data depends on the number of tests. The calculation error is proportional to $\sqrt{D/N}$, where D is a constant and N is number of tests. This means that, to increase the statistical accuracy by 10 times, N must be increased by 100 times [7]. Depending on the complexity of the modeled system, it may take several hours or even several days to simulate the system in a high-performance computer before producing meaningful statistical results [8, 9].

The navigation error analysis method based on Monte Carlo simulation is the most commonly used method for analyzing the error of navigation system. This method obtains the statistical results of error covariance of navigation states through the statistical analysis of the results produced in hundreds of simulation tests, so as to acquire the error information of navigation system. Its general structure is shown in Fig. 4.2.

In Fig. 4.2, the three states represent real state $\boldsymbol{x}$, nominal state $\bar{\boldsymbol{x}}$, and estimated state $\tilde{\boldsymbol{x}}$, respectively. Their mutual relations are shown in Fig. 4.3, where $\Delta\boldsymbol{x} = \boldsymbol{x} - \bar{\boldsymbol{x}}$, $\Delta\tilde{\boldsymbol{x}} = \tilde{\boldsymbol{x}} - \bar{\boldsymbol{x}}$, and $\boldsymbol{e} = \boldsymbol{x} - \tilde{\boldsymbol{x}}$ represent actual state deviation, navigation state deviation, and navigation error, respectively. The nominal state is determined by the pre-computed nominal orbit. The actual state of a spacecraft is unknown in the actual detection mission, but can be set in simulation to simulate observations. The estimated state is the spacecraft state estimated through navigation filtering by using the observation information and the orbit dynamics model.

From the results of N Monte Carlo simulation tests, the formulas for calculating the covariances of real state deviation, navigation state deviation, and navigation error can be derived as follows

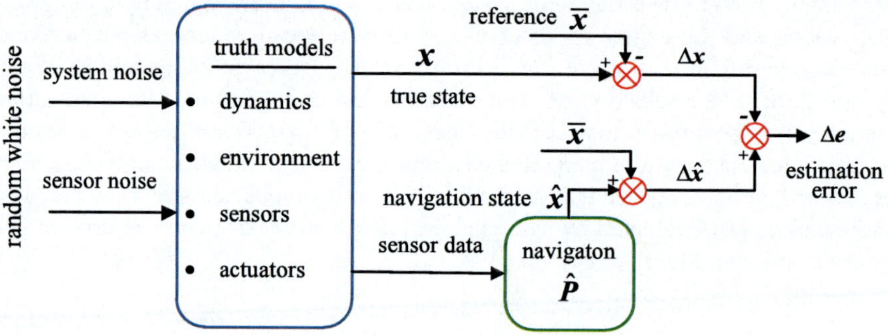

Fig. 4.2 Structure diagram Monte Carlo simulation [10]

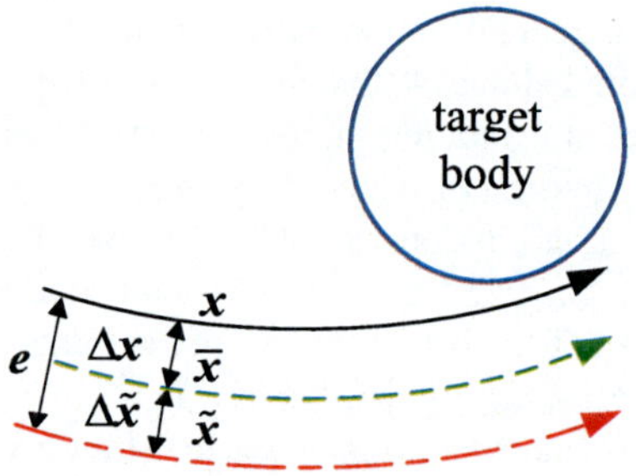

Fig. 4.3 Description of actual state deviation, navigation state deviation, and navigation error of onboard autonomous navigation system

$$D_{\text{true}} \approx \frac{1}{N-1}\sum_{i=1}^{N} \Delta x \Delta x^{\mathrm{T}}$$

$$D_{\text{nav}} \approx \frac{1}{N-1}\sum_{i=1}^{N} \Delta \tilde{x} \Delta \tilde{x}^{\mathrm{T}}$$

$$P_{\text{true}} \approx \frac{1}{N-1}\sum_{i=1}^{N} ee^{\mathrm{T}} \tag{4.44}$$

It can be seen from the above that the system error analysis based on Monte Carlo method is simple in principle and reliable in results, and free from the limitation by system type, but needs a complicated calculation.

4.5 Technique of Linear Covariance Analysis

Linear covariance analysis, sometimes referred to as covariance analysis, is a statistical analysis technique proposed by Maybeck [11] and Battin [12] separately. Linear covariance analysis is an effective method for the trade-off analysis of different designs. In the study of orbital rendezvous error and dispersion analysis problem, Geller improved the linear covariance analysis method as a general tool to analyze and design the onboard GNC system [10]. This tool was later used in the Altair mission for GNC system research, including the sensitivity analysis of inertial navigation and the trade-off analysis of relative terrain navigation [13].

By operating the linear covariance analysis method once, the statistical results similar to those of Monte Carlo simulation can be obtained, thus, greatly saving the simulation time. For the error analysis of autonomous navigation without orbit control, the influence of orbit control does not need to be considered. Therefore, the linear covariance analysis method proposed by Geller is simplified here by ignoring the orbit control.

The actual state model of navigation system as

$$\dot{x} = f(x, t) + w(t) \tag{4.45}$$

where $x(t) \in \mathbb{R}^n$, $w(t)$ is the Gaussian white noise with zero mean value, whose variance is $E\{w(t)w^T(t')\} = Q(t)\delta(t - t')$.

The observation obtained by non-inertial navigation sensor, such as navigation camera, at t_k is

$$z_k = h(x_k, t_k) + v_k \tag{4.46}$$

where $z_k \in \mathbb{R}^m$, v_k is the observation noise with zero mean value; $E\{v_k v_{k'}^T\} = R_v(t_k)\delta_{kk'}$.

The inertial sensors, including gyro and accelerometer, measurement equation is given by

$$\tilde{\boldsymbol{y}} = \boldsymbol{c}(\boldsymbol{x}, t) + \boldsymbol{\eta} \tag{4.47}$$

where $\boldsymbol{\eta}$ is the measurement noise with zero mean value; $E\left\{\boldsymbol{\eta}(t)\boldsymbol{\eta}^T(t')\right\} = \boldsymbol{S}_\eta(t)\delta(t-t')$.

Time update of system state and state error covariance applied in the autonomous navigation algorithm is

$$\dot{\tilde{\boldsymbol{x}}} = \tilde{\boldsymbol{f}}(\tilde{\boldsymbol{x}}, \tilde{\boldsymbol{y}}, t) \tag{4.48}$$

$$\dot{\tilde{\boldsymbol{P}}}(t) = [\tilde{\boldsymbol{A}}_{\tilde{x}} + \tilde{\boldsymbol{A}}_{\tilde{y}}\boldsymbol{C}_x]\tilde{\boldsymbol{P}}(t) + \tilde{\boldsymbol{P}}(t)[\tilde{\boldsymbol{A}}_{\tilde{x}} + \tilde{\boldsymbol{A}}_{\tilde{y}}\boldsymbol{C}_x]^T + \tilde{\boldsymbol{A}}_{\tilde{y}}\boldsymbol{S}_\eta\tilde{\boldsymbol{A}}_{\tilde{y}}^T + \boldsymbol{Q} \tag{4.49}$$

where $\tilde{\boldsymbol{A}}_{\tilde{x}}(t) = \frac{\partial \tilde{f}}{\partial \tilde{x}}$; $\tilde{\boldsymbol{A}}_{\tilde{y}} = \frac{\partial \tilde{f}}{\partial \tilde{y}}$; $\boldsymbol{C}_x = \frac{\partial C}{\partial x}$; $\boldsymbol{Q}$ is the variance matrix of model noise.

Measurement update of system state and state error covariance is

$$\tilde{\boldsymbol{x}}_k^+ = \tilde{\boldsymbol{x}}_k^- + \tilde{\boldsymbol{K}}(t_k)\left[z_k - \tilde{\boldsymbol{h}}(\tilde{\boldsymbol{x}}_k^-)\right] \tag{4.50}$$

$$\tilde{\boldsymbol{P}}(t_k^+) = \left[\boldsymbol{I}_n - \tilde{\boldsymbol{K}}(t_k)\tilde{\boldsymbol{H}}_{\tilde{x}}(t_k)\right]\tilde{\boldsymbol{P}}(t_k^-)\left[\boldsymbol{I}_n - \tilde{\boldsymbol{K}}(t_k)\tilde{\boldsymbol{H}}_{\tilde{x}}(t_k)\right]^T + \tilde{\boldsymbol{K}}(t_k)\boldsymbol{R}_v(t_k)\tilde{\boldsymbol{K}}^T(t_k) \tag{4.51}$$

where z_k is the observation of navigation sensor at t_k; $\tilde{\boldsymbol{h}}(\tilde{\boldsymbol{x}}_k^-)$ is the observation equation of autonomous navigation system; $\tilde{\boldsymbol{H}}_{\tilde{x}}(t_k) = \frac{\partial \tilde{h}}{\partial \tilde{x}}$; $\tilde{\boldsymbol{K}}(t_k)$ is Kalman filter gain

$$\tilde{\boldsymbol{K}}(t_k) = \tilde{\boldsymbol{P}}(t_k^-)\tilde{\boldsymbol{H}}_{\tilde{x}}^{\mathrm{T}}(t_k)\left[\tilde{\boldsymbol{H}}_{\tilde{x}}(t_k)\tilde{\boldsymbol{P}}(t_k^-)\tilde{\boldsymbol{H}}_{\tilde{x}}^{\mathrm{T}}(t_k) + \boldsymbol{R}_v(t_k)\right]^{-1} \tag{4.52}$$

Linearize the above time update Eqs. (4.45) and (4.48) along the pre-determined nominal orbit $\bar{\boldsymbol{x}}(t)$, ignore higher-order terms, and then obtain

$$\Delta\,\dot{\boldsymbol{x}} = \boldsymbol{A}_x(t)\Delta\boldsymbol{x} + \boldsymbol{w}(t) \tag{4.53}$$

$$\Delta\,\dot{\tilde{\boldsymbol{x}}} = \tilde{\boldsymbol{A}}_{\tilde{x}}\Delta\,\tilde{\boldsymbol{x}} + \tilde{\boldsymbol{A}}_{\tilde{y}}\boldsymbol{C}_x\Delta\,\boldsymbol{x} + \tilde{\boldsymbol{A}}_{\tilde{y}}\boldsymbol{\eta} \tag{4.54}$$

where $\boldsymbol{A}_x(t) = \left.\frac{\partial f}{\partial x}\right|_{\boldsymbol{x}=\bar{\boldsymbol{x}}(t)}$; $\tilde{\boldsymbol{A}}_{\tilde{x}}(t) = \left.\frac{\partial \tilde{f}}{\partial \tilde{x}}\right|_{\tilde{\boldsymbol{x}}=\bar{\boldsymbol{x}}(t)}$; $\tilde{\boldsymbol{A}}_{\tilde{y}} = \left.\frac{\partial \tilde{f}}{\partial \tilde{y}}\right|_{\tilde{\boldsymbol{y}}=\boldsymbol{c}(\bar{\boldsymbol{x}},t)}$.

The actual state deviation of the system $\Delta\boldsymbol{x}(t)$ is not affected by measurement update, so

$$\Delta\boldsymbol{x}_k^+ = \Delta\boldsymbol{x}_k^- \tag{4.55}$$

Linearizing the measurement update Eqs. (4.46) and (4.50) of the system and ignoring higher-order terms, assuming that $\tilde{\boldsymbol{h}}(\bar{\boldsymbol{x}}, t) = \boldsymbol{h}(\bar{\boldsymbol{x}}, t)$, gives

$$\Delta\tilde{\boldsymbol{x}}^{+} = \left[\boldsymbol{I}_n - \tilde{\boldsymbol{K}}(t_k)\tilde{\boldsymbol{H}}_{\tilde{x}}(t)\right]\Delta\tilde{\boldsymbol{x}}^{-} + \tilde{\boldsymbol{K}}(t_k)\boldsymbol{H}_x(t)\Delta\boldsymbol{x}_k^{-} + \tilde{\boldsymbol{K}}(t_k)\boldsymbol{v}(t) \tag{4.56}$$

Denoting an augmented state as $\boldsymbol{x} = \begin{bmatrix} \Delta\boldsymbol{x}(t) \\ \Delta\tilde{\boldsymbol{x}}(t) \end{bmatrix}$, Eqs. (4.53)–(4.56) can be written as

$$\begin{aligned} \dot{\boldsymbol{x}} &= \mathcal{A}\boldsymbol{x} + \boldsymbol{\Gamma}\boldsymbol{\eta} + \boldsymbol{G}\boldsymbol{w} \\ \boldsymbol{x}_k^{+} &= \boldsymbol{J}\boldsymbol{x}_k^{-} + \boldsymbol{B}\boldsymbol{v}_k \end{aligned} \tag{4.57}$$

where

$$\mathcal{A} = \begin{bmatrix} \boldsymbol{A}_x(t) & \boldsymbol{0}_{n\times n} \\ \tilde{\boldsymbol{A}}_{\tilde{y}}\boldsymbol{C}_x & \tilde{\boldsymbol{A}}_{\tilde{x}}(t) \end{bmatrix}$$

$$\boldsymbol{\Gamma} = \begin{bmatrix} \boldsymbol{0}_{n\times n_y} \\ \tilde{\boldsymbol{A}}_{\tilde{y}} \end{bmatrix} \quad \boldsymbol{G} = \begin{bmatrix} \boldsymbol{I}_n \\ \boldsymbol{0}_{n\times n} \end{bmatrix}$$

$$\boldsymbol{B} = \begin{bmatrix} \boldsymbol{0}_{n\times m} \\ \tilde{\boldsymbol{K}}(t_k) \end{bmatrix}$$

The covariance of the augmented state is denoted by $\boldsymbol{C} = E\{\boldsymbol{x}(t)\boldsymbol{x}(t)^T\}$. Its propagation and update equations are written as

$$\dot{\boldsymbol{C}} = \boldsymbol{J}\boldsymbol{C} + \boldsymbol{C}\boldsymbol{J}^T + \boldsymbol{\Gamma}\boldsymbol{S}_\eta\boldsymbol{\Gamma}^T + \boldsymbol{G}\boldsymbol{Q}\boldsymbol{G}^T \tag{4.58}$$

$$\boldsymbol{C}(t_k^{+}) = \boldsymbol{J}\boldsymbol{C}(t_k^{-})\boldsymbol{J}^T + \boldsymbol{B}\boldsymbol{R}_v(t_k)\boldsymbol{B}^T \tag{4.59}$$

Then, the covariances of actual state deviation, navigation state deviation, and navigation error can be calculated as

$$\begin{aligned} \boldsymbol{D}_{\text{true}} &= E\{\Delta\boldsymbol{x}(t)\Delta\boldsymbol{x}(t)^T\} = \left[\boldsymbol{I}_n \; 0_{n\times n}\right]\boldsymbol{C}\left[\boldsymbol{I}_n \; 0_{n\times n}\right]^T \\ \boldsymbol{D}_{\text{nav}} &= E\{\Delta\tilde{\boldsymbol{x}}(t)\Delta\tilde{\boldsymbol{x}}(t)^T\} = \left[0_{n\times n} \; \boldsymbol{I}_n\right]\boldsymbol{C}\left[0_{n\times n} \; \boldsymbol{I}_n\right]^T \\ \boldsymbol{P}_{\text{true}} &= E\{(\Delta\boldsymbol{x} - \Delta\tilde{\boldsymbol{x}})(\Delta\boldsymbol{x} - \Delta\tilde{\boldsymbol{x}})^T\} = \left[\boldsymbol{I}_n \; -\boldsymbol{I}_n\right]\boldsymbol{C}\left[\boldsymbol{I}_n \; -\boldsymbol{I}_n\right]^T \end{aligned} \tag{4.60}$$

To sum up, Eqs. (4.52), (4.58), and (4.59) are the covariance time/measurement update equations in the linear covariance analysis method. It can be seen that, through a single run of the linear covariance analysis method, navigation accuracy analysis results similar to those of Monte Carlo simulation can be obtained.

4.6 Brief Summary

This chapter presents methods of observability analysis, Monte Carlo method, and linear covariance analysis method. On the basis of the introduced observability analysis concept for linear systems, the observability and degree of observability analysis methods for nonlinear systems are specially highlighted. In terms of observability analysis, the observability rank condition for nonlinear systems and the observability analysis method based on singular value decomposition are given. In terms of degree of observability analysis, the degree of observability analysis methods based on observability Gramian, and estimation error covariance are presented. The Monte Carlo method evaluates the navigation system performance by establishing a mathematical simulation model with random quantity, conducting a large number of random sampling simulations, and collecting the probability statistics for simulation results. However, in the linear covariance analysis method, the linear time/measurement update equations of system error covariance are established. Therefore, a single run of this method can produce statistical results similar to those of Monte Carlo simulation, thus, greatly saving the simulation time.

References

1. Crassidis, John L., and John L. Junkins. 2012. *Optimal Estimation of Dynamic Systems*. Boca Raton: CRC Press, FL.
2. Hermann, R., and A.J. Krener. 1977. Nonlinear Controllability and Observability. *IEEE Transactions on Automatic Control* 22 (5): 728–740.
3. Xiangyu, Huang, Cui Pingyuan, and Cui Gutao. 2006. Observability Analysis of Deep Space Autonomous Navigation System. *Journal of Astronautics* 27 (3): 332–337.
4. Liu, Ping, Wang Dayi, Huang Xiangyu. 2007. Analysis of Degree of Observability in Autonomous Celestial Navigation System of Lunar Orbiter. Chinese Space Science and Technology, 27 (6): 12–18.
5. Ham, F.M., and T.G. Brown. 1983. Observability, Eigenvalues, and Kalman Filtering[J]. *IEEE Transactions on Aerospace and Electronic Systems* 19 (2): 269–273.
6. Hammersley, J.M., and D.C. Handscomb. 1964. *Monte Carlo Methods*. New York: Wiley.
7. Sobol, I.M. 1994. *A Primer for the Monte Carlo Method[M]*. Boca Raton: CRC Press, FL.
8. Rubinstein, R.Y. 1981. *Simulation and the Monte Carlo Method[M]*. New York: Wiley.
9. Spanos, P.D., and B.A. Zeldin. 1998. Monte Carlo Treatment of Random Fields: A Broad Perspective. *Applied Mechanics Review* 51 (3): 219–237.
10. Geller, D.K. 2006. Linear Covariance Techniques for Orbital Rendezvous Analysis and Autonomous Onboard Mission Planning. *Journal of Guidance, Control and Dynamics* 29 (6): 1404–1414.
11. Maybeck, P.S. 1979. Stochastic Models, Estimation and control, vol. 141 of Mathematics in Science and Engineering. Academic Press, New York
12. Battin, R.H. 1987. *An Introduction to the Mathematics and Methods of Astrodynamics*. Reston, VA: American Institute of Aeronautics and Astronautics.
13. Geller, D.K., and D.P. Christensen. 2012. Linear Covariance Analysis for Powered Lunar Descent and Landing. *Journal of Spacecraft & Rockets* 46 (6): 1231–1248.

Chapter 5
Time and Coordinate Systems

Time and coordinate systems are two basic elements for navigation system. For a navigation system, the position and velocity of a spacecraft should be described in a special time and coordinate system. In addition, the establishment of orbit dynamics model and the generation of navigation measurement data often involve the calculation of celestial ephemeris. This chapter presents the description of time system, definitions, and transformation of reference coordinate systems, and ephemeris calculation for navigation celestial bodies.

5.1 Time Systems

Time system is defined by the starting point of timing and the unit time interval. Since the spacecraft must measure its pointing direction and position relative to the earth, the sun, planets, stars, or asteroids, the use of astronomical time scale is needed. Interplanetary ephemeris information also involves the time systems. The time systems involved in this chapter will be detailed in this section.

5.1.1 Definition of Time System

Current time systems basically consist of Sidereal Time (ST), Universal Time (UT), Ephemeris Time (ET), International Atomic Time (TAI), and Terrestrial Dynamical Time (TDT) [1–3]. ST and UT are determined by the rotation of the Earth, ET by the motion of the earth, moon, and planets, and TAI by the electromagnetic oscillation of atoms.

1. Sidereal time

D. Wang et al., *Spacecraft Autonomous Navigation Technologies Based on Multi-source Information Fusion*, Space Science and Technologies, https://doi.org/10.1007/978-981-15-4879-6_5

Sidereal Time is measured by the position of the vernal equinox (see Sect. 5.2.1 for the definition of the vernal equinox) with respect to celestial meridian. The time interval between two successive upper meridians crossing of the vernal equinox is a sidereal day. A sidereal day is divided into 24 sidereal hours, each sidereal hour is divided into 60 sidereal minutes and each sidereal minute is divided into 60 sidereal seconds.

2. Universal time

Universal Time is a solar time standard reflecting the rotation of the earth. A solar day is the time period that the earth completes a full rotation related to the sun. UT has several different definitions: UT0, UT1, and UT2. UT0 is directly measured by astronomical observation, corresponding to the meridian of the instantaneous pole. UT1 is UT0 corrected for polar motion. UT1 is widely used in celestial navigation. UT2 is UT1 corrected for the seasonal variations of earth rotation.

3. Ephemeris time

Ephemeris Time (ET) is a uniform time scale under the framework of solar system barycenter and with the revolution of celestial bodies in the solar system as the reference. Since 1984, ET has been completely replaced by atomic time due to its low accuracy and slow update frequency.

4. International atomic time

The International System of Units (SI) defines a second as the time that the two ultra-fine energy levels in the ground state of a Cesium-133 atom at sea level to transition, radiate, and oscillate in a zero magnetic field for 9,192,631,770 cycles. The time system determined by this unit is called international atomic time. To meet the requirements for both of UT and TAI, the Coordinated Universal Time (UTC) was internationally specified as time standard.

5. Terrestrial dynamical time

In 1991, International Astronomical Union changed the original Terrestrial Dynamical Time (TDT) into Terrestrial Time (TT). TT is defined as differing from geocentric coordinate time (TCG), the time based on geocentric coordinate reference system, by one proportional constant. The unit of TT is the SI second on the geoid. In order to connect with the original ET, the 00:00:00 (TAI) on January 1, 1977 was read by TT as 0:00:32.184, January 1, 1977. TCG is the time based on geocentric coordinate reference system. The zero hour of TCG on January 1, 1977 is consistent with that of TT. The relationship between TCG and TT is

$$\mathrm{TCG} = \mathrm{TT} + L_G(\mathrm{JD} - 2443144.5) \times 86400\mathrm{s} \tag{5.1}$$

where JD is the Julian date at the time-converting moment, which will be defined in Sect. 5.1.2; $L_G = 6.969290134 \times 10^{-10}$ is the scale factor of conversion between

TCG and TT [3]. To describe the motion of solar system objects with the system barycenter as the origin under the framework of non-rotating relativity, the barycentric coordinate time (TCB) for solar system has been defined. TCB matched TCG at 0:00 (TAI) on January 1, 1977. The conversion formula between TCB and TCG is

$$\mathrm{TCB} = \mathrm{TCG} + L_C(\mathrm{JD} - 2443144.5) \times 86400\mathrm{s} + P \tag{5.2}$$

where

$$L_C = 1.48082686741 \times 10^{-8}$$

$$\begin{aligned} P \approx\ & 0.0016568\mathrm{s} \cdot \sin(35999.37°\mathrm{T} + 357.5°) \\ & + 0.0000224\mathrm{s} \cdot \sin(32964.5°\mathrm{T} + 246°) \\ & + 0.0000138\mathrm{s} \cdot \sin(71998.7°\mathrm{T} + 355°) \\ & + 0.0000048\mathrm{s} \cdot \sin(3034.9°\mathrm{T} + 25°) \\ & + 0.0000047\mathrm{s} \cdot \sin(34777.3°\mathrm{T} + 230°) \end{aligned}$$

$$T = (\mathrm{JD} - 2452545.0) / 36525$$

TCB supersedes a time scale known as barycentric dynamical time (TDB), which was introduced by the IAU in 1976 and defined to differ from TT by periodic terms, only. The conversion formula between TCB and TDB is

$$\mathrm{TDB} = \mathrm{TCB} - L_B(\mathrm{JD} - 2443144.5003725) \times 86400\mathrm{s} + \mathrm{TDB}_0 \tag{5.3}$$

where TDB_0 is chosen to provide consistency with the TDB–TT formula. Because of the same second length as TT, TDB is often used to analyze the small differences between TDB and TT and is frequently adopted by solar system dynamics.

In the orbit calculation, time is an independent variable. However, when calculating different physical quantities, different time systems are used. For example, UT is used for calculating the spacecraft's subsatellite point, ET is used for calculating the ephemeris of the sun, moon, planets, and asteroids, and UTC is used for sample time of measurements.

5.1.2 *Definition and Conversion of Julian Date*

In addition to the above time scales, Julian date (JD) is also commonly used for time system in the space activities.

A Julian year is defined as 365 mean solar days with one leap year (366 days) every four years, so the average length of Julian years is 365.25 mean solar days, and the length of the corresponding Julian century (100 years) is 36,525 mean solar days. Julian date, as an astronomical method of recording the days in a long time, is used to calculate the number of days between two dates that are several years apart. It dates back to Greenwich mean noon on January 1, 4713 BC. For example, the Julian date of 00:00 on February 1, 1992 is 2,448,653.5.

According to the new standard epoch (epoch is the marked time corresponding to the data in astronomical research) introduced in 1984, J2000.0 is January 1.5, 2000 TDB, whose Julian date is 2,451,545.0. The interval between the Julian year beginning of each year and the standard epoch is a multiple of Julian year 365.25. For example, the Julian year beginning in 1992 is on January 1.5, denoted as J1992.0; while the Julian year beginning in 1993 is on January 0.25, denoted as J1993.0.

In the spaceflights, Julian date is a very convenient way of time expression because it, just like year and day, doesn't need any complicated logic. However, to obtain a high-precision time expression, more digits are needed. 7 digits for the accuracy to day and 9 digits for the accuracy to millisecond. As Julian date is a large number with the first two digits unchanged in general applications and its starting point is noon (inconsistent with daily habits), it is often replaced by modified Julian date (MJD). MJD is defined as

$$\mathrm{MJD} = \mathrm{JD} - 2400000.5 \tag{5.4}$$

The conversion between Gregorian calendar date and Julian date is common in orbit calculation. The specific conversion methods are as follows:

1. Converting a Gregorian calendar date into a Julian date

Denote the year, month and day (including the decimal part of day) of a Gregorian calendar date as Y, M, D respectively. The corresponding Julian date will be

$$\begin{aligned} \mathrm{JD} = D - 32075 &+ \left[1461 \times \left(Y + 4800 + \left[\frac{M-14}{12}\right]\right) \div 4\right] \\ &+ \left[367 \times \left(M - 2 - \left[\frac{M-14}{12}\right] \times 12\right) \div 12\right] \\ &- \left[3 \times \left[\left(Y + 4900 + \left[\frac{M-14}{12}\right]\right) \div 100\right] \div 4\right] - 0.5 \end{aligned} \tag{5.5}$$

where $[X]$ is the integer part of X, whose decimal point is ignored.

2. Converting a Julian date into a Gregorian calendar date

Denote the Julian date for a given moment as JD (including the decimal part of day), and the year, month, and day of the corresponding Gregorian calendar date as Y, M, D (including the decimal part of day), respectively. Then,

$$\begin{cases} J = [\mathrm{JD} + 0.5] \\ N = \left[\frac{4(J+68569)}{146097}\right] \\ L_1 = J + 68569 - \left[\frac{N\times 146097+3}{4}\right] \\ Y_1 = \left[\frac{4000(L_1+1)}{1461001}\right] \\ L_2 = L_1 - \left[\frac{1461\times Y_1}{4}\right] + 31 \\ M_1 = \left[\frac{80\times L_2}{2447}\right] \\ L_3 = \left[\frac{M_1}{11}\right] \\ Y = 100(N-49) + Y_1 + L_3, \\ M = M_1 + 2 - 12L_3 \\ D = L_2 - \left[\frac{2447\times M_1}{80}\right] \end{cases} \quad (5.6)$$

5.2 Coordinate Frames

In navigation system, the description of attitude and orbit dynamics models should be based on an appropriate reference frame. This section gives the definitions of the reference frames commonly used in the research of autonomous spacecraft navigation and describes coordinate transformation between different frames.

5.2.1 Definition of Reference Coordinate System

To describe a reference coordinate system, the position of the coordinate origin, the orientation of reference plane (X-Y plane), the main direction (the direction of X-axis), and the direction of Z-axis must be specified. Since the Z-axis must be perpendicular to the reference plane, only its positive direction shall be indicated. The Y-axis direction completes the right-handed orthogonal set.

1. Barycentric celestial reference system ($O_{\mathrm{ssb}}X_{\mathrm{ssb}}Y_{\mathrm{ssb}}Z_{\mathrm{ssb}}$)

In the barycentric celestial reference system (BCRS), the solar system barycenter (SSB) is the origin, the reference plane is J2000.0 mean equatorial plane, the X-axis points to J2000.0 mean equinox, and the Z-axis is in the same direction as the angular velocity of earth rotation.

2. Heliocentric ecliptic coordinate system ($O_{\mathrm{s}}X_{\mathrm{si}}Y_{\mathrm{si}}Z_{\mathrm{si}}$)

In the heliocentric ecliptic coordinate system, sun center is the origin, and the reference plane is ecliptic plane (the orbit plane where the earth moves around the sun), as shown in Fig. 5.1. The intersection line between the ecliptic plane and the earth's equatorial plane determines the X-axis direction, which is called vernal equinox

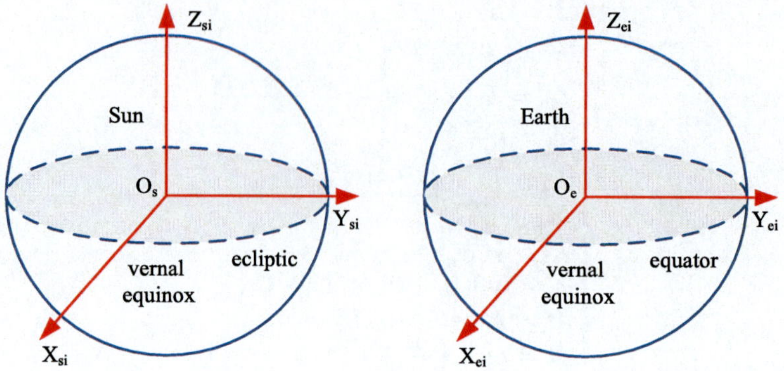

Fig. 5.1 Heliocentric ecliptic coordinate system and geocentric equatorial coordinate system

direction. The Z-axis is perpendicular to the ecliptic plane and is consistent with the angular velocity vector of earth revolution. Due to the slow drift of the intersection line between ecliptic plane and equatorial plane caused by the slow drift of the direction of the earth's spin axis, the heliocentric ecliptic coordinate system is not actually an inertial coordinate system. To establish an inertial coordinate system, it is necessary to point out according to the vernal equinox direction of which moment (epoch) the adopted coordinate system has been established. This book adopts the J2000.0 heliocentric ecliptic coordinate system, whose reference plane and main direction are J2000.0 mean ecliptic and mean equinox, respectively.

3. Geocentric coordinate systems

In geocentric coordinate systems, the origin is the earth's center.

(a) Geocentric equatorial coordinate system ($O_eX_{ei}Y_{ei}Z_{ei}$)
The reference plane of this frame is the earth equatorial plane. The X-axis points to the vernal equinox and the Z-axis points to the North Pole, as shown in Fig. 5.1. It should be noted that the geocentric equatorial coordinate system is not fixed on the earth and rotates with it. This book adopts the J2000.0 geocentric equatorial coordinate system, whose main direction is J2000.0 mean vernal equinox, and whose reference plane is mean equatorial plane.
(b) Geocentric ecliptic coordinate system ($O_eX_{si}Y_{si}Z_{si}$)
The reference plane of this frame is ecliptic plane. The X-axis points to vernal equinox and the Z-axis points to the North Pole. This coordinate system is formed by the translation of the origin of heliocentric ecliptic coordinate system to earth center.

4. Celestial coordinate systems

In celestial coordinate systems, the origin is the center of the targeted celestial body.

(a) Celestial inertial coordinate system ($O_t X_{ei} Y_{ei} Z_{ei}$)
The X-axis, Y-axis, and Z-axis are in the same direction as the X-axis, Y-axis, and Z-axis of J2000.0 geocentric equatorial coordinate system, respectively.
(b) Celestial fixed coordinate system ($O_t X_{tf} Y_{tf} Z_{tf}$)
The Z-axis is the rotation axis of the target, the X-axis points to a defined point in the equatorial plane of the target, and the Y-axis is selected to form a right-handed coordinate system.

5. Orbital coordinate system ($O_c X_o Y_o Z_o$)

The origin of an orbital coordinate system is nominal spacecraft barycenter or a special point (mainly because the actual spacecraft barycenter may change in the flight). The Z-axis points to the center of central celestial body along the direction of spacecraft flight; the X-axis is perpendicular to the Z-axis in the instantaneous orbital plane, and points to the velocity direction of the spacecraft; the Y-axis is parallel to the normal of instantaneous orbital plane, forming a right-handed coordinate system.

6. Spacecraft body coordinate system ($O_c X_b Y_b Z_b$)

The spacecraft body coordinate system takes nominal spacecraft barycenter or a special point on the spacecraft as the origin. The three axes are fixed with the body and usually point to the three principal axes of the spacecraft.

7. Sensor coordinate system ($O_{fc} X_{bs} Y_{bs} Z_{bs}$)

The origin of a sensor frame is a special point on the sensor, and the three axes are fixed with the sensor. For an optical imaging sensor frame, the focal plane center of optical imaging sensor is the origin. The Z-axis is its optical axis direction. The X-axis is the reference basis set in the focal plane. The Y-axis constitutes a right-handed coordinate system.

5.2.2 Coordinate Transformation

Coordinate transformation is sure to involve coordinate rotation. Therefore, the expression of coordinate transformation matrix corresponding to coordinate rotation should be defined first. If a vector is denoted by $\boldsymbol{r}$ in the original coordinate system and by $\boldsymbol{r}'$ in the new coordinate system after rotation, the rotation of Y-Z, Z-X, and X-Y planes about the X-axis, Y-axis, and Z-axis, respectively, by the angle θ (counterclockwise rotation is "+") is given by

$$\begin{aligned} \boldsymbol{r}' &= \boldsymbol{C}_{\mathrm{X}}(\theta)\boldsymbol{r} \\ \boldsymbol{r}' &= \boldsymbol{C}_{\mathrm{Y}}(\theta)\boldsymbol{r} \\ \boldsymbol{r}' &= \boldsymbol{C}_{\mathrm{Z}}(\theta)\boldsymbol{r} \end{aligned} \tag{5.7}$$

where

$$C_X(\theta)=\begin{bmatrix}1 & 0 & 0\\ 0 & \cos(\theta) & \sin(\theta)\\ 0 & -\sin(\theta) & \cos(\theta)\end{bmatrix}$$

$$C_Y(\theta)=\begin{bmatrix}\cos(\theta) & 0 & -\sin(\theta)\\ 0 & 1 & 0\\ \sin(\theta) & 0 & \cos(\theta)\end{bmatrix}$$

$$C_Z(\theta)=\begin{bmatrix}\cos(\theta) & \sin(\theta) & 0\\ -\sin(\theta) & \cos(\theta) & 0\\ 0 & 0 & 1\end{bmatrix}$$

and the coordinate transformation matrix $\boldsymbol{C}(\theta)$ has the following nature:

$$\boldsymbol{C}^{-1}(\theta)=\boldsymbol{C}^{T}(\theta)=\boldsymbol{C}(-\theta) \tag{5.8}$$

1. Heliocentric ecliptic coordinate system ⟶ geocentric ecliptic coordinate system ⟶ geocentric equatorial coordinate system

The transformation between epoch-specific heliocentric ecliptic coordinate system and epoch-specific geocentric equatorial coordinate system is composed of translation and rotation. The translation corresponds to a transitional epoch-specific geocentric ecliptic coordinate system. Denote the position vectors of epoch-specific geocentric equatorial coordinate system, epoch-specific geocentric ecliptic coordinate system, and epoch-specific heliocentric ecliptic coordinate system as $\boldsymbol{r}_{\text{ei}}$, $\boldsymbol{r}_{\text{e,si}}$, and $\boldsymbol{r}_{\text{si}}$, respectively. Then,

$$\begin{aligned}\boldsymbol{r}_{\text{e,si}}&=\boldsymbol{r}_{\text{si}}+\boldsymbol{r}_{\text{se,si}}\\ \boldsymbol{r}_{\text{ei}}&=\boldsymbol{C}_X(-\bar{\varepsilon})\boldsymbol{r}_{\text{e,si}}\end{aligned} \tag{5.9}$$

where $\boldsymbol{r}_{\text{se,si}}$ is the position vector of the sun (sun center) in the geocentric ecliptic coordinate system,

$$\begin{aligned}\bar{\varepsilon}&=\varepsilon-\Delta\varepsilon\\ \varepsilon&=23^\circ 26' 21''.448-46''.8150T-0''.00059T^2+0''.001813T^3\end{aligned} \tag{5.10}$$

where $\bar{\varepsilon}$ is mean ecliptic obliquity, $\Delta\varepsilon$ is small correction for the nutation in obliquity, $T=\frac{\text{JD}(t)-\text{JD(J2000.0)}}{36525.0}$ is the time in Julian centuries from the epoch J2000.0.

2. Celestial inertial coordinate system ⟶ celestial fixed coordinate system

As shown in Fig. 5.2, the orientation of celestial fixed coordinate system relative to Celestial inertial coordinate system can be described by three angles, namely φ, ψ, θ.

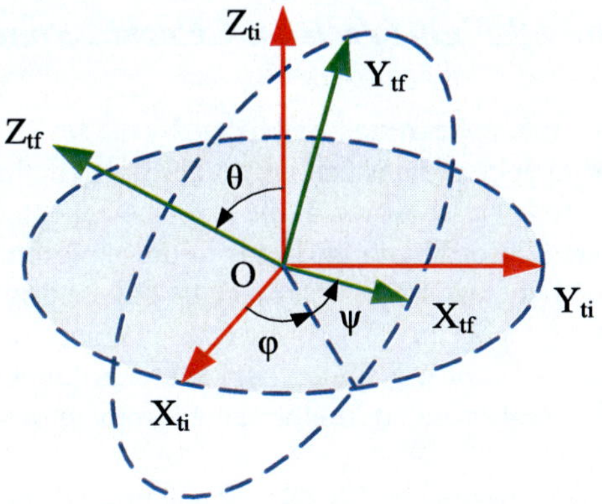

Fig. 5.2 Transformation relation between celestial inertial coordinate system and celestial fixed coordinate system

Denote the position vectors of celestial inertial coordinate system and celestial fixed coordinate system as $\boldsymbol{r}_{\mathrm{ti}}$ and $\boldsymbol{r}_{\mathrm{tf}}$, respectively. Then,

$$\boldsymbol{r}_{\mathrm{tf}} = \boldsymbol{C}_{\mathrm{Z}}(\psi)\boldsymbol{C}_{\mathrm{X}}(\theta)\boldsymbol{C}_{\mathrm{Z}}(\varphi)\boldsymbol{r}_{\mathrm{ti}} \tag{5.11}$$

3. Heliocentric ecliptic coordinate system ⟶ heliocentric orbital coordinate system

$$\boldsymbol{r}_{\mathrm{o}} = \boldsymbol{C}_{\mathrm{Z}}(\omega)\boldsymbol{C}_{\mathrm{X}}(i)\boldsymbol{C}_{\mathrm{Z}}(\Omega)\boldsymbol{r}_{\mathrm{si}} \tag{5.12}$$

where Ω is the longitude of ascending node; i is orbital inclination; ω is the argument of perigee; $\boldsymbol{r}_{\mathrm{o}}$ is the position of the spacecraft in the heliocentric orbital coordinate system; and $\boldsymbol{r}_{\mathrm{si}}$ is the position of the spacecraft in the heliocentric ecliptic coordinate system.

5.3 Ephemeris of Navigational Celestial Bodies

Optical autonomous navigation needs celestial ephemeris. Orbital propagation and measurement update algorithms require the calculation of celestial ephemeris. In this section, the commonly used celestial ephemeris algorithms will be presented.

5.3.1 Calculation of High-Precision Celestial Ephemerides

The DE planetary and lunar ephemerides published by the Jet Propulsion Laboratory (JPL) are high-precision ephemerides that are authoritative in the astronomical field [4–6]. Their main versions are shown in Table 5.1. The ephemeris with appropriate accuracy can be selected according to the needs of the exploration mission.

The ephemerides of planetary satellites are usually obtained by theoretical solution or numerical integration. Jacobson [7] describes the ephemeris origin of the satellites of Mars, Jupiter, Saturn, Uranus, Neptune, and Pluto. Although the ephemeris of each satellite has a different origin, it has the same format and is consistent with the planetary ephemeris.

The JPL ephemerides provide the Chebyshev polynomial coefficients and orders of celestial body position and nutation in the time interval. The time interval with 32 days is called big interval. For the fast-changing celestial bodies, the big interval is divided into several small subintervals to ensure the fitting accuracy.

In the ephemeris, each variable corresponds to a code, as shown in Table 5.2.

Table 5.1 Some DE ephemerides published by JPL

Ephemeris	Reference system	Date of publication (month, year)	Time span (day, month, year)
DE405	ICRF	May 1995	A. D. December 9, 1599–A. D. February 20, 2201
DE406	ICRF	May 1997	B. D. February 4, 3002–A. D. May 6, 3000
DE410	ICRF	April 2003	A. D. December 10, 1959–December 15, 2019
DE413	ICRF	November 2004	A. D. December 4, 1899–A. D. March 7, 2050
DE414	ICRF	May 2005	A. D. December 4, 1899–A. D. March 7, 2050
DE418	ICRF	May 2005	A. D. July 29, 1899–A. D. January 21, 2051
DE421	ICRF	February 2008	A. D. July 29, 1899–A. D. October 9, 2053
DE422	ICRF	September 2009	A. D. July 29, 1899–A. D. October 9, 2053
DE423	ICRF2	February 2010	A. D. December 26, 1799–A. D. February 2, 2200
DE424	ICRF2	2011	December 21, 3001 B. C.–A. D. January 30, 3000
DE431	ICRF2	2013	13201 B.C.–A. D. 17191
DE432	ICRF2	April 2014	A. D. December 21, 1549–A. D. January 25, 2650

Table 5.2 Code table of JPL ephemerides

Code	Definition	Code	Definition
1	Mercury	9	Pluto
2	Venus	10	Moon
3	Earth	11	Sun
4	Mars	12	Solar system barycenter
5	Jupiter	13	Earth-moon barycenter
6	Saturn	14	Nutation in longitude and nutation in ecliptic obliquity
7	Uranus	15	Physical libration and its variability of the moon
8	Neptune		

For each component (the X, Y, Z components of the celestial body or the nutation in longitude and nutation in ecliptic obliquity), each interval recorded in the ephemeris data contains NC Chebyshev coefficients. The required ephemeris components are calculated by a Chebyshev expansion of degree NC − 1

$$f(t_c) = \sum_{k=0}^{\mathrm{NC}-1} C_k P_k(t_c) \tag{5.13}$$

where $t_c \in [-1, 1]$ is the normalized Chebyshev time, C_k is the Chebyshev coefficient for position calculation, and is read from the ephemeris documentation, and P_k is a Chebyshev polynomial given by

$$P_k = \cos(k \arccos t_c) \tag{5.14}$$

From the Eq. (5.14), it can be seen that

$$\begin{cases} P_0(t_c) = 1 \\ P_1(t_c) = t_c \end{cases} \tag{5.15}$$

Using the trigonometric identity

$$\cos(n+1)\theta = 2\cos\theta\cos n\theta - \cos(n-1)\theta \quad n \geq 1 \tag{5.16}$$

and $t_c = \cos\theta$, the following iteration relation can be obtained

$$P_{k+1}(t_c) = 2t_c P_k(t_c) - 2P_{k-1}(t_c) \quad k \geq 2 \tag{5.17}$$

Nutation and celestial body position can be obtained by interpolation in Eq. (5.13) and the velocity of celestial body by the differentiation of Eq. (5.13)

$$\frac{\mathrm{d}f(t_c)}{\mathrm{d}t} = \sum_{k=1}^{NC-1} C_k V_k(t_c) \tag{5.18}$$

where

$$V_k(t_c) = \frac{\mathrm{d}P_k(t_c)}{\mathrm{d}t} \tag{5.19}$$

V_k can be obtained by the differentiation of Eq. (5.17)

$$\begin{cases} V_1(t_c) = 1 \\ V_2(t_c) = 4t_c \\ V_{k+1}(t_c) = 2t_c V_k(t_c) + 2P_k(t_c) - V_{k-1}(t_c) \end{cases} \tag{5.20}$$

The interpolation operation can work out the position and velocity of the planet, earth-moon barycenter or sun relative to solar system barycenter, as well as the position and velocity of the moon relative to earth barycenter. All of these position and velocity vectors are given in the J2000.0 geocentric equatorial coordinate system. If the position vector of the planet or sun relative to the target planet or spacecraft is calculated, the corresponding conversion will be needed.

The common star catalogues include Hipparcos catalogue, Tycho-2 catalogue, Gaia DR1 catalogue and Gaia DR2 catalogue. Table 5.3 shows the released years and the reference epochs of star catalogues. The Hipparcos [8] catalogue contains 118 218 stars. The Tycho-2 [9] catalogue contains positions and proper motions as well as two-colour photometric data for the 2.5 million brightest stars in the sky. The Gaia DR1 [10] consists of a primary astrometric data set which contains the positions, parallaxes, and mean proper motions for about 2 million of the brightest stars in common with the Hipparcos and Tycho-2 catalogues, and a secondary astrometric data set containing the positions for an additional 1.1 billion sources. The Gaia DR2 [11] contains celestial positions and the apparent brightness in G for approximately 1.7 billion sources, for 1.3 billion of those sources, parallaxes, and proper motions in addition available. The need for precision epoch conversion depends on navigation accuracy requirement.

Table 5.3 Released years and the reference epochs of star catalogues

Star catalogue	Released year	Positions and proper motions reference epoch
Hipparcos	1997	J1991.25
Tycho-2	2000	J2000.0
Gaia DR1	2016	J2015.0
Gaia DR2	2018	J2015.5

Table 5.4 Six mean orbital elements for 8 major planets and Pluto in the solar system (J2000.0 heliocentric ecliptic coordinate system)

Planet	Orbital element					
	a (AU)	e	i (°)	Ω (°)	ω (°)	f (°)
Mercury	0.38709893	0.20563069	7.00487	48.33167	77.45645	252.25084
Venus	0.72333199	0.00677323	3.39471	76.68069	131.53298	181.97973
Earth	1.00000011	0.01671022	0.00005	−11.26064	102.94719	100.46435
Mars	1.52366231	0.09341233	1.85061	49.57854	336.04084	355.45332
Jupiter	5.20336301	0.04839266	1.30530	100.55615	14.75385	34.40438
Saturn	9.53707032	0.05415060	2.48446	113.71504	92.43194	49.94432
Uranus	19.19126393	0.04716771	0.76986	74.22988	170.96424	313.23218
Neptune	30.06896348	0.00858587	1.76917	131.72169	44.97135	304.88003
Pluto	39.48168677	0.24880766	17.14175	110.30347	224.06676	238.92881

5.3.2 *Calculation of Simple Celestial Ephemerides*

Some ephemerides that do not require high accuracy (such as the celestial ephemeris required for the establishment of a celestial gravitational perturbation model) can be established with the method of simple ephemeris calculation. Due to low accuracy requirement, such a celestial ephemeris can be calculated by using the six mean orbital elements near the time interval of the exploration mission.

Table 5.4 shows the six mean orbital elements for 8 major planets and Pluto in the solar system. By using this table, the position and velocity vectors of the major planets in the J2000.0 heliocentric ecliptic coordinate system of an epoch can be calculated [12].

The problem of lunar orbit is complicated [13]. Because the magnitude of solar perturbation is as large as 10^{-2} when the moon moves around the earth, the lunar orbit changes very fast. Therefore, the lunar orbit can be treated as an invariant ellipsoid (or circle) only when the accuracy requirement is not high. The change of lunar orbit is usually taken into account when calculating the influence of the moon on spacecraft motion. The method of mean elements can give the approximate solution of lunar orbit variation at an accuracy of 10^{-3} to 10^{-4} and can calculate the lunar ephemeris at this accuracy. In the J2000.0 geocentric ecliptic coordinate system, this solution can be expressed as

$$\begin{cases} a = 384747.981\text{km} \\ \bar{e} = 0.054879905 \\ \bar{i} = 5.129835017^\circ \\ \bar{\Omega} = 125.044555556^\circ - 1934.1361850^\circ\text{T} + 0.0020767^\circ\text{T}^2 \\ \bar{\omega} = 138.308686110^\circ + 6003.1498961^\circ\text{T} - 0.0124003^\circ\text{T}^2 \\ \bar{M} = 134.963413889^\circ + 13.0649931553^\circ\text{d} + 0.0089939^\circ\text{d}^2 \end{cases} \tag{5.21}$$

For the calculation of simple asteroid ephemerides, the asteroid ephemeris documents provided on the Web site (ftp://lowell.edu/pub/elgb/astorb.dat.gz) can be used. In those documents, parameters (1) is the asteroid number, parameters (12) through (18) are the epoch of osculation and heliocentric osculating elements. By using these parameters, the position and velocity of asteroids of an epoch in the J2000.0 heliocentric ecliptic coordinate system can be calculated.

5.4 Brief Summary

This chapter addresses the time systems, reference coordinate systems, navigational celestial ephemerides, and other basic knowledge needed for autonomous navigation. The five-time systems in common use and Julian date are described. Several reference coordinate systems and their coordinate transformation are also introduced such that spacecraft state and observations can be described properly. Finally, calculation methods of celestial ephemeris are provided, which can be used for orbit propagation and filter update.

References

1. Montenbruck, O., and E. Gill. 2000. *Satellite orbits—Models, methods and applications*. Berlin: Springer Press.
2. Soffel, M., S.A. Klioner, G. Petit, et al. 2003. The IAU 2000 resolutions for astrometry, celestial mechanics, and metrology in the relativistic framework: Explanatory supplement. *The Astronomical Journal* 126 (6): 2687–2706.
3. Luzum, B., N. Capitaine, A. Fienga, et al. 2011. The IAU 2009 system of astronomical constants: The report of The IAU working group on numerical standards for fundamental astronomy. *Celestial Mechanics and Dynamical Astronomy* 110: 293–304.
4. Falkner, W.M., J.G. Williams, D.H. Boggs, et al. 2014. The planetary and lunar ephemerides DE430 and DE431. *Interplanetary Network Progress Report* 196: 1–81.
5. Yang, Yongzhang, Li Jinling, Ping Jinsong, et al. 2017. Development and comparison of the JPL ephemerides in deep space exploration. Journal of Deep Space Exploration 4 (1): 89–98.
6. Weiwei, Lei, Li Kai, and Zhang Hanwei. 2016. Structure, calculation and comparison of development ephemerides. *Journal of Spacecraft TT & C Technology* 35 (5): 375–384.
7. Jacobson, R. A. 2003. Natural satellite ephemerides at JPL: A report to the IAU commission 20 working group on natural satellites. Jet Propulsion Laboratory
8. Perryman, M.A.C., L. Lindegren, J. Kovalevsky, et al. 1997. The hipparcos catalogue. *Astronomy & Astrophysics* 323: 49–52.
9. Høg, E., C. Fabricius, V. V. Makarov, et al. 2000. The Tycho-2 catalogue of the 2.5 million brightest stars. Astronomy and Astrophysics 355: 27–30.
10. Collaboration, Gaia. 2016. Gaia data release 1 summary of the astrometric, photometric, and survey properties. *Astronomy & Astrophysics* 595 (A2): 1–23.
11. Collaboration, Gaia. 2018. Gaia data release 2 summary of the contents and survey properties. *Astronomy & Astrophysics* 616 (A1): 1–22.

12. Dayi, Wang, Huang Xiangyu, and Wei Chunling. 2012. *Principle and technology of autonomous control of deep space exploration based on optical imaging measurement*. Beijing: China Aerospace Publishing House.
13. Lin, Liu. 2000. *Orbit theory of spacecraft*. Beijing: National Defense Industry Press.

Chapter 6
Dynamic Models and Environment Models

In the design and performance evaluation of a navigation system, the orbit dynamics model, attitude kinematics model and environment model should be established. This chapter first addresses orbital perturbation model, establishes orbit dynamic equations, and then elaborates on four typical parametric forms for attitude description, the conversion relations between attitude parameters, and the attitude kinematics model corresponding to each parameter. Finally, Mars and asteroids are taken as examples to introduce the modeling methods of celestial environment, including celestial body shape modeling and gravitational field modeling.

6.1 Orbit Dynamics Model

An accurate orbit dynamics model is established to constitute the basis of autonomous navigation research.

6.1.1 Orbital Perturbation Model

In addition to the gravity of central body and the orbit control force, various perturbative forces act on the spacecraft. These perturbative forces mainly include additional gravitation generated by the non-spherical shape and non-uniform mass distribution of the central body, the gravitational force of other celestial bodies, the solar radiation pressure (SRP), and possible atmospheric resistance, as well as the disturbing force caused by attitude control.

D. Wang et al., *Spacecraft Autonomous Navigation Technologies Based on Multi-source Information Fusion*, Space Science and Technologies, https://doi.org/10.1007/978-981-15-4879-6_6

1. **Gravitational potential functions**

In the analysis of the celestial body's gravitational acting on the spacecraft, gravitational potential function U is often used. The gravitation force acting on the unit mass of the spacecraft at a point in space is given by

$$F = \text{grad}\, U \tag{6.1}$$

where grad is gradient function. This potential function is easy to use, as it is independent of the selection of coordinate system. If the mass M of the celestial body is concentrated at a single point, its potential function is

$$U_0 = \frac{GM}{r} = \frac{\mu}{r} \tag{6.2}$$

where $\mu = GM$ is the standard gravitational parameter of the central body and G is the gravitational constant of the central body, r is the distance from the barycenter of central body to the point. The potential function of a spherical celestial body with uniform mass to its external points is the same as that of the whole sphere when its mass is concentrated at its center. In addition, the gradient direction of this function always points to the sphere center. When considering the perturbation of celestial body shape, the potential function includes the following two parts:

$$U = U_0 + R \tag{6.3}$$

where R is the potential function of perturbative force, also called perturbation function.

When considering the perturbation caused by the shape of large planets, the moon, and other celestial bodies whose shape is similar to sphere, their gravitational potential functions can usually be expressed as the expansion of spherical harmonic term. For some ellipsoidal bodies, such as asteroids and comets, their gravitational potential functions can usually be expressed as the expansion of ellipsoidal harmonic term. For some celestial bodies with very special shapes, such as asteroids, their gravitational potential functions can be calculated with a polyhedral model.

The gravitational potential function expressed by the expansion of spherical harmonic term is

$$U = \frac{GM}{r}\sum_{n=0}^{\infty}\sum_{m=0}^{n}\left(\frac{r_0}{r}\right)^n P_{nm}(\sin\phi) \times \left[\overline{C}_{nm}\cos(m\lambda) + \overline{S}_{nm}\sin(m\lambda)\right] \tag{6.4}$$

where P_{nm} is Legendre polynomial function with degree n and order m, r_0 is the reference radius of the celestial body, r is the distance from the spacecraft to the center of the celestial body, ϕ and λ are the latitude and longitude of the celestial body, respectively, and $\overline{C}_{nm}$ and $\overline{S}_{nm}$ are normalized spherical harmonic coefficients.

The relationship between normalized and unnormalized coefficients is

$$(\overline{C}_{nm}; \overline{S}_{nm}) = \left[\frac{(n+m)!}{(2-\delta_{0m})(2n+1)(n-m)!}\right]^{\frac{1}{2}}(C_{nm}; S_{nm}) \tag{6.5}$$

where δ_{0m} is Kronecker symbol function.

The Legendre polynomial function is

$$P_{nm}(x) = \left(1-x^2\right)^{\frac{m}{2}}\frac{\mathrm{d}^m}{\mathrm{d}x^m}P_n(x) \tag{6.6}$$

The Legendre polynomial is

$$P_n(x) = \frac{1}{2^n n!}\frac{\mathrm{d}^n}{\mathrm{d}x^n}\left(x^2-1\right)^n \tag{6.7}$$

The gravitational potential function expressed by the expansion of ellipsoidal harmonic term [1, 2] is

$$U = GM\sum_{n=0}^{N\max}\sum_{p=1}^{2n+1}\bar{a}_n^p\frac{F_n^p(\lambda_1)}{F_n^p(a)}\overline{E}_n^p(\lambda_2)\overline{E}_n^p(\lambda_3) \tag{6.8}$$

where $\bar{a}_n^p$ is normalized ellipsoidal harmonic coefficient. Considering the shape and density of the celestial body, this coefficient satisfies

$$\bar{a}_n^p = \int_0^h\int_h^k\frac{U(\lambda_1=a,\lambda_2,\lambda_3)}{GM}\overline{E}_n^p(\lambda_2)\overline{E}_n^p(\lambda_3)\mathrm{d}s \tag{6.9}$$

This surface integral makes use of the potential function generated by the Brillouin ellipsoid corresponding to the celestial body. Figure 6.1 is the schematic diagram of Brillouin sphere and Brillouin ellipsoid. In Eq. (6.9), $\overline{E}_n^p(\lambda_2)\overline{E}_n^p(\lambda_3)$ satisfies the following relationship

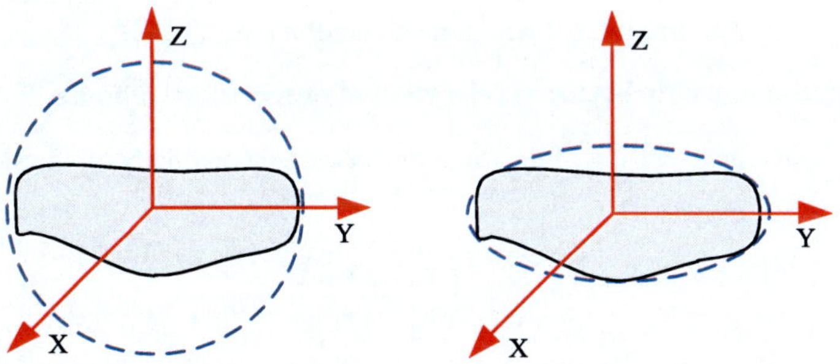

Fig. 6.1 Brillouin sphere (left) and Brillouin ellipsoid (right)

$$\overline{E}_n^p(\lambda_2)\overline{E}_n^p(\lambda_3) = \frac{E_n^p(\lambda_2)E_n^p(\lambda_3)}{\sqrt{\gamma_n^p}} \tag{6.10}$$

where E_n^p is the first kind of Lamé function (can be K_n^p, L_n^p, M_n^p or N_n^p); n is the number of dimension of the function; p is eigenvalue. $F_n^p(\lambda_1)$ and E_n^p have the following relationship

$$F_n^p(\lambda_1) = (2n+1)E_n^p(\lambda_1)\int_{\lambda_1}^{\infty} \frac{\mathrm{d}s}{\left(E_n^p\right)^2(s)\sqrt{\left(s^2-h^2\right)\left(s^2-k^2\right)}} \tag{6.11}$$

where s, h and k are the parameters of the ellipsoid equation that satisfies

$$\frac{x^2}{s^2} + \frac{y^2}{s^2-h^2} + \frac{z^2}{s^2-k^2} = 1 \tag{6.12}$$

For the given x, y, and z, the s^2 in Eq. (6.12) has three real roots $(\lambda_i)_{i=1,2,3}$, which satisfy the following constraints

$$\lambda_1^2 \in \left[k^2, +\infty\right), \lambda_2^2 \in \left[h^2, k^2\right], \lambda_3^2 \in \left[0, h^2\right] \tag{6.13}$$

The gravitational potential calculated with a polyhedral model [3] is

$$U = \sum_{i\in\text{cubes}} \left(\frac{G\rho_i}{2} \sum_{e\in\text{edges}} \boldsymbol{r}_e^{\mathrm{T}} \boldsymbol{E}_e \boldsymbol{r}_e \cdot L_e - \frac{G\rho_i}{2} \sum_{f\in\text{faces}} \boldsymbol{r}_f^{\mathrm{T}} \boldsymbol{F}_f \boldsymbol{r}_f \cdot \omega_f \right) \tag{6.14}$$

where ρ_i is the density of an ith finite element, $\boldsymbol{r}_e$ is the vector of gravitation calculation point pointing to any point on edge e or its infinite extension; $\boldsymbol{E}_e$ is the dyadic vector composed of the face/edge normal vectors related to each edge; L_e is the logarithmic term to express the one-dimensional linear potential; $\boldsymbol{r}_f$ is the vector of gravitation calculation point pointing to any point in the face plane; $\boldsymbol{F}_f$ is the outer product of face normal vectors; and ω_f is the signed solid angle subtended by planar region, when viewed from the gravitation calculation point.

2. **Gravitational perturbation acceleration of other celestial bodies**

The perturbation acceleration of the ith perturbative celestial body on the spacecraft is:

$$\boldsymbol{a}_i = \mu_i \left(\frac{\boldsymbol{r}_{\mathrm{r}i}}{r_{\mathrm{r}i}^3} - \frac{\boldsymbol{r}_{\mathrm{p}i}}{r_{\mathrm{p}i}^3} \right) \tag{6.15}$$

where μ_i is the gravitational constant of the ith perturbative celestial body, $\boldsymbol{r}_{\mathrm{p}i}$ is the position vector of the ith perturbative celestial body relative to central body, and

$r_{pi} = \|\boldsymbol{r}_{pi}\|$, $\boldsymbol{r}_{ri} = \boldsymbol{r}_{pi} - \boldsymbol{r}$ is the position vector of the ith perturbative celestial body relative to the spacecraft, where $\boldsymbol{r}$ is the position vector of the spacecraft relative to the center of the celestial body and $r_{ri} = \|\boldsymbol{r}_{ri}\|$.

3. **Perturbation acceleration caused by solar radiation pressure**

When a spacecraft is under sunlight, one part of the solar radiation energy will be absorbed, and the other part reflected. Through this energy conversion, a force will be applied on the spacecraft, which is called solar radiation pressure. The reflection of sunlight on spacecraft surface is very complex, including specular reflection and diffuse reflection. When studying the influence of solar radiation pressure on spacecraft orbit, the direction of radiation pressure can be considered consistent with the incidence direction of sunlight. The radiation pressure acting on the unit mass of the spacecraft can be expressed as

$$\boldsymbol{a}_s = -\frac{A_s E}{m r_{rs}^3} \boldsymbol{r}_{rs} \tag{6.16}$$

where A_s is the spacecraft section area perpendicular to the sunlight direction; m is spacecraft mass; $E = k' p_0 \Delta_0^2$ is solar flux constant, k' is comprehensive absorption coefficient; Δ_0 is the distance from the sun to the earth surface; p_0 is the solar radiation pressure intensity on the earth surface; $\boldsymbol{r}_{rs} = \boldsymbol{r}_{ps} - \boldsymbol{r}$ is the position vector of the sun relative to the spacecraft, and $r_{rs} = \|\boldsymbol{r}_{rs}\|$.

4. **Perturbation acceleration caused by atmospheric drag**

The drag acceleration of the spacecraft in the atmosphere is

$$\boldsymbol{a}_d = -\frac{1}{2} c_d \rho \frac{A_d}{m} v_a \boldsymbol{v}_a \tag{6.17}$$

where c_d is drag coefficient; ρ is atmospheric density; A_d is windward area, i.e., the projected area of the spacecraft in the velocity direction; $\boldsymbol{v}_a$ is the velocity of the spacecraft relative to the rotating atmosphere, and $v_a = \|\boldsymbol{v}_a\|$.

6.1.2 Spacecraft Orbit Dynamics Model

1. **Orbital dynamics model for heliocentric transfer phase**

For the heliocentric transfer phase, sun is the central body, and solar gravitation is the central gravitation. Other perturbative forces mainly include the third-body gravitational perturbation from other planets and solar radiation pressure. In the J2000.0 heliocentric ecliptic inertial coordinate system, the orbit dynamics equation of the spacecraft is given by

$$\dot{\boldsymbol{r}} = \boldsymbol{v}$$
$$\dot{\boldsymbol{v}} = -\frac{\mu_s}{r^3}\boldsymbol{r} + \sum_{i=1}^{n_p} \mu_i \left[\frac{\boldsymbol{r}_{ri}}{r_{ri}^3} - \frac{\boldsymbol{r}_{pi}}{r_{pi}^3} \right] - \frac{A_s E}{m r^3}\boldsymbol{r} + \frac{\boldsymbol{T}}{m} + \boldsymbol{a} \tag{6.18}$$

where $\boldsymbol{r}$ and $\boldsymbol{v}$ are the position and velocity vectors of the spacecraft in the heliocentric ecliptic coordinate system, respectively, $r = \|\boldsymbol{r}\|$; μ_s is solar gravitational constant; n_p is the number of perturbative planets; $\boldsymbol{T}$ is thrust vector; $\boldsymbol{a}$ is the total of other perturbation acceleration vectors.

2. **Orbital dynamics model for target-approaching phase**

The target-approaching phase refers to the orbit phase where the spacecraft has entered the targeted celestial body's gravitational field, and the forces on the spacecraft are dominated by the gravitation of the target. This phase mainly includes approaching phase, orbiting phase, and landing phase.

Because the spacecraft is close to the targeted celestial body, the gravitation of the target is the central force. Considering the perturbations caused by irregular target shape, solar gravitation, and radiation pressure, the orbital dynamics equation of the spacecraft in the target inertial coordinate system can be written as

$$\dot{\boldsymbol{r}}_i = \boldsymbol{v}_i$$
$$\dot{\boldsymbol{v}}_i = \frac{\partial V(\boldsymbol{r}_i)}{\partial \boldsymbol{r}_i} + \boldsymbol{a}_i \tag{6.19}$$

where $\boldsymbol{r}_i$ and $\boldsymbol{v}_i$ are the position and velocity vectors of the spacecraft, respectively; $\boldsymbol{a}_i$ are other accelerations to be considered; V is potential function, which is expressed as

$$V(\boldsymbol{r}_i) = U(\boldsymbol{r}_{tf}) + \frac{\beta \boldsymbol{d} \cdot \boldsymbol{r}_i}{d^3} - \frac{\mu_s}{2d^3}\left[\boldsymbol{r}_i \cdot \boldsymbol{r}_i - 3\left(\frac{\boldsymbol{d} \cdot \boldsymbol{r}_i}{d} \right)^2 \right] \tag{6.20}$$

where the first term is the gravitational potential function of the celestial body; the second term is the perturbation potential function of solar radiation pressure; the third term is the perturbation potential function of solar gravitation; $\boldsymbol{r}_{tf}$ is the position vector of the spacecraft relative to the center of the celestial body in the fixed coordinate system; β is solar radiation pressure; $\boldsymbol{d}$ is the position vector of the celestial body relative to the sun, $d = \|\boldsymbol{d}\|$.

6.2 Attitude Kinematics Model

6.2.1 Description of Attitude

Attitude can be expressed in various forms, including direction cosine matrix, Euler angles, Euler axis/angle, and quaternion. These attitude parameters can be converted to each other [4, 5].

1. **Direction cosine matrix**

Consider the transformation between body frame and reference frame. Define direction cosines as the cosines of the angles between the axes, which can be written as

$$\begin{cases} \cos(\boldsymbol{i}_\mathrm{b}, \boldsymbol{i}_\mathrm{r}) = \boldsymbol{i}_\mathrm{b} \cdot \boldsymbol{i}_\mathrm{r} \triangleq C_{xx} \\ \cos(\boldsymbol{i}_\mathrm{b}, \boldsymbol{j}_\mathrm{r}) = \boldsymbol{i}_\mathrm{b} \cdot \boldsymbol{j}_\mathrm{r} \triangleq C_{xy} \\ \cdots \\ \cos(\boldsymbol{k}_\mathrm{b}, \boldsymbol{k}_\mathrm{r}) = \boldsymbol{k}_\mathrm{b} \cdot \boldsymbol{k}_\mathrm{r} \triangleq C_{zz} \end{cases} \tag{6.21}$$

where $\boldsymbol{i}$,$\boldsymbol{j}$, and $\boldsymbol{k}$ are unit vectors along the three orthogonal axes of a frame, subscripts b and r represent body frame (denoted as $\mathcal{F}_\mathrm{b}$) and reference frame (denoted as $\mathcal{F}_\mathrm{r}$), respectively, and "·" is the dot product of two vectors.

Using these direction cosines, $\boldsymbol{i}_\mathrm{b}$ can be expressed in the reference frame as

$$\begin{aligned} \boldsymbol{i}_\mathrm{b} &= (\boldsymbol{i}_\mathrm{b} \cdot \boldsymbol{i}_\mathrm{r})\boldsymbol{i}_\mathrm{r} + (\boldsymbol{i}_\mathrm{b} \cdot \boldsymbol{j}_\mathrm{r})\boldsymbol{j}_\mathrm{r} + (\boldsymbol{i}_\mathrm{b} \cdot \boldsymbol{k}_\mathrm{r})\boldsymbol{k}_\mathrm{r} \\ &= C_{xx}\boldsymbol{i}_\mathrm{r} + C_{xy}\boldsymbol{j}_\mathrm{r} + C_{xz}\boldsymbol{k}_\mathrm{r} \end{aligned} \tag{6.22}$$

Combine the direction cosines of Eq. (6.21) together into a matrix, namely direction cosine matrix

$$\boldsymbol{C} = \begin{bmatrix} C_{xx} & C_{xy} & C_{xz} \\ C_{yx} & C_{yy} & C_{yz} \\ C_{zx} & C_{zy} & C_{zz} \end{bmatrix} \tag{6.23}$$

According to Eq. (6.22), the geometric direction of $\mathcal{F}_\mathrm{b}$ in $\mathcal{F}_\mathrm{r}$ can be determined as

$$\begin{bmatrix} \boldsymbol{i}_\mathrm{b} & \boldsymbol{j}_\mathrm{b} & \boldsymbol{k}_\mathrm{b} \end{bmatrix} = \boldsymbol{C}\begin{bmatrix} \boldsymbol{i}_\mathrm{r} & \boldsymbol{j}_\mathrm{r} & \boldsymbol{k}_\mathrm{r} \end{bmatrix} \tag{6.24}$$

Because the direction cosine matrix $\boldsymbol{C}$ completely determines the state of spacecraft attitude in the reference system and each of its elements is an attitude parameter, it is called attitude matrix. Since both $\mathcal{F}_\mathrm{r}$ and $\mathcal{F}_\mathrm{b}$ are orthogonal coordinates, the nine elements also satisfy six constraint equations. Three of them are derived by the norm constraint of unit vectors:

$$\begin{cases} \boldsymbol{i}_b \cdot \boldsymbol{i}_b = C_{xx}^2 + C_{xy}^2 + C_{xz}^2 = 1 \\ \boldsymbol{j}_b \cdot \boldsymbol{j}_b = C_{yx}^2 + C_{yy}^2 + C_{yz}^2 = 1 \\ \boldsymbol{k}_b \cdot \boldsymbol{k}_b = C_{zx}^2 + C_{zy}^2 + C_{zz}^2 = 1 \end{cases} \tag{6.25}$$

Three of them are derived from the orthogonal property of body coordinate axes:

$$\begin{cases} \boldsymbol{i}_b \cdot \boldsymbol{j}_b = C_{xx}C_{yx} + C_{xy}C_{yy} + C_{xz}C_{yz} = 0 \\ \boldsymbol{i}_b \cdot \boldsymbol{k}_b = C_{xx}C_{zx} + C_{xy}C_{zy} + C_{xz}C_{zz} = 0 \\ \boldsymbol{j}_b \cdot \boldsymbol{k}_b = C_{yx}C_{zx} + C_{yy}C_{zy} + C_{yz}C_{zz} = 0 \end{cases} \tag{6.26}$$

Therefore, only three attitude parameters are independent. In other words, the state of the three-axis attitude of a spacecraft in $\mathcal{F}_r$ can be described with only three independent parameters.

It can be inferred from the above six constraint equations that the attitude matrix satisfies:

$$\boldsymbol{C}\boldsymbol{C}^{\mathrm{T}} = \boldsymbol{I}_3$$

The above equation shows that $\boldsymbol{C}$ is an orthogonal matrix. In fact, attitude matrix is just the coordinate transformation matrix between $\mathcal{F}_r$ and $\mathcal{F}_b$. A unit reference vector $\boldsymbol{V}$ can be expressed in $\mathcal{F}_b$ (denoted as $\boldsymbol{V}^b$) and $\mathcal{F}_r$ (denoted as $\boldsymbol{V}^r$) as

$$\begin{cases} \boldsymbol{V}^b = V_x^b \boldsymbol{i}_b + V_x^b \boldsymbol{j}_b + V_x^b \boldsymbol{k}_b \\ \boldsymbol{V}^r = V_x^r \boldsymbol{i}_r + V_x^r \boldsymbol{j}_r + V_x^r \boldsymbol{k}_r \end{cases} \tag{6.27}$$

The dot products of the above equation with $\boldsymbol{i}_b$, $\boldsymbol{j}_b$, and $\boldsymbol{k}_b$, respectively, give

$$\boldsymbol{V}^b = \boldsymbol{C}\boldsymbol{V}^r \tag{6.28}$$

The attitude matrix expressed by direction cosines matrix is usually used to describe the attitude. Through the matrix multiplication, the attitude matrix relative to different reference systems can be obtained. Denote the attitude matrix of $\mathcal{F}_b$ relative to $\mathcal{F}_r$ as $\boldsymbol{C}_r^b$ and the coordinate transformation matrix of $\mathcal{F}_r$ relative to another reference coordinate system $\mathcal{F}_f$ as $\boldsymbol{C}_f^r$, then the attitude matrix $\boldsymbol{C}_f^b$ of $\mathcal{F}_b$ relative to $\mathcal{F}_f$ is given by

$$\boldsymbol{C}_f^b = \boldsymbol{C}_r^b \boldsymbol{C}_f^r \tag{6.29}$$

2. **Euler Angles**

In engineering practice, it is expected that the three attitude parameters have simpler and physical meaning, so that they can be directly measured by an attitude sensor, and their attitude equations can be more conveniently solved. In this case, Euler angles are good choices to represent attitude. According to the Euler's theorem, the rotation of a rigid body around a fixed point can also be treated as a combination of finite rotations around that point. In the Euler rotation, body coordinate system can be obtained by rotating the reference coordinate system for three times. In the three rotations, the axis of each rotation is an axis of the rotated coordinate system, and the angle of each rotation is just Euler angle. Therefore, the attitude matrix determined by Euler angles is the product of three coordinate transformation matrices, whose standard forms are as follows:

$$\boldsymbol{C}_x(\theta) = \begin{bmatrix} 1 & 0 & 0 \\ 0 & \cos\theta & \sin\theta \\ 0 & -\sin\theta & \cos\theta \end{bmatrix} \tag{6.30}$$

$$\boldsymbol{C}_y(\theta) = \begin{bmatrix} \cos\theta & 0 & -\sin\theta \\ 0 & 1 & 0 \\ \sin\theta & 0 & \cos\theta \end{bmatrix} \tag{6.31}$$

$$\boldsymbol{C}_z(\theta) = \begin{bmatrix} \cos\theta & \sin\theta & 0 \\ -\sin\theta & \cos\theta & 0 \\ 0 & 0 & 1 \end{bmatrix} \tag{6.32}$$

where the subscript x represents the rotation axis x; θ is rotation angle.

Take $\boldsymbol{C}_x(\theta)$ as an example. Suppose that there is a point B in space, which is (X, Y, Z) in the coordinate system O-xyz. By rotating this orthogonal coordinate system around the x-axis by θ, a new coordinate system $O{-}x'y'z'$ is obtained, as shown in Fig. 6.2.

Obviously, the x'-axis and x-axis are coincident. Denote the position of the point B in the new coordinate system as $\left(X', Y', Z'\right)$. Then,

$$\begin{cases} X' = X \\ Y' = Y\cos\theta + Z\sin\theta \\ Z' = -Y\sin\theta + Z\cos\theta \end{cases}$$

whose matrix form is

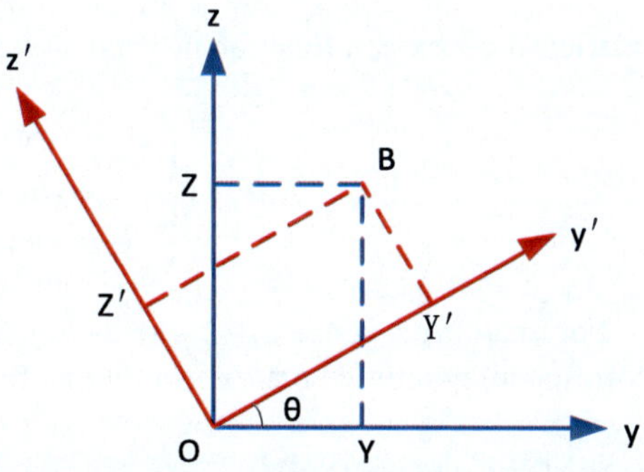

Fig. 6.2 Coordinate system rotation around the x-axis

$$\begin{bmatrix} X' \\ Y' \\ Z' \end{bmatrix} = \boldsymbol{C}_x(\theta) \begin{bmatrix} X \\ Y \\ Z \end{bmatrix}$$

If the first rotation is around the x-axis by φ, the second rotation is around the y'-axis by θ in the new coordinate system, and the third rotation is around the z''-axis by ψ in the new coordinate system obtained after the second rotation, and then, the coordinates of point B in the final coordinate system $O{-}x'''y'''z'''$ can be written as

$$\begin{bmatrix} X''' \\ Y''' \\ Z''' \end{bmatrix} = \boldsymbol{C}_z(\psi)\boldsymbol{C}_y(\theta)\boldsymbol{C}_x(\varphi) \begin{bmatrix} X \\ Y \\ Z \end{bmatrix} \tag{6.33}$$

Clearly, the attitude matrix also depends on the sequence of the three rotations, which may have two options. In the first option, the first and the third rotations are about the same axis, and the second rotation is about one of the other two axes. In the second option, each rotation is about a different axis.

If the numbers 1, 2, and 3 represent, respectively, the coordinate axes x, y, and z of various coordinate systems, then 12 Euler rotation sequences can be expressed as 1-2-1, 1-3-1, 2-1-2, 2-3-2, 3-1-3, 3-2-3, 1-2-3, 1-3-2, 2-1-3, 2-3-1, 3-1-2, and 3-2-1.

The most commonly used Euler angles are obtained by the rotation sequence 3-1-3 or 3-1-2, in which the Euler angles are denoted as (ψ, θ, φ) or (ψ, φ, θ).

Consider a 3-1-3 rotation sequence from a reference frame $\mathcal{F}_r$ to a body frame $\mathcal{F}_b$. First is a rotation around the z-axis by ψ to an intermediate frame $\mathcal{F}_{r'}$. Second is a rotation around the x-axis of $\mathcal{F}_{r'}$ by θ to a another intermediate frame $\mathcal{F}_{r''}$. Finally, it is a rotation around the z-axis of $\mathcal{F}_{r''}$ by φ to $\mathcal{F}_b$. The attitude matrix from $\mathcal{F}_r$ to $\mathcal{F}_b$ expressed by Euler angles can be written as

$$\begin{aligned} \boldsymbol{C}_r^b &= \boldsymbol{C}_z(\varphi)\boldsymbol{C}_x(\theta)\boldsymbol{C}_z(\psi) \\ &= \begin{bmatrix} \cos\varphi\cos\psi - \cos\theta\sin\varphi\sin\psi & \cos\varphi\sin\psi + \cos\theta\sin\varphi\cos\psi & \sin\theta\sin\varphi \\ -\sin\varphi\cos\psi - \cos\theta\cos\varphi\sin\psi & \sin\varphi\sin\psi + \cos\theta\cos\varphi\cos\psi & \sin\theta\cos\varphi \\ \sin\theta\sin\psi & -\sin\theta\cos\psi & \cos\theta \end{bmatrix} \end{aligned} \tag{6.34}$$

Through the comparison with direction cosine matrix, it is found that the relationship between Euler angles and direction cosine is

$$\begin{cases} \psi = -\arctan\dfrac{C_{zx}}{C_{zy}} \\ \theta = \arccos C_{zz} \\ \varphi = \arctan\dfrac{C_{xz}}{C_{yz}} \end{cases}$$

For rotation sequence 3-1-2 with the rotation angles ψ, θ and φ successively. The corresponding attitude matrix from initial frame to the final frame can be written as

$$\boldsymbol{C} = \boldsymbol{C}_y(\theta)\boldsymbol{C}_x(\varphi)\boldsymbol{C}_z(\psi)$$

$$= \begin{bmatrix} \cos\theta\cos\psi - \sin\varphi\sin\theta\sin\psi & \cos\varphi\sin\psi + \sin\varphi\sin\theta\cos\psi & -\cos\varphi\sin\theta \\ -\cos\theta\sin\psi & \cos\varphi\cos\psi & \sin\varphi \\ \sin\theta\cos\psi - \cos\theta\sin\varphi\sin\psi & \sin\theta\sin\psi - \sin\varphi\cos\theta\cos\psi & \cos\varphi\cos\theta \end{bmatrix} \tag{6.35}$$

If ψ, θ and φ are all small angles, and Eq. (6.35) can be approximated as

$$\boldsymbol{C} = \begin{bmatrix} 1 & \psi & -\theta \\ -\psi & 1 & \varphi \\ \theta & -\varphi & 1 \end{bmatrix} \tag{6.36}$$

3. Euler Axis/Angle

Euler axis/angle is another way to represent attitude. According to Euler's theorem, an arbitrary rotation of a rigid body around a fixed point can be obtained by rotating an angle around an axis that passes through this point, which originates from the following property of orthogonal matrix $\boldsymbol{C}$: a real orthogonal matrix has at least one eigenvector with the eigenvalue of 1, namely a unit vector $\boldsymbol{e}$ that satisfies the following equation:

$$\boldsymbol{e} = \boldsymbol{C}\boldsymbol{e} \tag{6.37}$$

This equation shows that, for the vector representing the direction of rigid-body rotation axis, its components in the body coordinate system are the same as those in the reference coordinate system. In addition, any attitude rotation corresponds to a transformation matrix. In this way, the rotation axis $\boldsymbol{e}$ and rotation angle $\boldsymbol{\Phi}$ are called Euler axis/angle of the rotation. The attitude matrix corresponding to the Euler axis/angle is given by

$$\begin{aligned} \boldsymbol{C}(\boldsymbol{\Phi}) &= \begin{bmatrix} \cos\Phi + e_x^2(1-\cos\Phi) & e_x e_y(1-\cos\Phi) + e_z\sin\Phi & e_x e_z(1-\cos\Phi) - e_y\sin\Phi \\ e_x e_y(1-\cos\Phi) - e_z\sin\Phi & \cos\Phi + e_y^2(1-\cos\Phi) & e_y e_z(1-\cos\Phi) + e_x\sin\Phi \\ e_x e_z(1-\cos\Phi) + e_y\sin\Phi & e_y e_z(1-\cos\Phi) - e_x\sin\Phi & \cos\Phi + e_z^2(1-\cos\Phi) \end{bmatrix} \\ &= \cos\Phi\boldsymbol{I} + (1-\cos\Phi)\boldsymbol{e}\boldsymbol{e}^{\mathrm{T}} - \sin\Phi[\boldsymbol{e}\times] \end{aligned} \tag{6.38}$$

where $[\boldsymbol{e}\times]$ is the cross product matrix of $\boldsymbol{e}$ defined by

$$[\boldsymbol{e}\times] \triangleq \begin{bmatrix} 0 & -e_z & e_y \\ e_z & 0 & -e_x \\ -e_y & e_x & 0 \end{bmatrix} \tag{6.39}$$

It should be mentioned that the Euler axis/angle representation only has three independent parameters, because of $\|\boldsymbol{e}\| = e_x^2 + e_y^2 + e_z^2 = 1$. From Eqs. (6.38), it can be seen that the direction cosines of the two coordinate systems can be expressed by Euler axis/angle parameters. Similarly, if direction cosines are known, Euler axis/angle parameters can also be calculated with the following equation:

$$\begin{cases} e_x = \frac{C_{yz}-C_{zy}}{2\sin\Phi} \\ e_y = \frac{C_{zx}-C_{xz}}{2\sin\Phi} \\ e_z = \frac{C_{xy}-C_{yx}}{2\sin\Phi} \\ \cos\Phi = \frac{\mathrm{Tr}[C]-1}{2} \end{cases} \tag{6.40}$$

where $\mathrm{Tr}[\boldsymbol{C}] = C_{xx} + C_{yy} + C_{zz}$ is the trace of attitude matrix. In the rotation by the same angle Φ around any axis, the trace of attitude matrix remains unchanged.

Let us continue to analyze the relationship between Euler angle Φ and the two sets of coordinates. Suppose Φ_x, Φ_y, and Φ_z are the included angles between the coordinate axes in the reference coordinate system and the body coordinate system. Obviously, the diagonal elements in the attitude matrix can be expressed as

$$C_{mm} = \cos\Phi_m = \cos\Phi + e_m^2(1-\cos\Phi) \quad (m = x, y, z) \tag{6.41}$$

Using the identity of $\cos 2\alpha = \cos^2\alpha - \sin^2\alpha$, the following equation can be obtained

$$\cos\Phi_m = \cos^2\frac{\Phi_m}{2} - \sin^2\frac{\Phi_m}{2} = 1 - 2\sin^2\frac{\Phi_m}{2}$$
$$\cos\Phi + e_m^2(1-\cos\Phi) = \left(1-e_m^2\right)\cos\Phi + e_m^2$$

Substituting the above equation into Eq. (6.41) gives

$$\begin{aligned} -2\sin^2\frac{\Phi_m}{2} &= \left(1-e_m^2\right)\cos\Phi - \left(1-e_m^2\right) \\ &= \left(1-e_m^2\right)(\cos\Phi - 1) \\ &= \left(1-e_m^2\right)\left(1 - 2\sin^2\frac{\Phi}{2} - 1\right) \\ &= -2\left(1-e_m^2\right)\sin^2\frac{\Phi}{2} \end{aligned}$$

According to the relationship in the above equation, the addition of the equations corresponding to $m = x, y, z$ can obtain

$$\sin^2\frac{\Phi}{2} = \frac{1}{2}\left(\sin^2\frac{\Phi_x}{2} + \sin^2\frac{\Phi_y}{2} + \sin^2\frac{\Phi_z}{2}\right) \tag{6.42}$$

The above equation shows the relationship between the deviation angle of the corresponding coordinate axis and the Euler angles. If the deviation angle is small, then

$$\Phi = \sqrt{\frac{\Phi_x^2 + \Phi_y^2 + \Phi_z^2}{2}} \tag{6.43}$$

This formula can be used to evaluate the error of attitude determination.

4. **Quaternion**

To facilitate the matrix operation of attitude matrix, Euler axis/angle parameters constitute the other four attitude parameters. The first three represent the directions of Euler rotation axes, and the fourth represents Euler rotation angle. The quaternion $\boldsymbol{q}$ is defined as the combination of three-dimensional vectors and scalar

$$\boldsymbol{q} = \begin{bmatrix} \boldsymbol{q}_{\mathrm{v}} \\ q_4 \end{bmatrix} = \begin{bmatrix} q_1 \\ q_2 \\ q_3 \\ q_4 \end{bmatrix} = \begin{bmatrix} e_x \sin\frac{\Phi}{2} \\ e_y \sin\frac{\Phi}{2} \\ e_z \sin\frac{\Phi}{2} \\ \cos\frac{\Phi}{2} \end{bmatrix} \tag{6.44}$$

where $\boldsymbol{q}_{\mathrm{v}} = \begin{bmatrix} q_1 & q_2 & q_3 \end{bmatrix}^{\mathrm{T}}$. This quaternion satisfies the constraint equation

$$q_1^2 + q_2^2 + q_3^2 + q_4^2 = 1 \tag{6.45}$$

The quaternion $\boldsymbol{q}$ can be expressed in a plural form

$$\boldsymbol{q} = q_1\boldsymbol{i} + q_2\boldsymbol{j} + q_3\boldsymbol{k} + q_4 \tag{6.46}$$

where $\boldsymbol{i}, \boldsymbol{j}$ and $\boldsymbol{k}$ are the unit vectors of reference axes.

With $\cos\Phi = 2\cos^2\frac{\Phi}{2} - 1$ and $\sin\Phi = 2\sin\frac{\Phi}{2}\cos\frac{\Phi}{2}$, the relationship between attitude matrix and quaternion can be obtained

$$\begin{aligned} \boldsymbol{C}(\boldsymbol{q}) &= \begin{bmatrix} q_1^2 - q_2^2 - q_3^2 + q_4^2 & 2(q_1q_2 + q_3q_4) & 2(q_1q_3 - q_2q_4) \\ 2(q_1q_2 - q_3q_4) & -q_1^2 + q_2^2 - q_3^2 + q_4^2 & 2(q_2q_3 + q_1q_4) \\ 2(q_1q_3 + q_2q_4) & 2(q_2q_3 - q_1q_4) & -q_1^2 - q_2^2 + q_3^2 + q_4^2 \end{bmatrix} \\ &= (q_4^2 - \boldsymbol{q}_{\mathrm{v}}^{\mathrm{T}}\boldsymbol{q}_{\mathrm{v}})\boldsymbol{I} + 2\boldsymbol{q}_{\mathrm{v}}\boldsymbol{q}_{\mathrm{v}}^{\mathrm{T}} - 2q_4[\boldsymbol{q}_{\mathrm{v}}\times] \end{aligned} \tag{6.47}$$

The relationship between quaternion and direction cosines is:

$$\begin{cases} \boldsymbol{q}_{\mathrm{v}} = \frac{1}{4q_4}\begin{bmatrix} C_{yz} - C_{zy} \\ C_{zx} - C_{xz} \\ C_{xy} - C_{yx} \end{bmatrix} \\ q_4 = \pm\frac{1}{2}\sqrt{\mathrm{Tr}(\boldsymbol{C}) + 1} \end{cases} \tag{6.48}$$

Here, the sign of q_4 can be "+" or "−." "+" means the rotation around a fixed axis by Φ, and "−" means the rotation by $2\pi + \Phi$. If $1 + C_{xx} + C_{yy} + C_{zz} \approx 0$, the quaternion components calculated by Eq. (6.48) may have a large error due to the use of q_4 as denominator. To avoid this situation, the components of quaternion can be calculated by the following formula according to the relationship between the quaternion given by Eq. (6.47) and the direction cosine matrix:

$$\begin{cases} q_2 = \pm\frac{1}{2}\sqrt{1 - C_{11} + C_{22} - C_{33}} \\ q_1 = \frac{1}{4q_2}(C_{12} - C_{21}) \\ q_3 = \frac{1}{4q_2}(C_{32} - C_{23}) \\ q_4 = \frac{1}{4q_2}(C_{31} - C_{13}) \end{cases} \tag{6.49}$$

Furthermore, if $q_2 \approx 0$, the same method can be used to solve q_1 or q_3 first and then to solve other components.

Similar to direction cosine matrix, the quaternion obtained after two consecutive rotations is equal to the product of the quaternion obtained after each rotation. Let the quaternion obtained after the first rotation be $\boldsymbol{q}$, the quaternion obtained after the second rotation be $\boldsymbol{q}'$, and the quaternion obtained after two consecutive rotations be $\boldsymbol{q}''$. If the result after two rotations is represented by an attitude matrix, then

$$\boldsymbol{C}(\boldsymbol{q}'') = \boldsymbol{C}(\boldsymbol{q}')\boldsymbol{C}(\boldsymbol{q}), \tag{6.50}$$

i.e.,

$$\begin{bmatrix} q_1'' \\ q_2'' \\ q_3'' \\ q_4'' \end{bmatrix} = \begin{bmatrix} q_4' & q_3' & -q_2' & q_1' \\ -q_3' & q_4' & q_1' & q_2' \\ q_2' & -q_1' & q_4' & q_3' \\ -q_1' & -q_2' & -q_3' & q_4' \end{bmatrix} \begin{bmatrix} q_1 \\ q_2 \\ q_3 \\ q_4 \end{bmatrix} \tag{6.51}$$

or

$$\begin{bmatrix} q_1'' \\ q_2'' \\ q_3'' \\ q_4'' \end{bmatrix} = \begin{bmatrix} q_4 & -q_3 & q_2 & q_1 \\ q_3 & q_4 & -q_1 & q_2 \\ -q_2 & q_1 & q_4 & q_3 \\ -q_1 & -q_2 & -q_3 & q_4 \end{bmatrix} \begin{bmatrix} q_1' \\ q_2' \\ q_3' \\ q_4' \end{bmatrix} \tag{6.52}$$

Equation (6.52) can also be expressed by a simple quaternion product, namely

$$\begin{aligned} \boldsymbol{q}'' &= \boldsymbol{q}\boldsymbol{q}' \\ &= (q_1\boldsymbol{i} + q_2\boldsymbol{j} + q_3\boldsymbol{k} + q_4)\left(q_1'\boldsymbol{i} + q_2'\boldsymbol{j} + q_3'\boldsymbol{k} + q_4'\right) \end{aligned} \tag{6.53}$$

where

$$\begin{cases} \boldsymbol{ii} = \boldsymbol{jj} = \boldsymbol{kk} = -1 \\ \boldsymbol{ij} = -\boldsymbol{ji} = \boldsymbol{k} \\ \boldsymbol{jk} = -\boldsymbol{kj} = \boldsymbol{i} \\ \boldsymbol{ki} = -\boldsymbol{ik} = \boldsymbol{j} \end{cases}$$

Comparison between Eqs. (6.50) and (6.53) shows that the multiplying sequence of quaternions is opposite to that of attitude matrices.

6.2.2 Attitude Kinematics Equation

Attitude kinematics equation describes the change of attitude parameters during the attitude maneuver. The angular velocity of $\mathcal{F}_b$ with respect to $\mathcal{F}_r$ is denoted as $\boldsymbol{\omega}$. Then, $\boldsymbol{\omega} = \omega \boldsymbol{e}$, where $\omega = \|\boldsymbol{\omega}\|$ and $\boldsymbol{e}$ is the rotation axis. Denote $\boldsymbol{C}(t)$ and $\boldsymbol{C}(t + \Delta t)$ as the attitude matrices at t and $t + \Delta t$, respectively, then

$$\boldsymbol{C}(t + \Delta t) = \boldsymbol{C}'\boldsymbol{C}(t) \tag{6.54}$$

$\boldsymbol{C}'$ is the rotation matrix obtained after rotating around the axis $\boldsymbol{e}$ by the angle $\Phi = \omega\Delta t$. According to the Euler axis/angle equation, this rotation matrix can be written as

$$\boldsymbol{C}' = \cos\Phi \boldsymbol{I} + (1 - \cos\Phi)\boldsymbol{e}\boldsymbol{e}^{\mathrm{T}} - \sin\Phi[\boldsymbol{e}\times] \tag{6.55}$$

For small angle approximations, $\sin\Phi \approx \omega\Delta t$ and $\cos\Phi \approx 1$, then

$$\boldsymbol{C}' = \boldsymbol{I} - [\boldsymbol{\omega}\times]\Delta t \tag{6.56}$$

Therefore, the attitude matrix at $t + \Delta t$ can be expanded as

$$\boldsymbol{C}(t + \Delta t) = \boldsymbol{C}(t) - \Delta t[\boldsymbol{\omega}\times]\boldsymbol{C}(t), \tag{6.57}$$

so that the attitude kinematics equation in the form of attitude matrix can be obtained

$$\frac{\mathrm{d}\boldsymbol{C}(t)}{\mathrm{d}t} = \lim_{\Delta t \to 0} \frac{\boldsymbol{C}(t + \Delta t) - \boldsymbol{C}(t)}{\Delta t} = -[\boldsymbol{\omega}\times]\boldsymbol{C}(t) \tag{6.58}$$

By using the relationship between quaternion and direction cosines, the equation of quaternion change can be obtained. Take q_4 as an example. Differentiating Eq. (6.48) can obtain

$$\dot{q}_4 = \frac{1}{8q_4}(\dot{C}_{xx} + \dot{C}_{yy} + \dot{C}_{zz}) \tag{6.59}$$

According to Eq. (6.58), the derivatives of matrix elements can be listed as

$$\begin{cases} \dot{C}_{xx} = \omega_z C_{yx} - \omega_y C_{zx} \\ \dot{C}_{yy} = -\omega_z C_{xy} + \omega_x C_{zy} \\ \dot{C}_{zz} = \omega_y C_{xz} - \omega_x C_{yz} \end{cases}$$

By substituting the equation of $\dot{q}_4$ and using Eq. (6.48), the following equation can be obtained

$$\dot{q}_4 = -\frac{1}{2}\left(q_1\omega_x + q_2\omega_y + q_3\omega_z\right) \tag{6.60}$$

Similarly, the differential formula of other quaternions can be obtained and reduced to

$$\begin{bmatrix} \dot{q}_1 \\ \dot{q}_2 \\ \dot{q}_3 \\ \dot{q}_4 \end{bmatrix} = \frac{1}{2}\begin{bmatrix} 0 & \omega_z & -\omega_y & \omega_x \\ -\omega_z & 0 & \omega_x & \omega_y \\ \omega_y & -\omega_x & 0 & \omega_z \\ -\omega_x & -\omega_y & -\omega_z & 0 \end{bmatrix}\begin{bmatrix} q_1 \\ q_2 \\ q_3 \\ q_4 \end{bmatrix}$$

$$= \frac{1}{2}\begin{bmatrix} q_4 & -q_3 & q_2 \\ q_3 & q_4 & -q_1 \\ -q_2 & q_1 & q_4 \\ -q_1 & -q_2 & -q_3 \end{bmatrix}\begin{bmatrix} \omega_x \\ \omega_y \\ \omega_z \end{bmatrix}$$

The above equation is the kinematics equation of quaternion. It can be simplified as

$$\dot{\boldsymbol{q}} = \frac{1}{2}\boldsymbol{\Omega}(\boldsymbol{\omega})\boldsymbol{q} \tag{6.61}$$

where

$$\boldsymbol{\Omega}(\boldsymbol{\omega}) = \begin{bmatrix} 0 & \omega_z & -\omega_y & \omega_x \\ -\omega_z & 0 & \omega_x & \omega_y \\ \omega_y & -\omega_x & 0 & \omega_z \\ -\omega_x & -\omega_y & -\omega_z & 0 \end{bmatrix} \tag{6.62}$$

Equation (6.61) is a linear differential equation without trigonometric functions or singularities. Its solution must satisfy the constraint Eq. (6.45).

The kinematics equation of attitude Euler angles can be obtained from the rotation sequence of Euler angles. In the body coordinate system, the rotation speed of attitude relative to reference coordinate system can be expressed as

$$\boldsymbol{\omega} = \omega_x \boldsymbol{x}_b + \omega_y \boldsymbol{y}_b + \omega_z \boldsymbol{z}_b$$

This rotation speed can be treated as a synthesis of three Euler rotations. For example, for the sequences 3-1-3 and 3-1-2,

$$\boldsymbol{\omega}_{313} = C_3(\varphi)\left\{\begin{bmatrix} 0 \\ 0 \\ \dot{\varphi} \end{bmatrix} + C_1(\theta)\left[\begin{bmatrix} 0 \\ \dot{\theta} \\ 0 \end{bmatrix} + C_3(\psi)\begin{bmatrix} 0 \\ 0 \\ \dot{\psi} \end{bmatrix}\right]\right\},$$

i.e.,

$$\begin{bmatrix} \omega_x \\ \omega_y \\ \omega_z \end{bmatrix}_{313} = \begin{bmatrix} \dot{\psi} \sin\theta \sin\varphi + \dot{\theta} \cos\varphi \\ \dot{\psi} \sin\theta \cos\varphi - \dot{\theta} \sin\varphi \\ \dot{\psi} \cos\theta + \dot{\varphi} \end{bmatrix} \tag{6.63}$$

and

$$\boldsymbol{\omega}_{312} = C_2(\theta)\left\{\begin{bmatrix} 0 \\ \dot{\theta} \\ 0 \end{bmatrix} + C_1(\varphi)\left[\begin{bmatrix} \dot{\varphi} \\ 0 \\ 0 \end{bmatrix} + C_3(\psi)\begin{bmatrix} 0 \\ 0 \\ \dot{\psi} \end{bmatrix}\right]\right\},$$

i.e.,

$$\begin{bmatrix} \omega_x \\ \omega_y \\ \omega_z \end{bmatrix}_{312} = \begin{bmatrix} -\dot{\psi} \sin\theta \cos\varphi + \dot{\varphi} \cos\theta \\ \dot{\psi} \sin\varphi - \dot{\theta} \\ \dot{\psi} \cos\theta \cos\varphi + \dot{\varphi} \sin\theta \end{bmatrix} \tag{6.64}$$

For the sequence 3-1-3, the kinematic equation of attitude angle can be derived from Eq. (6.63):

$$\begin{bmatrix} \dot{\psi} \\ \dot{\theta} \\ \dot{\varphi} \end{bmatrix}_{313} = \frac{1}{\sin\theta}\begin{bmatrix} \omega_x \sin\varphi + \omega_y \cos\varphi \\ \omega_x \cos\varphi \sin\theta - \omega_y \sin\varphi \sin\theta \\ -\omega_x \sin\varphi \cos\theta - \omega_y \cos\varphi \cos\theta + \omega_z \sin\theta \end{bmatrix} \tag{6.65}$$

For the sequence 3-1-2, the kinematic equation of attitude angle can be derived from Eq. (6.64):

$$\begin{bmatrix} \dot{\psi} \\ \dot{\varphi} \\ \dot{\theta} \end{bmatrix}_{312} = \frac{1}{\cos\varphi}\begin{bmatrix} -\omega_x \sin\theta + \omega_z \cos\theta \\ \omega_x \cos\varphi \cos\theta - \omega_z \cos\varphi \sin\theta \\ \omega_x \sin\varphi \sin\theta + \omega_y \cos\varphi - \omega_z \sin\varphi \cos\theta \end{bmatrix} \tag{6.66}$$

The above two sets of equations are nonlinear equations. It can be seen that Eq. (6.65) is singular when $\theta = 0°$ or $\theta = 180°$ and Eq. (6.66) is singular when $\varphi = 90°$ or $\varphi = 270°$.

6.3 Mars Environment Model

6.3.1 Mars Ellipsoid Model

To describe the topography of Martian surface, Mars ellipsoid model needs to be defined first. Currently, two Mars ellipsoid models are commonly used, including ellipsoid model and spheroid model. Their specific parameters are slightly different,

Table 6.1 Typical Mars ellipsoid models

Type of ellipsoid	Semi-major axis (km)	Semi-minor axis (km)	Remarks
Ellipsoid	3396.19	3376.20	Less practical application
Spheroid	3396.0	3396.0	Topographic data from HRSC
Spheroid	3396.19	3396.19	Topographic data from MOLA

as given in Table 6.1. Due to the relatively complex application of ellipsoid model in map projection calculation, the ellipsoid IAU_MARS_2000 defined by IAU has not gained popularity. Actual Mars exploration missions often use a spheroid for terrain data processing. For example, a spheroid with the radius of 3396.0 km is used for processing the topographic data obtained by the high-resolution stereo camera (HRSC) of "Mars Express," and a spheroid with the radius of 3396.19 km for processing the topographic data obtained by the Mars orbiter laser altimeter (MOLA) of "Mars Global Surveyor" [6].

6.3.2 *Mars Gravitation Field Model*

Mars gravitation field model was established with the method of spherical harmonic function. In 1995, NASA released the 50th-order Mars gravitation field model "Masr50c," which had been developed on the basis of the Doppler tracking data from "Viking 1," "Viking 2," and "Mariner 9" probes [7]. In 2001, NASA released the 80th-order model "GMM-2B" (Table 6.2). This model had been developed using the X-band tracking data obtained by "Mars Global Surveyor" from October 1997 to February 2000 and the ranging data obtained by MOLA from March 1997 to December 1997 [8]. In 2016, NASA released the 120th-order model "MRO120D" based on years of tracking data from Mars reconnaissance orbiter [9].

6.4 Asteroid Environment Model

6.4.1 *Asteroid 3D Model*

Due to a relatively small mass, the gravitation of asteroids is not enough to overcome solid stress and shape them into almost a spheroid, so they often present various shapes. In addition, their surface is often densely filled with the craters caused by meteorite impacts [10], as shown on the surface of Toutatis asteroid in Fig. 6.3.

Table 6.2 First four orders of spherical harmonic coefficients in the Mars gravitation field model "GMM-2B"

Order, n	Degree, m	$\overline{C}_{nm}$	$\overline{S}_{nm}$
2	0	$-8.745054708 \times 10^{-4}$	0
3	0	$-1.188691065 \times 10^{-5}$	0
4	0	$5.125798718 \times 10^{-6}$	0
2	1	$1.3938449166 \times 10^{-10}$	$1.704428064 \times 10^{-10}$
2	2	$-8.417751981 \times 10^{-5}$	$4.960534884 \times 10^{-5}$
3	1	$3.905344232 \times 10^{-6}$	$2.513932404 \times 10^{-5}$
3	2	$-1.586341103 \times 10^{-5}$	$8.485798716 \times 10^{-6}$
3	3	$3.533854114 \times 10^{-5}$	$2.511398426 \times 10^{-5}$
4	1	$4.227157505 \times 10^{-6}$	$3.741321503 \times 10^{-6}$
4	2	$-1.025388411 \times 10^{-6}$	$-8.962295163 \times 10^{-6}$
4	3	$6.446128873 \times 10^{-6}$	$-2.729779031 \times 10^{-7}$
4	4	$9.638433482 \times 10^{-8}$	$-1.286136169 \times 10^{-5}$

Note $\overline{C}_{nm}$ and $\overline{S}_{nm}$ are the normalized spherical harmonic coefficients in Eq. (6.4)

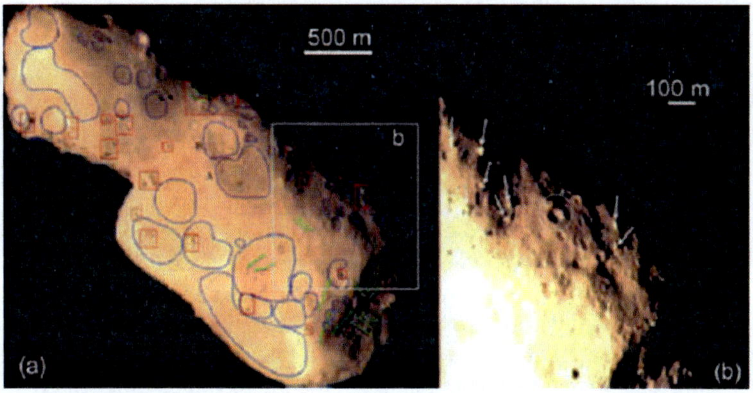

Fig. 6.3 Toutatis's appearance photographed by Chang'E-2

To describe the shape of an asteroid as accurately as possible, a polyhedron whose surface is composed of a series of triangles can be used to approximate the asteroid. This polyhedron is just the polyhedral model of the asteroid. Taking Eros433 as an example, a polyhedron composed of 5184 faces can approximate the asteroid shape very well. The polyhedral models of asteroids are available from NASA's Planetary Data System. The contents of the polyhedral model package provided by NASA are divided into two parts. The first part lists the positions of the points on polyhedral surface. For example, the first point in the data package is "1 8.33272E+00 −4.80902E+00 −4.78454E+00," where 1 is the number of this point, and the three numbers (in km) behind represent the position components of this point in the asteroid-fixed coordinate system. The second part gives the triangle index of the polyhedron. For example, one triangle is written as "1 1 105 101," where the first 1 is the number of that triangle and the three numbers behind are the numbers of its three vertexes. By querying the first part, the location of that triangle in the asteroid-fixed coordinate system can be determined [11].

6.4.2 Asteroid Gravitation Field Model

Studies show that, because of long-term spatial weathering, the internal structure of most asteroids is not a whole dense rock, but a heaped structure composed of many gravels gathered by gravitation [12]. That is to say, the internal density distribution of an asteroid is not uniform. This, together with the irregularity of asteroid shape, makes it difficult to establish an accurate asteroid gravitation field model.

The method of spherical harmonic function can approximate the gravitational potential function of an asteroid through an infinite series. However, its main problem is the lack of orbit data to determine the spherical harmonic coefficient and the dependence of its convergence upon the calculated position. When the calculated position is close to the effective radius, namely Brillouin ellipsoid, convergence is very slow. When the calculated position is within the effective radius, the calculation result will have a large error and even diverge completely. Therefore, this method is not suitable for the surface gravitation field modeling of small celestial bodies.

The polyhedron model method uses a polyhedral model to approximate the shape of an irregular asteroid and then deals with the volume fraction of gravitational field through integral transformation to obtain the gravitational potential of any polyhedron. Therefore, it can be used for the gravitational field modeling of any small celestial body that is irregular and homogeneous. Now this method has been widely used in the analysis of the gravitational field characteristics of irregular small celestial bodies and in the related studies of their orbital dynamics [13, 14]. The formula for calculating the gravitational potential of polyhedral model is shown in Eq. (6.14).

6.5 Brief Summary

This chapter addresses basic knowledge on the orbit dynamics model, attitude kinematics model, and celestial environment model of spacecrafts. The spacecraft orbit perturbations mainly include the perturbations caused by the gravitation and shape of central body, the third-body gravitation, the solar radiation pressure, and the atmospheric drag. Then, the orbit dynamics equations are established for different phases of deep-space exploration missions. Spacecraft attitude is usually expressed in four parametric forms, namely direction cosine matrix, Euler angles, Euler axis/angle, and quaternion. The attitude of a spacecraft at a certain moment is uniquely determined, so various forms of attitude parameters can be converted to each other. Attitude matrix and quaternion are more commonly used in the research of navigation algorithm. The former is used for the conversion of various vectors between different coordinate systems, while the latter is for describing the state of spacecraft attitude in the inertial system without producing singularities. In terms of environment modeling, the shape of large celestial bodies (such as Mars) can be effectively described with a simple revolving ellipsoid model owing to their shape regularity. Gravitational field model can be accurately calculated through spherical harmonic function. The shape of small celestial bodies (such as asteroids) is usually irregular and thus described with a polyhedral model pieced together by many triangles, which is also used to calculate their gravitational field.

References

1. Roman, G., and B. Jean-Pierre. 2001. Ellipsoidal harmonic expansions of the gravitational potential: Theory and application. *Celestial Mechanics & Dynamical Astronomy* 79 (4): 235–275.
2. Casotto, S., and S. Musotto. 2013. Methods for computing the potential of an irregular, homogeneous, solid body and its gradient. In *Astrodynamics specialist conference*, 82–96.
3. Park, R.S., R.A. Werner, and S. Bhaskaran. 2010. Estimating small-body gravity field from shape model and navigation data. *Journal of Guidance, Control and Dynamics* 33 (1): 212–221.
4. Zhang, Renwei. 1998. *Attitude dynamics and control of satellite orbit*. Beijing: Beihang University Press.
5. Lu, Zhenduo, and Yongjun Lei. 2013. *Satellite attitude measurement and determination*. Beijing: National Defense Industry Press.
6. Xu, Qing, Xun Geng, Chaozhen Lan, et al. 2014. Review of mars topographic mapping. *Journal of Deep Space Exploration* 1 (1): 28–35.
7. Konopliv, A.S., and W.L. Sjogren (1995) The JPL Mars gravity field, Mars50c, based upon Viking and Mariner 9 Doppler tracking data. NASA STI/RECON Technical Report N.
8. Lemoine, F.G., D.E. Smith, D.D. Rowlands, et al. 2001. An improved solution of the gravity field of Mars (GMM-2B) from Mars global surveyor. *Journal of Geophysical Research: Planets* 106 (E10): 23359–23376.
9. Konopliv, A.S., R.S. Park, and W.M. Folkner. 2016. An improved JPL Mars gravity field and orientation from Mars orbiter and lander tracking data. *Icarus* 274: 253–260.
10. Li, Junfeng, Hexi Baoyin, and Fanghua Jiang. 2014. *Dynamics and control of interplanetary flight*. Beijing: Tsinghua University Press.
11. Cui, Hutao, Zhenjiang Zhang, and Meng Yu. 2012. Computing and analysis of gravity field of Eros433 using polyhedron model. *Journal of Haerbin Institute of Technology* 44 (3): 17–22.
12. Richardson, D.C., Z.M. Leinhardt, H.J. Melosh, et al. 2002. Gravitational aggregates: evidence and evolution. In *Asteroids III*, ed. F. William, et al., 501–515. Tucson, AZ: University of Arizona Press.
13. Xiao, Yao, Xiaogang Ruan, and Ruoyan Wei. 2016. Research on the accuracy and operation time of polyhedron gravity model base on 433 Eros. *Journal of Deep Space Exploration* 3 (1): 41–46.
14. Jiang, Yu, and Hengnian Li. 2017. *Asteroid detector orbital mechanics*, 18. Beijing: China Aerospace Publishing House.

Chapter 7
Inertial Autonomous Navigation Technology

In this chapter, the measurement equations of gyroscope and accelerometer and the differential equations of strapdown inertial navigation system (SINS) are presented at first. For high accurate application, in order to reduce the influence of output noise on the calculation accuracy of the system and make full use of the output information, the outputs of gyroscope and accelerometer are all increments, namely the angle increment output by gyroscope and the velocity increment output by accelerometer. The core of strapdown inertial navigation algorithm is the attitude, velocity, and position update equation. Among them, attitude update resolving is the most important. Updating equations are generally obtained by solving the integrals of differential equations. When the spacecraft is in angular and linear vibration or maneuvering, the coning and sculling effects will be caused in the attitude and velocity resolving, respectively. The multi-sample algorithm can effectively compensate for these effects [1]. The installation error and parameter error of gyroscope and accelerometer are important factors that cause the navigation error. The error of inertial navigation will accumulate with time, so the calibration and error compensation of inertial devices are very meaningful as an important way to improve the precision of inertial navigation. Two methods of calibrating the gyro installation error, scale factor error, and constant bias of a spacecraft are also presented in this chapter. Finally, lunar soft landing is taken as an example to describe the application of inertial navigation technique to space missions.

7.1 Measurement Equation

7.1.1 Gyroscope Measurement Equation

Take optical gyro module as an example for the illustration of gyro measurement equation. Without loss of generality, assume that gyroscopes are installed along each of the three axes of $\mathcal{F}_b$. Then, without considering the dynamic response, the

D. Wang et al., *Spacecraft Autonomous Navigation Technologies Based on Multi-source Information Fusion*, Space Science and Technologies, https://doi.org/10.1007/978-981-15-4879-6_7

measurement equation of the gyros is given by [2]

$$\tilde{\boldsymbol{\omega}}_{bi}^{b} = (\boldsymbol{I}_3 + \boldsymbol{\Delta}_{\mathrm{G}} + [\boldsymbol{\Psi}_{\mathrm{G}}\times])\boldsymbol{\omega}_{bi}^{b} + \boldsymbol{b}_{\mathrm{G}} + \boldsymbol{d}_{\mathrm{G}} + \boldsymbol{n}_{\mathrm{G}} \tag{7.1}$$

$$\dot{\boldsymbol{b}}_{\mathrm{G}} = 0 \tag{7.2}$$

$$\dot{\boldsymbol{d}}_{\mathrm{G}} = \boldsymbol{n}_{\mathrm{d}} \tag{7.3}$$

$\tilde{\boldsymbol{\omega}}_{bi}^{b}$ and $\boldsymbol{\omega}_{bi}^{b}$ are measured and actual values of angular velocity of the $\mathcal{F}_b$ with respect to $\mathcal{F}_i$ expressed in $\mathcal{F}_b$, $\boldsymbol{\Delta}_{\mathrm{G}} \in \mathbb{R}^{3\times3}$ is the scale factor error matrix, $\boldsymbol{\Psi}_{\mathrm{G}} \in \mathbb{R}^3$ is the installation angle error matrix; $\boldsymbol{b}_{\mathrm{G}}$ is static gyroscope bias, $\boldsymbol{d}_{\mathrm{G}}$ is the angular rate random walk of gyro, $\boldsymbol{n}_{\mathrm{G}}$ is the angle random walk of gyro, and both of $\boldsymbol{n}_{\mathrm{G}}$ and $\boldsymbol{n}_{\mathrm{d}}$ are zero-mean Gaussian white noises.

The measurement model of optical gyro is shown in Fig. 7.1. In the figure, ω_{n} is the natural oscillation frequency of the second-order model, and $\boldsymbol{\omega}_{\mathrm{out}}$ is the actual output of gyros.

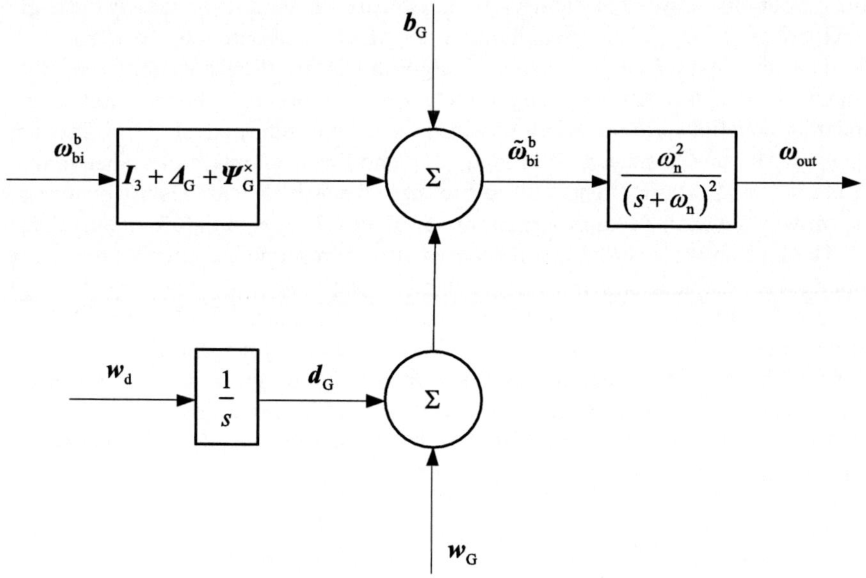

Fig. 7.1 Measurement model of optical gyro

7.1.2 Accelerometer Measurement Equation

Take quartz accelerometer module as an example for the illustration of accelerometer measurement equation. Without loss of generality, assume that accelerometers are installed along each of the three axes of spacecraft coordinate system. Then, without considering the dynamic responses of accelerometer, the measurement equation of accelerometers is given by

$$\tilde{\boldsymbol{a}}^b = (\boldsymbol{I}_3 + \boldsymbol{\Delta}_{\mathrm{A}} + [\boldsymbol{\Psi}_{\mathrm{A}}\times])\boldsymbol{a}^b + \boldsymbol{b}_{\mathrm{A}} + \boldsymbol{n}_{\mathrm{A}} \tag{7.4}$$

$$\dot{\boldsymbol{b}}_{\mathrm{A}} = 0 \tag{7.5}$$

where $\tilde{\boldsymbol{a}}^b$ and $\boldsymbol{a}^b$ are measured and actual values of specific force, $\boldsymbol{\Delta}_{\mathrm{A}} \in \mathbb{R}^{3\times3}$ is the scale factor error matrix, $\boldsymbol{\Psi}_{\mathrm{A}} \in \mathbb{R}^3$ is the installation angle error, $\boldsymbol{b}_{\mathrm{A}}$ is the zero bias, and $\boldsymbol{n}_{\mathrm{A}}$ is zero-mean Gaussian white noise.

The measurement model of accelerometer is shown in Fig. 7.2.

7.2 Differential Equation of Strapdown Inertial Navigation

When applying the inertial navigation technique to a spacecraft, the position and velocity of the spacecraft are calculated directly in an inertial frame $\mathcal{F}_i$. Denote the attitude quaternion of $\mathcal{F}_b$ with respect to $\mathcal{F}_i$ as $\boldsymbol{q} = \begin{bmatrix} q_1 & q_2 & q_3 & q_4 \end{bmatrix}^{\mathrm{T}}$. Then, the attitude kinematics equation is given by Eq. (6.61), and rewritten here for the sake of convenience.

$$\dot{\boldsymbol{q}} = \frac{1}{2}\boldsymbol{\Omega}(\boldsymbol{\omega}_{bi}^b)\boldsymbol{q} \tag{7.6}$$

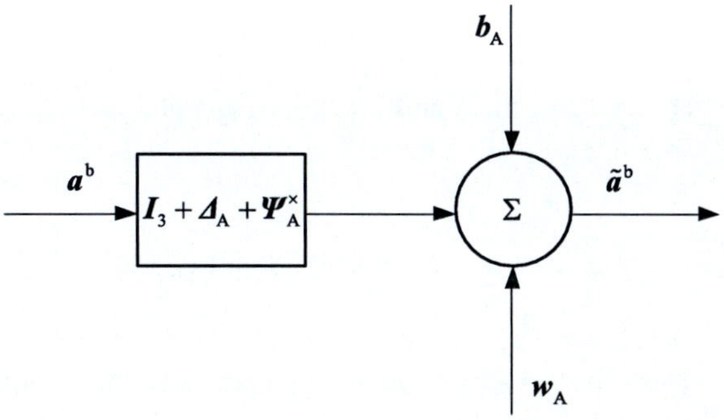

Fig. 7.2 Measurement model of accelerometer

where $\boldsymbol{\omega}_{bi}^{b}$ is the angular velocity of the spacecraft, which can be measured by gyros. Equation (7.6) is used to update the attitude quaternion.

Denote the position and velocity vectors of the spacecraft in $\mathcal{F}_i$ as $\boldsymbol{r}^i$ and $\boldsymbol{v}^i$. Then, the differential equations of position and velocity in $\mathcal{F}_i$ are

$$\dot{\boldsymbol{r}}^i = \boldsymbol{v}^i \tag{7.7}$$

$$\dot{\boldsymbol{v}}^i = \boldsymbol{C}_b^i \boldsymbol{a}^b + \boldsymbol{g}^i \tag{7.8}$$

where $\boldsymbol{C}_b^i$ is the attitude matrix of from $\mathcal{F}_b$ to $\mathcal{F}_i$, $\boldsymbol{g}^i$ is the gravitational acceleration expressed in $\mathcal{F}_i$. In addition to the gravitation of the central body, the third-body gravitation should be taken into account in the accurate calculation of gravitational acceleration. However, for some applications, such as the accuracy requirement is not high or the navigation time is short, the gravitational acceleration may be modeled as a two-body model:

$$\boldsymbol{g}^i = -\frac{\mu}{r^3} \boldsymbol{r}^i \tag{7.9}$$

where μ is the gravitational constant of the central body, and $r = \left\| \boldsymbol{r}^i \right\|$.

7.3 Strapdown Inertial Navigation Update Equations

7.3.1 *Attitude Update Equation*

Denote the angle increment of the spacecraft measured by gyro in a navigation update period as $\Delta\boldsymbol{\theta}$, and the updating equation for Eq. (7.6) is given by [1]

$$\boldsymbol{q}_{k+1} = \boldsymbol{e}^{\frac{1}{2}\int_{t_k}^{t_{k+1}} \boldsymbol{\Omega}(\boldsymbol{\omega}_{bi}^{b})\mathrm{d}\tau} \boldsymbol{q}_k \tag{7.10}$$

where the subscript k and $k+1$ correspond to a quantity at t_k and t_{k+1}, respectively.

Let

$$\Delta\boldsymbol{\Theta} = \int_{t_k}^{t_{k+1}} \boldsymbol{\Omega}\left(\boldsymbol{\omega}_{bi}^{b}\right)\mathrm{d}\tau \approx \boldsymbol{\Omega}(\Delta\boldsymbol{\theta})$$

Then, the Taylor expansion of exponential function for Eq. (7.10) gives

$$q_{k+1} = \left[I + \frac{\frac{\Delta\boldsymbol{\Theta}}{2}}{1!} + \frac{\left(\frac{\Delta\boldsymbol{\Theta}}{2}\right)^2}{2!} + \cdots \right] q_k \tag{7.11}$$

Usually, the spacecraft attitude update is calculated by using the first two terms of the above Taylor expansion. This is called single-sample algorithm.

However, the angle increment in a navigation cycle is not infinitesimal, so when the rigid body is in finite rotation, its angular position in space is related to the rotation sequence, that is, the non-commutative error of rotation is caused. But this error is ignored by single-sample algorithm. Therefore, in the case of high dynamic motion, the accuracy will decrease. This single-sample algorithm is only applicable to the low-dynamic environment where the angular motion is almost unchanged.

A solution for high dynamic motion is using the rotation vector algorithm for attitude updating and to compensate for the non-commutative error in an attitude updating period by using more than two subsamples, so as to improve the attitude accuracy under dynamic conditions. The rotation vector defines an axis of rotation and magnitude for rotation about the axis, and describes the state change required to transition from one attitude to another. The subsample refers to a number of gyro measurement outputs in a cycle of attitude updating. For example, if gyro measurement period is T and attitude updating period is NT, then the number of subsamples will be N. A bigger N means higher accuracy, but a heavier calculation burden.

The steps of updating the attitude quaternion with the rotation vector algorithm are:

1. Calculate the rotation vector

The single-sample algorithm of rotation vector is

$$\boldsymbol{\Phi} = \Delta\boldsymbol{\theta} \tag{7.12}$$

where $\Delta\boldsymbol{\theta}$ is the angle increment collected in the navigation cycle.

In the single-sample algorithm, the formula is simple and the sampling frequency of gyro is low, but the calculation accuracy under coning motion is also low. To calculate the rotation vector more accurately, multi-sample algorithm can be used, which will be presented in 7.4.1.

2. Construct the quaternion of attitude rotation with rotation vector

Use the rotation vector to calculate the attitude quaternion corresponding to attitude rotation

$$q' = \begin{bmatrix} \frac{\boldsymbol{\Phi}}{\Phi} \sin\frac{\Phi}{2} \\ \cos\frac{\Phi}{2} \end{bmatrix} \tag{7.13}$$

where $\Phi = \|\boldsymbol{\Phi}\|$.

3. Attitude quaternion updating

Update the attitude quaternion through the multiplication operation

$$\boldsymbol{q}_{k+1} = \boldsymbol{q}_k \otimes \boldsymbol{q}' \tag{7.14}$$

Once $\boldsymbol{q}$ is updated, the attitude transformation matrix $\boldsymbol{C}_i^b$ can be calculated by

$$\boldsymbol{C}_i^b = \begin{bmatrix} 1-2\left(q_2^2+q_3^2\right) & 2(q_1q_2+q_3q_4) & 2(q_1q_3-q_2q_4) \\ 2(q_1q_2-q_3q_4) & 1-2\left(q_1^2+q_3^2\right) & 2(q_2q_3+q_1q_4) \\ 2(q_1q_3+q_2q_4) & 2(q_2q_3-q_1q_4) & 1-2\left(q_1^2+q_2^2\right) \end{bmatrix} \tag{7.15}$$

In a highly dynamic environment, multi-sample rotation vector algorithm is more accurate than single-sample algorithm when updating the attitude. With the navigation period unchanged, more subsamples will signify higher attitude update accuracy, but also a higher gyro-sampling frequency and a heavier calculation burden. In the system design and algorithm selection, the requirements for real time and accuracy shall be fully weighed to make a compromise.

7.3.2 Inertial Velocity Update Equation

Through the integration of Eq. (7.8) from t_k to t_{k+1}, the velocity update equation can be written as [1, 3]

$$\boldsymbol{v}_{k+1}^i = \boldsymbol{v}_k^i + \boldsymbol{C}_{b,k}^i \int_{t_k}^{t_{k+1}} \boldsymbol{C}_{b,\tau}^{b,t_k} \boldsymbol{a}^b \mathrm{d}\tau + \int_{t_k}^{t_{k+1}} \boldsymbol{g}^i \mathrm{d}\tau \tag{7.16}$$

where $\boldsymbol{C}_b^i = \left(\boldsymbol{C}_i^b\right)^{\mathrm{T}}$, $\boldsymbol{C}_{b,\tau}^{b,t_k}$ is the coordinate transformation matrix within $\mathcal{F}_b$ from t_k to τ; In this period, the attitude change is represented by a small angle $\Delta\boldsymbol{\theta}$. Then

$$\boldsymbol{C}_{b,\tau}^{b,t_k} \approx \boldsymbol{I} + [\Delta\boldsymbol{\theta}\times] \tag{7.17}$$

and

$$\begin{aligned} \boldsymbol{v}_{k+1}^i &= \boldsymbol{v}_k^i + \boldsymbol{C}_{b,k}^i \int_{t_k}^{t_{k+1}} \left(\boldsymbol{a}^b + \Delta\boldsymbol{\theta} \times \boldsymbol{a}^b\right) \mathrm{d}\tau + \int_{t_k}^{t_{k+1}} \boldsymbol{g}^i \mathrm{d}\tau \\ &= \boldsymbol{v}_k^i + \boldsymbol{C}_{b,k}^i \left(\Delta\boldsymbol{v}(k+1) + \int_{t_k}^{t_{k+1}} \Delta\boldsymbol{\theta} \times \boldsymbol{a}^b \mathrm{d}\tau \right) + \int_{t_k}^{t_{k+1}} \boldsymbol{g}^i \mathrm{d}\tau \end{aligned} \tag{7.18}$$

where $\Delta \boldsymbol{v}(k+1) = \int_{t_k}^{t_{k+1}} \boldsymbol{a}^b \mathrm{d}\tau$. Then, the velocity compensation term caused by the specific force is

$$\begin{aligned}\Delta \boldsymbol{v}_{sf,k+1} &= \Delta \boldsymbol{v}(k+1) + \int_{t_k}^{t_{k+1}} \Delta \boldsymbol{\theta} \times \boldsymbol{a}^b \mathrm{d}\tau \\ &= \Delta \boldsymbol{v}(k+1) + \frac{1}{2}\Delta \boldsymbol{\theta}(k+1) \times \Delta \boldsymbol{v}(k+1) \\ &+ \frac{1}{2}\int_{t_k}^{t_{k+1}} \left(\Delta \boldsymbol{\theta} \times \boldsymbol{a}^b + \Delta \boldsymbol{v} \times \boldsymbol{\omega}_{bi}^b\right)\mathrm{d}\tau \end{aligned} \tag{7.19}$$

where $\Delta \boldsymbol{\theta}(k+1) = \int_{t_k}^{t_{k+1}} \boldsymbol{\omega}_{bi}^b \mathrm{d}\tau$; $\Delta \boldsymbol{v} = \int_{t_k}^{t} \boldsymbol{a}^b \mathrm{d}\tau$; The second and third terms on the right of Eq. (7.19) are the rotation compensation term and sculling effect compensation term of the velocity, respectively

$$\Delta \boldsymbol{v}_{\mathrm{rot},k+1} = \frac{1}{2}\Delta \boldsymbol{\theta}(k+1) \times \Delta \boldsymbol{v}(k+1) \tag{7.20}$$

$$\Delta \boldsymbol{v}_{\mathrm{scul},k+1} = \frac{1}{2}\int_{t_k}^{t_{k+1}} \left(\Delta \boldsymbol{\theta} \times \boldsymbol{a}^b + \Delta \boldsymbol{v} \times \boldsymbol{\omega}_{bi}^b\right)\mathrm{d}\tau \tag{7.21}$$

Then, the velocity update equation be written as

$$\boldsymbol{v}_{k+1}^i = \boldsymbol{v}_k^i + \boldsymbol{C}_{b,k}^i\left[\Delta \boldsymbol{v}(k+1) + \Delta \boldsymbol{v}_{\mathrm{rot},k+1} + \Delta \boldsymbol{v}_{\mathrm{scul},k+1}\right] + \int_{t_k}^{t_{k+1}} \boldsymbol{g}^i \mathrm{d}\tau \tag{7.22}$$

The multi-sample calculation formula of sculling effect compensation term will be presented in 7.4.2. When the spacecraft is free from linear or angular vibration, this compensation term is zero. The simplified calculation formula for the integral term of gravitational acceleration is

$$\int_{t_k}^{t_{k+1}} \boldsymbol{g}^i \mathrm{d}\tau \approx \boldsymbol{g}_k^i T \tag{7.23}$$

$$\boldsymbol{g}_k^i \approx -\frac{\mu}{r_k^3}\boldsymbol{r}_k^i \tag{7.24}$$

where $T = t_{k+1} - t_k$ is navigation update period.

7.3.3 *Inertial Position Update Equation*

Through the integration of Eq. (7.7) from t_k to t_{k+1}, the inertial position update equation can be written as

$$\boldsymbol{r}_{k+1}^{i} = \boldsymbol{r}_{k}^{i} + \int\limits_{t_k}^{t_{k+1}} \boldsymbol{v}^{i}\mathrm{d}t \tag{7.25}$$

The integral term in Eq. (7.25) cannot be solved accurately, but can be approximately solved with the trapezoidal method as

$$\int\limits_{t_k}^{t_{k+1}} \boldsymbol{v}\mathrm{d}t \approx \frac{(\boldsymbol{v}_{k+1} + \boldsymbol{v}_k)}{2}T \tag{7.26}$$

Substituting Eq. (7.26) into Eq. (7.25) gives

$$\boldsymbol{r}_{k+1} \approx \boldsymbol{r}_k + \frac{(\boldsymbol{v}_{k+1} + \boldsymbol{v}_k)}{2}T \tag{7.27}$$

7.4 Compensation for Coning and Sculling Effects

Note that inertial navigation states are updated by discretizing the corresponding differential equations. High dynamic motion of spacecraft will bring errors into the discretization process. These errors include the attitude coning error, velocity sculling error, and position scroll error [1]. The scroll error has a small weight in the total position error of strapdown inertial navigation, so it will not be discussed in this book. This section will present multi-sample algorithms to compensate for the errors mentioned above, by dividing the navigation update interval into several sensor-data intervals.

7.4.1 *Coning Effect Compensation*

Among various forms of dynamics or vibration, coning motion has the worst effect on the accuracy of attitude solution. Coning motion refers to the coning or nearly coning motion of an axis of gyro module in inertial space, caused by slight angular vibration of the other two orthogonal axes in the equilibrium position, as shown in Fig. 7.3. Coning motion will induce the serious drift of the mathematical platform

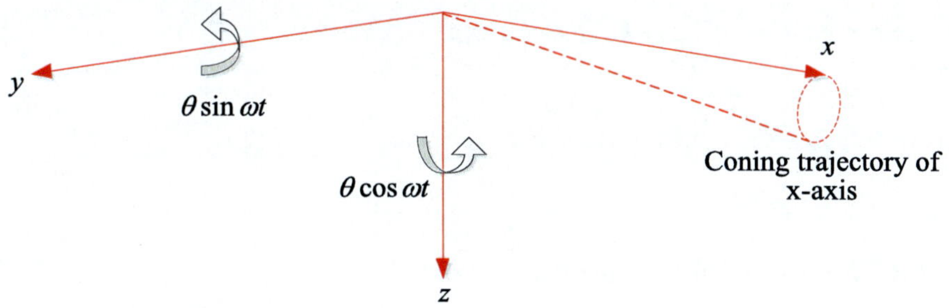

Fig. 7.3 Coning motion [4]

of strapdown inertial navigation.

In the typical coning motion, angular vibrations with the phase difference of 90° will be found on the two axes of the spacecraft, while a constant angular velocity related to vibration angle and amplitude will appear on the third axis. Assume that the angular vibrations on *y*-axis and *z*-axis have the same frequency but different phases. Then, the rotation vector of typical coning motion can be expressed as

$$\boldsymbol{\Phi} = \begin{bmatrix} 0 & A\sin\omega t & A\cos\omega t \end{bmatrix}^{\mathrm{T}} \tag{7.28}$$

where A is the amplitude of coning motion; ω is the frequency of coning motion. In this motion, a constant angular velocity will be sensed from the *x*-axis of the spacecraft. The output $\omega_{\mathrm{x}} = -2\omega\sin^2\frac{A}{2}$ is related to the amplitude and frequency of coning motion [5].

The coefficient of multi-sample algorithm is obtained by optimizing and minimizing the calculation error under the condition of typical coning motion. If the vibration frequency is lower than the frequency band of gyro, the calculation error can be reduced by optimizing the coefficient given the limited sampling frequency. Denote m as the number of samples during an attitude update period. Then, a m-sample algorithm of rotation vector can be written as

$$\boldsymbol{\Phi} = \sum_{i=1}^{m}\Delta\boldsymbol{\theta}_i + \sum_{i=1}^{m-1}\sum_{j=i+1}^{m}k_{ij}\Delta\boldsymbol{\theta}_i \times \Delta\boldsymbol{\theta}_j \tag{7.29}$$

where $\Delta\boldsymbol{\theta}_i$ is the angle increment over the ith interval within a navigation update cycle, k_{ij} is constant coefficient. The second term of the above equation is the coning correction term, different k_{ij} values can be obtained depending on the choice of optimal criteria, for example, the four-sample algorithm for the Chang'E-3 lunar lander [6] is given by

$$\begin{aligned}\boldsymbol{\Phi} &= \Delta\boldsymbol{\theta}_1 + \Delta\boldsymbol{\theta}_2 + \Delta\boldsymbol{\theta}_3 + \Delta\boldsymbol{\theta}_4 \\ &\quad + \frac{214}{315}(\Delta\boldsymbol{\theta}_1 \times \Delta\boldsymbol{\theta}_2 + \Delta\boldsymbol{\theta}_2 \times \Delta\boldsymbol{\theta}_3 + \Delta\boldsymbol{\theta}_3 \times \Delta\boldsymbol{\theta}_4)\end{aligned}$$

$$+\frac{46}{105}(\Delta\boldsymbol{\theta}_1 \times \Delta\boldsymbol{\theta}_3 + \Delta\boldsymbol{\theta}_2 \times \Delta\boldsymbol{\theta}_4)$$
$$+\frac{54}{105}\Delta\boldsymbol{\theta}_1 \times \Delta\boldsymbol{\theta}_4 \tag{7.30}$$

7.4.2 Sculling Effect Compensation

The sculling motion of the spacecraft appears when high-frequency angular and linear vibrations occur simultaneously on two or more orthogonal axes of the spacecraft. In this case, sculling effect compensation is required. Since the outputs of gyro and accelerometer are angle increment and velocity increment, the compensation term of sculling effect cannot be accurately calculated from Eq. (7.21). Furthermore, in some applications, the changes of angular velocity and acceleration are very complex due to high-frequency vibration and randomness, so they cannot be accurately described with definite functions. One solution is using typical sculling motion functions to approximate the motion, and thus, multi-sample methods can be applied to calculate the compensation term of sculling effect. In the sculling environment, the optimization algorithm of each subsample can be derived by taking the drift rate of an algorithm as accuracy criterion. The m-sample algorithm [7] for the compensation term of sculling effect is

$$\Delta\boldsymbol{v}_{\text{scul}} = \sum_{i=1}^{m-1}\sum_{j=i+1}^{m} k_{ij}\left(\Delta\boldsymbol{\theta}_i \times \Delta\boldsymbol{v}_j + \Delta\boldsymbol{v}_i \times \Delta\boldsymbol{\theta}_j\right) \tag{7.31}$$

where $\Delta\boldsymbol{v}_i$ is the velocity increment over the ith interval within a navigation update cycle, and k_{ij} is constant coefficient.

7.5 Calibration and Error Compensation of Inertial Devices

Because of the presence of initial alignment error as well as the measurement error and installation error of IMU (gyro and accelerometer), the inertial navigation error will be accumulated continuously. For soft landing missions, to increase INS's attitude accuracy, onboard calibration for gyros systematic errors is necessary. Li [8] proposed a two-layer onboard calibration algorithm for systematic errors of a 3-axis orthogonal gyroscope, including constant bias, scale factor error, and installation error. The first layer provides a relatively accurate angle velocity estimate for the second layer. Aided by star trackers, the second layer estimates the scale factor error, installation error, and constant bias of gyro via a UD decomposition-based Kalman

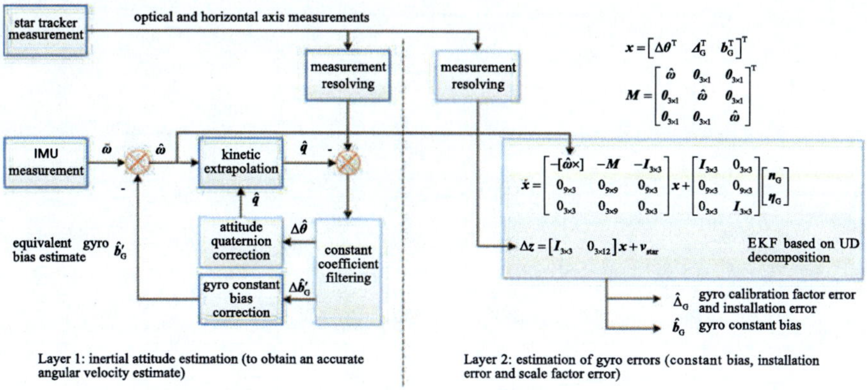

Fig. 7.4 Structure of layered filtering system

filter. To improve the filter observability, several appropriate attitude maneuvers are performed. The calibration algorithm has been successfully applied to the Chang'E-3 lunar soft landing. The structure of layered filtering system is shown in Fig. 7.4.

Yu [9] extended the calibration algorithm to the case of non-orthogonal gyroscopes, propose a lightweight calibration algorithm based on least square method, which achieved high accuracy calibration on-orbit within a limited amount of computing capacity. The ground mathematic simulation and the results of test validation show that this method has the similar accuracy to EKF, but the computational complexity is much less than EKF. This algorithm has already been used successfully in actual flight of circumlunar return and reentry spacecraft of third phase of China lunar exploration program, and thus achieved initial attitude alignment of two-sets of inertial navigating platforms with a high degree of accuracy.

7.6 Numerical Simulation

In this section, numerical simulations of the inertial navigation for the lunar soft landing are presented. The landing attitude and trajectory under the explicit guidance of constant thrust are adopted, the initial descending altitude is 15 km and the final altitude is 100 m. The initial position/velocity errors of navigation system in radial, normal, and forward directions are 1000 m and 1 m/s respectively. The initial attitude estimation error is 0.1° in each axis. The installation error of the IMU is 0.3°. In the inertial measurement unit, the bias estimation error of gyro is 0.5°/h, the random walk coefficient of angular rate of gyro is $0.02°/\sqrt{\text{h}}$, and the scale factor error of gyro is 0.1%; the bias estimation error of accelerometer is 0.002 m/s^2, the measurement noise of accelerometer is 0.005 m/s^2, and the scale factor error of accelerometer is 0.1%. The navigation update period is 100 ms. Figures 7.5 and 7.6 show the position/velocity estimation error of inertial navigation in a single run. It can

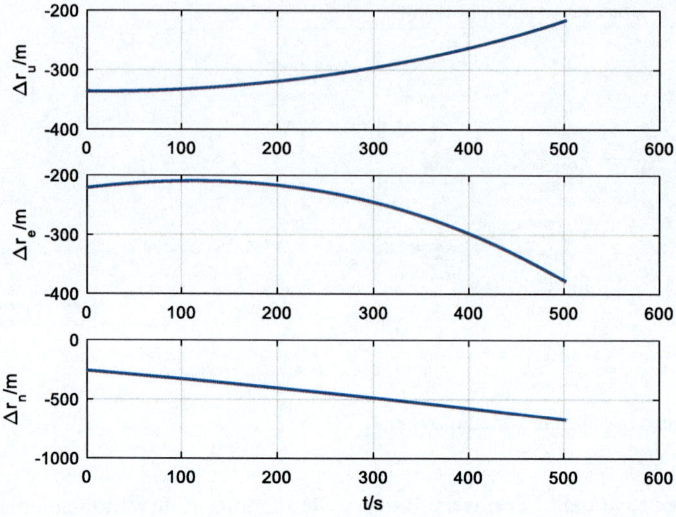

Fig. 7.5 Position estimation error of the autonomous navigation based on inertial measurement unit in lunar soft landing

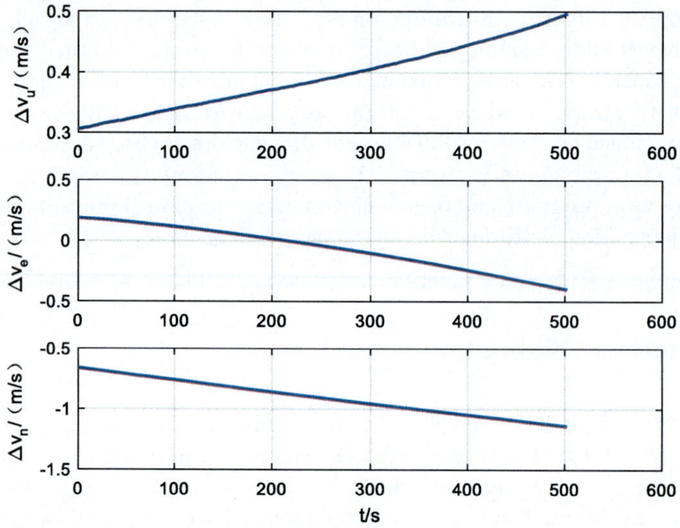

Fig. 7.6 Velocity estimation error of the autonomous navigation based on inertial measurement unit in lunar soft landing

be seen that the position/velocity estimation errors of inertial navigation accumulate and increase with time. Therefore, for lunar soft landing missions, inertial navigation alone cannot meet the requirement for navigation accuracy and other types of measurements shall be introduced to correct the position/velocity estimation errors.

7.7 Brief Summary

In this chapter, the measurement equations of gyro and accelerometer are firstly addressed, followed by the differential equation, INS update equation, and multi-sample compensation algorithm of strapdown inertial navigation system, and the calibration and error compensation methods of inertial devices. Finally, an example is given to illustrate the application of inertial autonomous navigation technique to spacecrafts. Inertial navigation is strongly autonomous and independent of external measurement information, and is also capable of long-time continuous working. It has become the core of navigation system in the deep-space exploration missions such as landing on the surface of an extraterrestrial body. The error in inertial navigation will accumulate over time, so it needs to be corrected by other types of measurements. In Chap. 9, the autonomous navigation technique based on inertial and ranging-velocity/optical fusion will be presented.

References

1. Qin, Yongyuan. 2006. *Inertial navigation.* Beijing: Science Press.
2. Lv, Zhenduo, and Yongjun Lei. 2013. *Satellite attitude measurement and determination.* Beijing: National Defense Industry Press.
3. Savage, P.G. 1998. Strapdown inertial navigation algorithm design, part 2: Velocity and position algorithms. *AIAA Journal of Guidance, Control and Dynamics* 21 (2): 208–221.
4. Titterton, D.H., and J.L. Weston. 2004. *Strapdown inertial navigation technology.* United Kingdom: The Institution of Electrical Engineers.
5. Qin, Yongyuan, and Shimiao Zhang. 2001. Research on petimal four-sample rotation vector algorithm for sins attitude updating. *Journal of Chinese Inertial Technology* 9 (4): 1–7.
6. Zhang, Honghua, Ji Li, Yifeng Guan, et al. 2014. Autonomous navigation for powered descent phase of Chang'E–3 lunar lander. *Control Theory & Applications* 31 (12): 1686–1694.
7. Ignagni, M.B. 1998. Duality of optimal strapdown sculling and coning algorithms. *Navigation* 45 (2): 85–95.
8. Li, Ji, Honghua Zhang, Yu Zhao, et al. 2014. In-flight calibration of the gyros of the Chang'E-3 lunar lander. *Scientia Sinica Technologica* 44 (6): 582–588.
9. Yu, Dan, Wenqiang Dong, Yong Wang, et al. 2015. Research and implementation of on-orbit self-calibration for gyroscope of circumlunar return and reentry spacecraft. *Scientia Sinica Technologica* 45 (2): 213–220.

Chapter 8 Optical Autonomous Navigation Technology

This chapter presents optical autonomous navigation technology. The principles of optical autonomous navigation and optical imaging sensors are introduced in 8.1 and Sect. 8.2, respectively. Section 8.3 addresses selection criteria of candidate navigation beacons. Section 8.4 discusses the measurement equation of optical autonomous navigation and classifies the observations into two categories: apparent radius information of large celestial bodies and LOS (line of sight) direction information. Section 8.5 studies the geometric observability of optical autonomous navigation and the navigation beacon planning algorithm based on geometric observability analysis. Section 8.6 is navigation filtering algorithm, presenting the implementation process of the optical autonomous navigation based on batch least squares and Kalman filtering algorithm. Section 8.7 simulates and demonstrates the optical autonomous navigation technologies based on different types of navigation beacons for deep-space transfer, approaching, and orbiting phases.

8.1 Principle of Optical Autonomous Navigation

In an optical autonomous navigation system, the spacecraft processes images of navigational celestial bodies (including the images of their surface) acquired by an onboard optical imaging sensor, extracts information of the navigational celestial bodies, and then combines such information with ephemeris information and the measurement information from other sensors (if required) to autonomously determine its own orbit and attitude through filtering algorithms.

In deep-space transfer phase, images of multiple navigation asteroids and background stars are often used for optical autonomous navigation, e.g., The "Deep Space 1" [1] takes the images of distant celestial bodies (such as planets and asteroids, with stars as the reference background) for navigation.

D. Wang et al., *Spacecraft Autonomous Navigation Technologies Based on Multi-source Information Fusion*, Space Science and Technologies, https://doi.org/10.1007/978-981-15-4879-6_8

When approaching a small celestial body, optical autonomous navigation is usually based on the center of the target body. First, the image captured by navigation camera is processed to obtain the target center information, and then, this image information is combined with the spacecraft attitude to estimate spacecraft's position and velocity information through navigation filter algorithms.

When flying around or flying by the targeted celestial body, optical autonomous navigation is usually based on the edge feature of the target. One method is to determine the spacecraft orbit by taking the extracted information of the target center and the apparent radius of the target as observations. Another method to determine the spacecraft's orbit is matching the acquired target image with the pre-processed target model.

In the descent phase of soft landing or the impact phase of impact mission, a large number of craters, rocks, and textures on the surface of the extraterrestrial body can be used as natural navigation beacons. By imaging these landmarks, processing their images and matching their features, and combining the obtained information with the information on landmark locations, the complete information on spacecraft position, velocity, and attitude can be estimated.

For autonomous navigation based on optical imaging, its important feature is to acquire the image data of celestial bodies through optical imaging sensors. Depending on the characteristics of different mission segments, the following types of navigation information can be obtained: center-point information, edge-point information, and feature-point information.

8.1.1 Center-Point Information

Through image processing, the center of Near-Earth Objects (NEOs) and the energy center of distant objects (stars) can be extracted from their images captured by optical imaging sensors. These center points are the most commonly used measurements for optical navigation in the mission segments such as transfer, approaching, orbiting/flyby, and impact. For example, the Deep Space 1 used optical imaging sensors to optically measure the asteroids and background stars and obtained their image information [2].

The use of the physical parameters of sensors can convert the center-point pixel information of NEOs and distant objects into LOS direction information, avoid complicated and inaccurate pixel calculation, and facilitate navigational positioning analysis by means of the concept of position plane in the astronomical navigation theory.

According to the LOS directions of celestial bodies, the LOS direction of a NEO and the angle between the LOS direction of the NEO and that of a distant object can be determined. This angle is called angular distance (the angle between the directions of two vectors at a certain time). For example, the angular between the centers of a star and an asteroid, and the angular between the Mars centers and Mars natural satellite. The biggest benefit of angular information is that it is independent of spacecraft attitude and remains unchanged in different coordinate systems. The angle measured in a body frame is equal to that in an inertial frame.

8.1.2 Edge-Point Information

Through image processing, edge-point information of a NEO can be extracted from its images captured by optical imaging sensors. Both the direction of a spherical NEO center and its apparent radius can be determined using edge-point information of a spherical NEO (or only part of its edge image when the spacecraft is too close to it), so as to provide the distance between the spacecraft and the NEO. Edge-point information is mainly when approaching and flying around/over large celestial objects or flying around small objects. For example, the spacecraft Clementine relies on lunar edge images to determine lunar apparent radius. When flying around a small celestial object with an irregular shape, the reference positions of its edge points can be determined by using its 3D model and matching its edge-point images with its image model.

8.1.3 Feature-Point Information

Through processing image, feature-point information of the target objects (including artificial targets) can be extracted from their surface images captured by optical imaging sensors. Then, this information can be matched with the stored model. The feature points are mainly used in attachment/landing, impact, and orbiting/flyby phase. For example, the MUSES-C mission [2] obtained the information on the visible landing sites and surface feature points of asteroids using optical imaging sensors.

8.2 Optical Imaging Sensor

Optical imaging sensor is the key component of optical autonomous navigation system. Depending on actual flight phases, technical specifications of optical imaging sensors may be different.

8.2.1 *Heliocentric Transfer Phase*

During the heliocentric transfer phase, the celestial bodies near the spacecraft are usually selected as observed objects, but they are still very far away from the spacecraft and very dim. Therefore, the sensor should have extremely high resolution and sensitivity to successfully obtain the navigation information meeting the accuracy requirements. The navigation sensor used by "Deep Space 1" is Miniature Integrated Camera and Spectrometer (MICAS) [1, 3], with a focal length of 677 mm. Its imaging element is a 1024 × 1024 CCD, where the field angle per pixel and total field angle are 13 μrad and 13.4 mrad (0.77°), respectively. The image code has 12 bits. The brightness ranges from 0 (dark) to 4095 (bright). Such a narrow field of view (FOV) and millions of pixels provide the optical imaging sensor of Deep Space 1 with very high resolution. The navigation sensor of Deep Space 1 is fixed with respect to the spacecraft and spacecraft attitude maneuvers are required when taking pictures.

8.2.2 *Approaching Phase*

During the approaching phase, the targeted celestial body is relatively close to the spacecraft, so the imaged target body is no longer a point, the size of which is constantly increasing. In this phase, a wider FOV and a shorter focal length optical sensor is required, so the sensor can have a smaller size and weight.

For example, an optical imaging sensor of Mars Reconnaissance Orbiter (MRO) is used to obtain the navigation images of Mars and its satellites (Phobos and Deimos) when approaching the Mars. The sensor is mounted at the aft end of the spacecraft, and its optical axis points toward the Mars during the approaching phase. The sensor's aperture is 6 cm and the FOV is 1.4 × 1.4°. In addition, the shutter times and exposure times can be adjusted [4].

To adapt to the rapid change of spacecraft direction relative to the target during and to meet the energy and other scientific exploration requirements, some optical imaging sensors contain a rotating mechanism or moving part that can change their FOV pointing. For example, the imaging sensor system of Stardust also includes a reflector. The installation of the sensor itself ensures that the optical axis is parallel to the $-Y$-axis of this spacecraft. The reflector is 45° to the sensor, and can rotate around the Y-axis of the spacecraft. In this way, the effective observation range of the sensor will cover the sky area from the $+X$-axis to the $-X$-axis in the $-Z$-axis direction of the spacecraft. For the optical sensor of Stardust, its design indices are as follows: focal length: 200 mm; Size of CCD: 1024 × 1024; FOV: 3.5°; the minimum shutter time: 5 ms [5].

8.2.3 Orbiting Phase

At the orbiting phase of a planet exploration mission, navigation measurement shall be performed by an optical imaging sensor with a large FOV, because the spacecraft is close to the target and the observation angle is big. The specific FOV size shall be determined according to the altitude of orbit and the size of the target, which is usually 100°–150°. The accuracy index of this sensor is determined according to the navigation accuracy requirement. If the navigation position accuracy is required to be 10 km, the imaging sensor accuracy indices are as follows: 3σ of random errors are 0.1° and systematic errors are about 0.01°.

In a small-body exploration mission, the spacecraft is very close to the target, but the target is small. Therefore, an optical imaging sensor with medium FOV can be configured. The specific FOV size is determined according to the altitude of orbiting orbit and the size of the target, which is usually 30°–60°. The sensor accuracy indices are determined according to the navigation accuracy requirement. If the navigation position accuracy is required to be 100 m, the imaging sensor accuracy is as follows (considering that the spacecraft is quite close, specifically 20–50 km, to the target): 3σ of random errors are 0.1° and systematic errors are about 0.01°.

8.2.4 Soft Landing Phase

The optical sensors used in the soft landing missions can be light detection and ranging (LIDAR) imaging sensor or optical camera, where LIDAR is an active sensor, and optimal camera is a passive sensor. The optical sensors are usually used for safe landing site selection. For example, an optical camera was configured at the bottom of the Chang'e-3 lander [6], where image size is 1024 × 1024 pixels and the field angle is 45.4°. When the lander was 15 km away from the lunar surface, the descending camera was activated. During the descent and landing, it acquired 4672 descending images of the lander at the speed of 10 frames/s. The "Curiosity" Mars rover carries two cameras with the same CCD sensor. One camera has a 34 mm f/8 lens, with the field angle of 15°, the resolution of 1600 × 1200 pixels, and the photograph resolution of 1200 × 1200 pixels. The other camera has a 100 mm f/10 lens, with the field angle of 5.1°.

8.2.5 Impact Phase

The optical sensor used at the impact phase is similar to that used in the approaching phase in terms of appearance and performance parameters. For example, the impactor used for deep impact is equipped with an optical imaging sensor [7], called Medium Resolution Imager (MRI). This sensor has an optical lens with a 12 cm aperture, a

1.2 m focal length, and a 10 mrad FOV. In addition, it has a 1024 × 1024 pixel CCD device, which can provide 14-bit digitization. The ITS provides navigation images and high-resolution scientific images captured during the impact.

8.3 Selection Criteria of Candidate Navigation Beacons

Beacons for spacecraft optical navigation are the natural celestial bodies, whose ephemeris information is known, as well as their surface feature points. Natural celestial bodies mainly include the sun, planets and their satellites, asteroids, and comets [8]. The selection and planning of navigation beacons is very important for optical autonomous navigation, which consists of the following three steps:

(1) Based on sensor characteristics and various constraints, select m navigation beacons from the database as candidates.
(2) According to observability, and choose $n(n \leq m)$ of m candidate navigation beacons to observe.
(3) Plan the observation sequence of n navigation beacons in a way that starting from the initial attitude, the observation of n navigation beacons is completed successively by maneuvering, and then the spacecraft returns to the initial attitude. The most cost-effective attitude maneuver must be considered in the whole process.

Step (3) can be formulated as a traveling salesman problem (TSP) and solved by an optimization algorithm (such as genetic algorithm). The specific implementation has been described in the literature [9], and will not be repeated here. Only the steps 1 and 2 will be discussed in this book.

8.3.1 Natural Celestial Bodies as Navigation Beacons

To design an autonomous navigation scheme based on optical imaging measurement in a deep-space exploration mission, candidate beacons should be selected according to some standards. The standards derived from the imaging constraints of navigation sensors include the standards for solar phase angle and apparent magnitude. The standards derived from the image processing algorithm include the standards for apparent motion and three-star probability. The standard derived from navigation accuracy is the distance standard between spacecraft and celestial body [10–12]. Next, the bases on which the first four standards are established will be detailed.

8.3.1.1 Solar Phase Angle

Similar with star trackers, stray light suppression should be considered when designing a optical navigation sensor. In addition to designing a special lens hood to suppress stray light as much as possible, the angle between the sensor optical axis and bright sources must also be constrained to prevent the stray light from entering the through-light hole of navigation sensor. For deep-space exploration, the main stray light source is the sun. As shown in Fig. 8.1, the solar phase angle α is defined as the angle between the sun and the spacecraft as seen from the navigational celestial body and the spacecraft, which can be calculated as follows

$$\alpha = \arccos\left[\frac{(\boldsymbol{r}_a - \boldsymbol{r}) \cdot \boldsymbol{r}_a}{\|\boldsymbol{r}_a - \boldsymbol{r}\| \cdot \|\boldsymbol{r}_a\|}\right] \tag{8.1}$$

where $\boldsymbol{r}_a$ is position of navigational celestial body in a heliocentric ecliptic frame and $\boldsymbol{r}$ is position of the spacecraft in the heliocentric ecliptic coordinate system.

Obviously, the larger the solar phase angle α is, the more the stray light will enter the through-light hole. In particular, sunlight will directly enter the hole when the solar phase angle is equal to 180°. Therefore, depending on the sunlight suppression ability of lends hood, the optical axis of navigation sensor for the imaging of the navigational celestial body will be constrained by the maximum solar phase angle $\alpha_{\max}$. If the solar phase angle α is $> \alpha_{\max}$, sunlight will interfere with imaging. That is to say, there is a spatial cone, with the navigational celestial body as its vertex, the line from spacecraft to the navigational celestial body as its axis, and the maximum

Fig. 8.1 Definition of solar phase angle

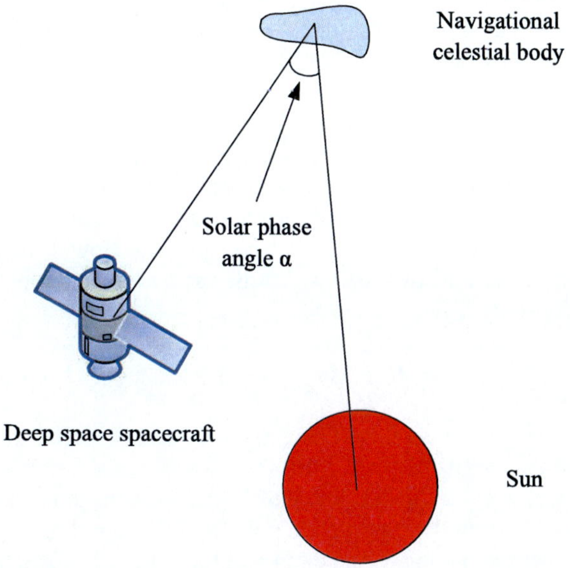

solar phase angle α_{max} as its semi-cone angle. The sun must be confined to this cone. The semi-cone angle of this cone is generally 40°–60°.

8.3.1.2 Apparent Magnitude

Apparent magnitude was first established by ancient Greek astronomer Hipparchus, who divided the 1022 stars in his catalogue into six brightness levels, namely from first magnitude to sixth magnitude. In 1850, British astronomer Pogson found that the stars of magnitude 1 were 100 times brighter than the stars of magnitude 6. According to this relationship, apparent magnitudes were redefined. A difference of 1 redefined magnitude is equal to a 2.512 times difference in brightness [13].

But the magnitudes 1--6 could not cover the brightness of all the celestial objects discovered at that time, so astronomers expanded the original magnitudes to include the concept of "negative magnitude." This apparent magnitude system is still in use today. For example, Altair is 0.77, Vega is 0.03, Siriusis −1.45, the sun is −26.7, the full moon is −12.8, and Venus is −4.89 at its brightest. The largest telescope on the ground can now see stars of magnitude 24, while Hubble can observe stars with magnitude 30.

However, apparent magnitude is the brightness measure of a star observed from the earth. In fact, it is only equivalent to illuminance in the optics. Because the distances between different stars and the earth are different, apparent magnitudes cannot indicate the luminous intensities of the stars themselves. Therefore, the concept of absolute magnitude is introduced. Absolute magnitude refers to the apparent magnitude displayed by a celestial body when it is placed at a specified distance, e.g., 10 parsecs (see Appendix A.1 for the value of 1 parsec). By this measure, the absolute magnitudes of Altair, Vega, Sirius, and the sun are 2.19, 0.5, 1.43, and 4.8, respectively.

But for non-stellar objects such as planets, asteroids, and comets, they can only be seen by reflecting sunlight and starlight and their brightness varies even when they are observed from a fixed distance, so the absolute magnitudes of planets, asteroids, and comets need to be defined separately.

Using some empirical formulas, the apparent magnitude can be calculated from the absolute magnitude. For example, the Bowell model [14] introduced by IAU in 1985, as shown below, can be used for calculating the apparent magnitudes of asteroids [15]:

$$m = H + 5\log_{10}(d \cdot r) - 2.5\log_{10}[(1 - G)\phi_1(\alpha) + G\phi_2(\alpha)] \quad (8.2)$$

where m is the apparent magnitude, H and G are the absolute magnitude and albedo, respectively, obtained from the asteroid ephemeris; r is the distance in AU (see Appendix A.1 for the value of 1 AU) from the sun to the asteroid; d is the distance in AU from the observation point to the asteroid; α is solar phase angle. The phase functions ϕ_1 and ϕ_2 are empirical formulas, as shown below [15]

$$\phi_1(\alpha) = \exp\left(-3.33 \cdot \tan^{0.63}\left(\frac{\alpha}{2}\right)\right)$$
$$\phi_2(\alpha) = \exp\left(-1.87 \cdot \tan^{1.22}\left(\frac{\alpha}{2}\right)\right) \tag{8.3}$$

Obviously, only the celestial body whose apparent magnitude is higher than the minimum sensitive magnitude of navigation sensor can be used for navigation observation.

8.3.1.3 Apparent Motion

Apparent motion is the angular velocity of a celestial body relative to the spacecraft. Generally, the distance between a star and solar system is much larger than the radius of the solar system, so the star position on the celestial sphere can be considered unchanged during the exploration mission. But the angular position of a celestial object in the solar system relative to the spacecraft cannot be assumed unchanged. During the imaging exposure, if the angular velocity of the navigational celestial body relative to the spacecraft is small, the shape of its image point will be the same as that of the background star so that its argument can be extracted more easily through the image processing algorithm. On the contrary, if the angular velocity is very big, the image-point trajectory of the star may be point-like, while the image-point trajectory of navigation object may be linear. This will bring great difficulties to image processing. Therefore, the apparent motion of the navigational celestial body, namely relative angular velocity, must be constrained. The apparent motion of the navigational celestial body can be calculated as

$$\omega_{\mathrm{LOS}} = \frac{\|(\boldsymbol{v}_a - \boldsymbol{v}) \times (\boldsymbol{r}_a - \boldsymbol{r})\|}{\|(\boldsymbol{r}_a - \boldsymbol{r})\|^2} \tag{8.4}$$

where $\boldsymbol{v}_a$ is the velocity of the celestial body in the heliocentric ecliptic coordinate system; $\boldsymbol{r}_a$ is the position of the celestial body in heliocentric ecliptic coordinate system; $\boldsymbol{v}$ is the velocity of the spacecraft in this coordinate system; $\boldsymbol{r}$ is the position of the spacecraft in heliocentric ecliptic coordinate system.

8.3.1.4 Three-Star Probability

In the transfer and approaching segments of a deep-space exploration mission, the navigation images usually contain background stars. From the ephemeris information, the inertial pointing information of the optical axis of navigation sensor at the imaging time can be obtained. The premise that image processing algorithm can identify the starry sky area to which the image belongs is that the image background must contain at least three stars. The upper and lower limits of background star magnitude are constrained by the exposure time and dynamic range of navigation sensor. Exposure time is mainly determined by the brightness of navigation object. Dynamic range is the brightness multiple difference between the brightest and darkest unsaturated image points in an image. It is determined by the performance of navigation sensor. If the star brightness is above the upper magnitude limit, the image will be saturated. If the star brightness is below the lower limit, the image will not be formed.

After defining the upper and lower magnitude limits of background stars, the number of stars in the FOV of navigation sensor still cannot be accurately predicted, because it is also closely related to the attitude of navigation sensor during imaging, and small changes in attitude may force the stars on the FOV boundary out of or into the FOV.

Chausson et al. [11] proposed a probability method to solve this problem. First, suppose that the average star density in the observed space area is ρ stars$/(°)^2$, and that in the given FOV $\Omega(°)^2$, the probability of containing k stars is subject to Poisson distribution

$$P(k) = \frac{1}{k!}(\rho\Omega)^k e^{-\rho\Omega} \tag{8.5}$$

Then, the three-star probability is

$$P(k \geq 3) = 1 - \sum_{k=0}^{2} P(k) \tag{8.6}$$

Take the $1° \times 1°$ FOV of sensor as an example. The relationship between three-star probability and average star density is shown in Fig. 8.2.

The three-star probability is analyzed by a star catalogue. Currently, the star catalogues in common use include SAO, Hipparcos, Tycho-2. A method to calculate the average star density ρ is counting the number of stars in a sufficiently large circular space centered on the celestial body, and then divide it by the area of that space.

8.3.2 *Surface Feature Points as Navigation Beacons*

If navigation beacons are the feature points on the surface of a natural celestial body, a landmark model database needs to be established. This model database is generally constructed by acquiring many optical images during the orbiting phase. An important step for autonomous navigation is selecting candidate suitable navigation beacons according to some screen criteria.

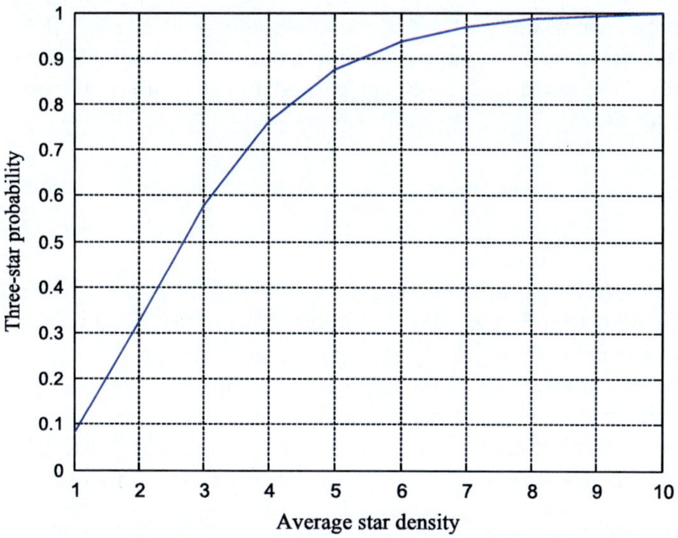

Fig. 8.2 Relationship between three-star probability and average star density

Referring to the screening criteria of navigation asteroids presented in Sect. 8.3.1, the screening criteria of surface feature points can be determined by considering the following factors:

- Visibility/identifiability (e.g., features, texture)
- Phase angle (e.g., the angle between spacecraft and the sun as seen from the landmard)
- Robustness (e.g., changing in scale)
- LOS angle between two navigation beacons.

To meet the imaging requirements of navigation sensor, landmarks should have a good visibility which are easy for image recognition and matching. The phase angle is as small as possible to prevent sunlight from entering the navigation sensor. A large LOS angle between two landmarks can improve the navigation accuracy. These criteria are all related to the parameters of navigation sensor.

When screening the navigation beacons, the landmarks should be evaluated against the above criteria separately. The evaluation results based on different criteria are different for the same landmark. It is impossible to screen out the best navigation beacon by using a single criterion. This requires finding a rule to screen out the best navigation beacon. The general rule is as follows. First, comprehensively evaluate all the candidate landmarks that meet the requirements for visibility/identifiability, phase angle, robustness, and LOS angle, and then rank them by the comprehensive navigation evaluation value, and finally, choose the one with the highest value as navigation beacon.

The comprehensive navigation evaluation value of a landmark is defined as the weighted sum of the values of various landmark parameters (visibility/identifiability, phase angle, robustness, and LOS angle) and the deviations of corresponding screening criteria, i.e.,

$$V = \sum_{i=1}^{3} w_i Er_i \tag{8.7}$$

where w_i is the weight of each parameter, depending on the influence of each parameter on navigation accuracy; Er_i is the absolute value of the deviation between each landmark parameter and the corresponding selection criterion.

As can be seen from Eq. (8.7), the larger the comprehensive navigation evaluation value of a landmark, the more suitable it is for this navigation. All the candidate landmarks that meet the criteria can be ranked by Eq. (8.7), and then the landmark with high values of V can be selected as navigation beacons. The specific steps are as follows:

Step 1: rank all the candidate landmarks that meet the criteria by the comprehensive navigation evaluation value.
Step 2: choose the landmark with the highest value from this rank, and cancel it.
Step 3: treat the landmark selected at the step 2 as the first navigation beacon.
Step 4: choose the landmark with the highest value from the current rank, and cancel it.
Step 5: if the LOS angle between the landmark selected at the step 4 and the existing navigation beacon is greater than θ (θ is a fixed angle), then select it as the next navigation beacon; otherwise, go to the step 4.
Step 6: complete the screening if the number of navigational landmarks is as required; otherwise, continue the step 4.

8.4 Measurement Equations for Optical Autonomous Navigation

Depending on the geometric characteristics of navigation beacons and their distance from the spacecraft, their images in the optical sensor may be area targets or point targets, as discussed in the following cases:

(1) If the navigation beacon is a large celestial body, its image will be a disc target when it is close to the spacecraft (for example, at the orbiting or approaching phase). If the disc edge is used for navigation, the navigation observation variable will be the apparent radius of the celestial body. If the disc center is used for navigation, the navigation observation variable will be the LOS direction pointing to the center of the celestial body.

(2) If the navigation beacon is far away from the spacecraft, its image will be a point target, and the navigation observation variable will be the LOS direction pointing to it.
(3) If the navigation beacon is a feature point on the surface of a natural celestial body, the navigation observation variable will be the LOS direction pointing to the landmark.

To sum up, the observation variables in optical autonomous navigation can be divided into two categories: apparent radius and LOS direction. The corresponding measurement equations will be described below.

8.4.1 Measurement Equations Based on the Apparent Radius Information of Large Celestial Bodies

Figure 8.3 shows geometric relationship based on the apparent radius observations of a large celestial body. The apparent radius ρ can be written as

$$\rho = \arcsin \frac{R_p}{r} + \nu \tag{8.8}$$

where R_p is the radius of the large celestial body, r is the distance from its center to the spacecraft, and ν is measurement noise. To implement navigation filter, it is usually required to calculate the measurement sensitivity matrix. If the state variables $\boldsymbol{x}$ consist of the position $\boldsymbol{r}$ and velocity $\boldsymbol{v}$ relative to the celestial body, then the measurement sensitivity matrix can be written as [16]

$$\begin{aligned} \boldsymbol{H} &= \left[\frac{\partial \rho}{\partial \boldsymbol{r}} \ \frac{\partial \rho}{\partial \boldsymbol{v}} \right] \\ &= \left[-\frac{R_p}{r^2\sqrt{r^2 - R_p^2}} \boldsymbol{r}^{\mathrm{T}} \ \boldsymbol{0}_{1\times 3} \right] \end{aligned} \tag{8.9}$$

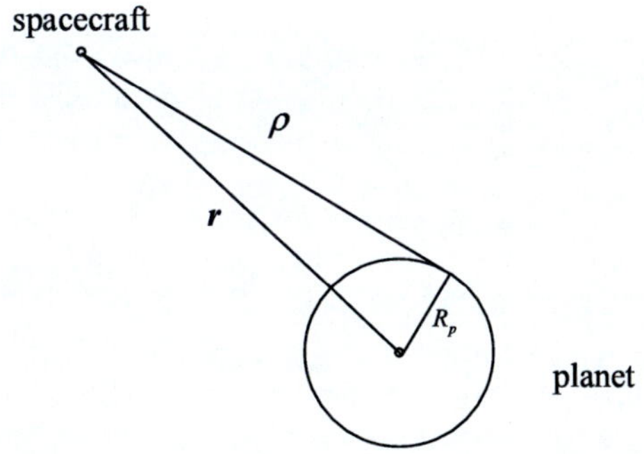

Fig. 8.3 Geometric relation diagram based on apparent radius observations [16]

8.4.2 Measurement Equations Based on LOS Direction Information

8.4.2.1 Unit Vector Form

Figure 8.4 shows the geometric relationship diagram of the optical autonomous navigation based on LOS direction information. Denote $\tilde{\boldsymbol{n}}^s$ and $\boldsymbol{n}^n$ as the LOS direction of navigation beacon expressed in the sensor frame $\mathcal{F}_s$ and the navigation frame $\mathcal{F}_n$, respectively. Then

$$\tilde{\boldsymbol{n}}^s = \boldsymbol{C}_n^s \boldsymbol{n}^n + \boldsymbol{v} \tag{8.10}$$

where $\boldsymbol{C}_n^s$ is the attitude transformation matrix from $\mathcal{F}_n$ to $\mathcal{F}_s$, which is obtained from the attitude determination system and assumed to be known in the navigation system; $\boldsymbol{v}$ is observation noise. Suppose that the FOV is small. The observation noise can be approximated to Gaussian noise [17], satisfying

$$\begin{aligned} &\mathrm{E}\{\boldsymbol{v}\} = \boldsymbol{0} \\ &\mathrm{E}\{\boldsymbol{v}\boldsymbol{v}^{\mathrm{T}}\} = \sigma^2[\boldsymbol{I} - (\boldsymbol{C}_n^s \boldsymbol{n}^n)(\boldsymbol{C}_n^s \boldsymbol{n}^n)^{\mathrm{T}}] \end{aligned} \tag{8.11}$$

Denote the position vector of navigation beacon relative to the origin of coordinates as $\boldsymbol{R}_b$, as shown in Fig. 8.4. Then

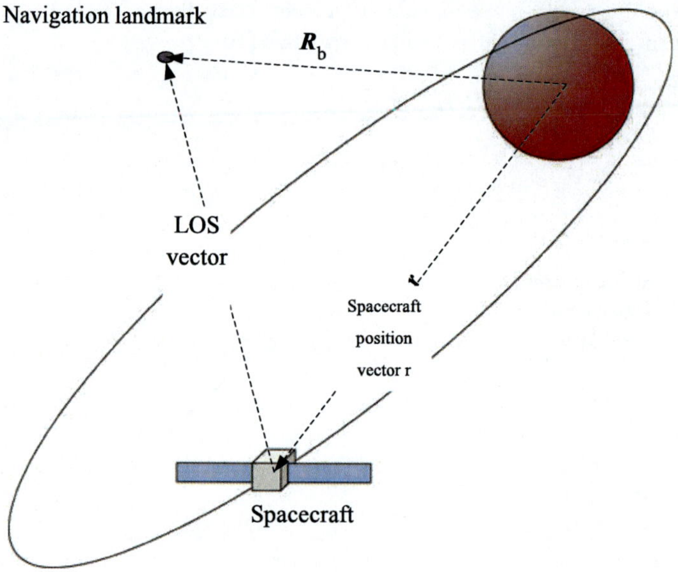

Fig. 8.4 Geometric relationship of the optical autonomous navigation based on LOS information

$$\boldsymbol{n}^n = \frac{\boldsymbol{R}_b^n - \boldsymbol{r}^n}{\left\|\boldsymbol{R}_b^n - \boldsymbol{r}^n\right\|} \tag{8.12}$$

Substituting Eq. (8.12) into Eq. (8.10), the measurement equation based on LOS direction information can be obtained:

$$\tilde{\boldsymbol{n}}^s = \boldsymbol{C}_n^s \frac{\boldsymbol{R}_b^n - \boldsymbol{r}^n}{\left\|\boldsymbol{R}_b^n - \boldsymbol{r}^n\right\|} + \boldsymbol{v} \tag{8.13}$$

The measurement sensitivity matrix corresponding to Eq. (8.13) can be written as

$$\begin{aligned} \boldsymbol{H} &= \left[\frac{\partial \tilde{\boldsymbol{n}}^s}{\partial \boldsymbol{r}} \; \frac{\partial \tilde{\boldsymbol{n}}^s}{\partial \boldsymbol{v}}\right] \\ &= \left[-\frac{1}{\left\|\boldsymbol{R}_b^n - \boldsymbol{r}^n\right\|}\boldsymbol{C}_n^s\left(\boldsymbol{I} - \frac{\boldsymbol{R}_b^n - \boldsymbol{r}^n}{\left\|\boldsymbol{R}_b^n - \boldsymbol{r}^n\right\|}\frac{(\boldsymbol{R}_b^n - \boldsymbol{r}^n)^{\mathrm{T}}}{\left\|\boldsymbol{R}_b^n - \boldsymbol{r}^n\right\|}\right) \; \boldsymbol{0}_{3\times 3}\right] \end{aligned} \tag{8.14}$$

For different types of optical navigation beacons, the measurements used in the autonomous navigation filtering algorithm are also different, as shown below:

(1) $R_b^n = 0$: In this case, navigation beacon is the targeted celestial body, whose the measurement is LOS direction. This usually occurs in the approaching and orbiting phases. In this case, Eq. (8.13) can be simplified as

$$\tilde{\boldsymbol{n}}^s = -\boldsymbol{C}_n^s \frac{\boldsymbol{r}^n}{\left\|\boldsymbol{r}^n\right\|} + \boldsymbol{v} \tag{8.15}$$

The measurement sensitivity matrix corresponding to Eq. (8.15) can be written as

$$\begin{aligned} \boldsymbol{H} &= \left[\frac{\partial \tilde{\boldsymbol{n}}^s}{\partial \boldsymbol{r}} \; \frac{\partial \tilde{\boldsymbol{n}}^s}{\partial \boldsymbol{v}}\right] \\ &= \left[-\frac{1}{\left\|\boldsymbol{r}^n\right\|}\boldsymbol{C}_n^s\left(\boldsymbol{I} - \frac{\boldsymbol{r}^n}{\left\|\boldsymbol{r}^n\right\|}\frac{(\boldsymbol{r}^n)^{\mathrm{T}}}{\left\|\boldsymbol{r}^n\right\|}\right) \; \boldsymbol{0}_{3\times 3}\right] \end{aligned} \tag{8.16}$$

(2) $0 < R_b^n < \infty$, $\boldsymbol{R}_b^n$ is known: In this case, navigation beacons are usually large distant celestial objects (such as the sun, earth, and moon), asteroids, and feature points on the surface of natural celestial objects. The corresponding flight phases may be deep-space transfer, approaching, orbiting and landing. Obviously, if Eq. (8.13) is taken as measurement equation in the navigation filtering, the attitude determination error will affect the navigation accuracy. If the navigation sensor has a FOV wide enough to observe multiple LOS directions simultaneously, then the angle between two LOS directions can be taken as the measurement to eliminate the influence of attitude determination error. In this case, the measurement equation can be written as

$$\begin{aligned}\theta_{12} &= \arccos(\boldsymbol{n}_1^n \cdot \boldsymbol{n}_2^n) + v \\ &= \arccos\left(\frac{\boldsymbol{R}_{b,1}^n - \boldsymbol{r}^n}{\left\|\boldsymbol{R}_{b,1}^n - \boldsymbol{r}^n\right\|} \cdot \frac{\boldsymbol{R}_{b,2}^n - \boldsymbol{r}^n}{\left\|\boldsymbol{R}_{b,2}^n - \boldsymbol{r}^n\right\|}\right) + v\end{aligned} \tag{8.17}$$

where θ_{12} is the angle between two LOS directions, $\boldsymbol{R}_{b,1}^n$ and $\boldsymbol{R}_{b,1}^n$ are, respectively, the position vectors of two navigation beacons expressed in $\mathcal{F}_n$. The corresponding measurement sensitivity matrix is

$$\begin{aligned}\boldsymbol{H} &= \left[\begin{matrix}\frac{\partial\theta_{12}}{\partial \boldsymbol{r}} & \frac{\partial\theta_{12}}{\partial \boldsymbol{v}}\end{matrix}\right] \\ &= \frac{1}{\sqrt{1-(\boldsymbol{n}_1^n \cdot \boldsymbol{n}_2^n)^2}} \\ &\left[\frac{1}{\left\|\boldsymbol{R}_{b,2}^n - \boldsymbol{r}^n\right\|}(\boldsymbol{n}_1^n)^{\mathrm{T}}\left(\boldsymbol{I} - \boldsymbol{n}_2^n(\boldsymbol{n}_2^n)^{\mathrm{T}}\right) + \frac{1}{\left\|\boldsymbol{R}_{b,1}^n - \boldsymbol{r}^n\right\|}(\boldsymbol{n}_2^n)^{\mathrm{T}}\left(\boldsymbol{I} - \boldsymbol{n}_1^n(\boldsymbol{n}_1^n)^{\mathrm{T}}\right) \; \boldsymbol{0}_{1\times 3}\right]\end{aligned} \tag{8.18}$$

where $\boldsymbol{n}_1^n = \frac{\boldsymbol{R}_{b,1}^n - \boldsymbol{r}^n}{\left\|\boldsymbol{R}_{b,1}^n - \boldsymbol{r}^n\right\|}$, $\boldsymbol{n}_2^n = \frac{\boldsymbol{R}_{b,2}^n - \boldsymbol{r}^n}{\left\|\boldsymbol{R}_{b,2}^n - \boldsymbol{r}^n\right\|}$.

Specially, if one of two navigation beacons is the central body, e.g., $\boldsymbol{R}_{b,2}^n = \boldsymbol{0}$, then Eq. (8.17) can be written as

$$\begin{aligned}\theta_{12} &= \arccos(\boldsymbol{n}_1^n \cdot \boldsymbol{n}_2^n) + v \\ &= \frac{\boldsymbol{R}_{b,1}^n - \boldsymbol{r}^n}{\left\|\boldsymbol{R}_{b,1}^n - \boldsymbol{r}^n\right\|} \cdot \frac{-\boldsymbol{r}^n}{\left\|\boldsymbol{r}^n\right\|} + v\end{aligned} \tag{8.19}$$

The corresponding measurement sensitivity matrix can be written as

$$\begin{aligned}\boldsymbol{H} &= \left[\begin{matrix}\frac{\partial\theta_{12}}{\partial \boldsymbol{r}} & \frac{\partial\theta_{12}}{\partial \boldsymbol{v}}\end{matrix}\right] \\ &= \frac{1}{\sqrt{1-(\boldsymbol{n}_1^n \cdot \boldsymbol{n}_2^n)^2}}\left[\frac{1}{\left\|\boldsymbol{r}^n\right\|}(\boldsymbol{n}_1^n)^{\mathrm{T}}\left(\boldsymbol{I} - \boldsymbol{n}_2^n(\boldsymbol{n}_2^n)^{\mathrm{T}}\right) + \frac{1}{\left\|\boldsymbol{R}_{b,1}^n - \boldsymbol{r}^n\right\|}(\boldsymbol{n}_2^n)^{\mathrm{T}}\left(\boldsymbol{I} - \boldsymbol{n}_1^n(\boldsymbol{n}_1^n)^{\mathrm{T}}\right) \; \boldsymbol{0}_{1\times 3}\right]\end{aligned} \tag{8.20}$$

(3) $0 < R_b^n < \infty$, $\boldsymbol{R}_b^n$ is unknown: In this case, navigation beacons are feature points on the surface of natural celestial objects. The corresponding flight phase is usually orbiting or landing. The optical autonomous navigation in this case will not be discussed here. Interested readers may refer to [18–20].

(4) $R_b^n \to \infty$: In this case, the navigation beacon is usually an infinitely distant star. If $R_b^n \to \infty$, Eq. (8.13) can be written as

$$\tilde{\boldsymbol{n}}^s = \boldsymbol{C}_n^s \frac{\boldsymbol{R}_b^n}{\left\|\boldsymbol{R}_b^n\right\|} + \boldsymbol{v} \tag{8.21}$$

Obviously, Eq. (8.21) is not a function of spacecraft position. Therefore, the autonomous navigation of a spacecraft cannot be achieved by simply observing stars. In practical applications, star observation is usually for determining the spacecraft attitude. Nevertheless, star observations can be combined with other observations to achieve autonomous navigation. In this case, the angle between a NEO (at the finite distance) and a distant object (at the infinite distance) can be taken as observation variable. Denote the LOS direction of the distant object as $\boldsymbol{n}_{star}$. The LOS direction of the NEO is determined by Eq. (8.13). The angle θ between the two LOS vectors is the measurement, which is given by

$$\begin{aligned}\theta &= \arccos(\boldsymbol{n}^n \cdot \boldsymbol{n}^n_{\text{star}}) + v \\ &= \frac{\boldsymbol{R}^n_b - \boldsymbol{r}^n}{\left\| \boldsymbol{R}^n_b - \boldsymbol{r}^n \right\|} \cdot \boldsymbol{n}^n_{\text{star}} + \boldsymbol{v}\end{aligned} \tag{8.22}$$

The measurement sensitivity matrix corresponding to Eq. (8.22) is [16]

$$\begin{aligned}\boldsymbol{H} &= \left[\frac{\partial \tilde{\boldsymbol{n}}^s}{\partial \boldsymbol{r}} \ \frac{\partial \tilde{\boldsymbol{n}}^s}{\partial \boldsymbol{v}} \right] \\ &= \left[\left(\left\| \boldsymbol{R}^n_b - \boldsymbol{r}^n \right\|^2 - (\boldsymbol{R}^n_b - \boldsymbol{r}^n) \cdot \boldsymbol{n}_{\text{star}} \right)^{-\frac{1}{2}} \left(\boldsymbol{n}^T_{\text{star}} - \frac{(\boldsymbol{R}^n_b - \boldsymbol{r}^n)^{\mathrm{T}} \boldsymbol{n}_{\text{star}}}{\left\| \boldsymbol{R}^n_b - \boldsymbol{r}^n \right\|^2} (\boldsymbol{R}^n_b - \boldsymbol{r}^n)^{\mathrm{T}} \right) \ \boldsymbol{0}_{1\times 3} \right]\end{aligned} \tag{8.23}$$

In addition, the time when the distant object is occulted by the NEO can also be used for navigation. This navigation mode is called occultation navigation [21–23]. If the NEO is surrounded by atmosphere, starlight refraction can be used for navigation [24–27], which are not discussed in this book.

8.4.2.2 Pixels Form

The optical imaging sensor images the navigation beacon and obtains two-dimensional pixels from the image plane, so its measurement equation can also be written in the form of pixels. Figure 8.5 shows the measurement principle of optical imaging sensor, where $x_c y_c$ is the imaging plane and z_c is the focal length direction. Denote the Cartesian coordinates of the imaging point on the focal plane as (x_c, y_c). Then, $\tilde{\boldsymbol{n}}^s$ expressed in the sensor coordinate system will be

$$\tilde{\boldsymbol{n}}^s = \frac{1}{\sqrt{x_c^2 + y_c^2 + f^2}} \begin{bmatrix} -x_c \\ -y_c \\ f \end{bmatrix} \tag{8.24}$$

where f is the focal length of the sensor. Write $\boldsymbol{R}^n_b$ and $\boldsymbol{r}^n$ as

$$\begin{aligned}\boldsymbol{R}^n_b &= [x_d, y_d, z_d]^{\mathrm{T}} \\ \boldsymbol{r}^n &= [x, y, z]^{\mathrm{T}}\end{aligned} \tag{8.25}$$

Therefore,

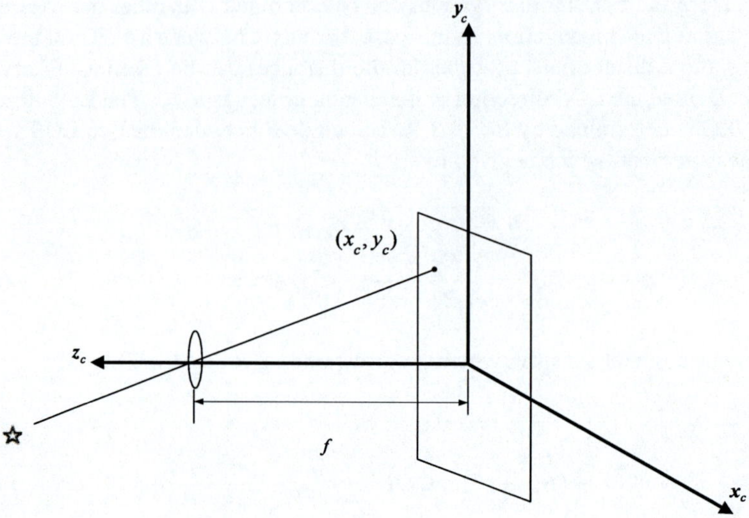

Fig. 8.5 Measurement principle of optical imaging sensor

$$\frac{\boldsymbol{R}_b^n - \boldsymbol{r}^n}{\left\| \boldsymbol{R}_b^n - \boldsymbol{r}^n \right\|} = \frac{1}{\sqrt{(x_d - x)^2 + (y_d - y)^2 + (z_d - z)^2}} \begin{bmatrix} x_d - x \\ y_d - y \\ z_d - z \end{bmatrix} \tag{8.26}$$

Without considering the measurement noise and substituting Eqs. (8.24) and (8.26) into Eq. (8.13) gives

$$\begin{aligned} x_c &= -f \frac{a_{11}(x_d - x) + a_{12}(y_d - y) + a_{13}(z_d - z)}{a_{31}(x_d - x) + a_{32}(y_d - y) + a_{33}(z_d - z)} \\ y_c &= -f \frac{a_{21}(x_d - x) + a_{22}(y_d - y) + a_{23}(z_d - z)}{a_{31}(x_d - x) + a_{32}(y_d - y) + a_{33}(z_d - z)} \end{aligned} \tag{8.27}$$

where $a_{ij}(i = 1, 2, 3; j = 1, 2, 3)$ are the corresponding elements in the direction cosine matrix $\boldsymbol{C}_n^s$. In practice, the projection coordinates of the imaging point on the focal plane are pixel coordinates, denoted as (p, l). Considering the distortion effect, (p, l) is given by

$$\begin{bmatrix} p \\ l \end{bmatrix} = \boldsymbol{k} \begin{bmatrix} x_c \\ y_c \\ x_c y_c \end{bmatrix} + \begin{bmatrix} p_0 \\ l_0 \end{bmatrix} + \begin{bmatrix} v_p \\ v_l \end{bmatrix} \tag{8.28}$$

where

$$\boldsymbol{k} = \begin{bmatrix} k_x & k_{xy} & k_{xxy} \\ k_{yx} & k_y & k_{yxy} \end{bmatrix} \tag{8.29}$$

is the conversion matrix from Cartesian coordinates to pixels; p_0 and l_0 are the image element and image line at the sensor center; v_p and v_l are the measurement noise. Because the values of k_{xy}, k_{yx}, k_{xxy} and k_{yxy} are very small, the distortion effect can be ignored to simplify the analysis. Let $\boldsymbol{k} = \begin{bmatrix} k_x & 0 & 0 \\ 0 & k_y & 0 \end{bmatrix}$. Then, Eq. (8.28) can be simplified as

$$\begin{bmatrix} p \\ l \end{bmatrix} = \begin{bmatrix} k_x x_c \\ k_y y_c \end{bmatrix} + \begin{bmatrix} p_0 \\ l_0 \end{bmatrix} + \begin{bmatrix} v_{p,i} \\ v_{l,i} \end{bmatrix} \tag{8.30}$$

Substituting Eq. (8.30) into Eq. (8.27) and considering the measurement noise, the following results can be obtained

$$\begin{aligned} p &= -k_x f \frac{a_{11}(x_d - x) + a_{12}(y_d - y) + a_{13}(z_d - z)}{a_{31}(x_d - x) + a_{32}(y_d - y) + a_{33}(z_d - z)} + p_0 + v_p \\ l &= -k_y f \frac{a_{21}(x_d - x) + a_{22}(y_d - y) + a_{23}(z_d - z)}{a_{31}(x_d - x) + a_{32}(y_d - y) + a_{33}(z_d - z)} + l_0 + v_l \end{aligned} \tag{8.31}$$

The measurement sensitivity matrix corresponding to Eq. (8.31) can be written as

$$\boldsymbol{H} = \frac{\partial z}{\partial \boldsymbol{x}} = \begin{bmatrix} \partial p/\partial x & \partial p/\partial y & \partial p/\partial z & \boldsymbol{0}_{2\times 3} \\ \partial l/\partial x & \partial l/\partial y & \partial l/\partial z & \end{bmatrix} \tag{8.32}$$

where

$$\begin{aligned} \partial p/\partial x &= -k_x f \frac{(a_{12}a_{31} - a_{11}a_{32})(y_d - y) + (a_{13}a_{31} - a_{11}a_{33})(z_d - z)}{[a_{31}(x_d - x) + a_{32}(y_d - y) + a_{33}(z_d - z)]^2} \\ \partial p/\partial y &= -k_x f \frac{(a_{11}a_{32} - a_{12}a_{31})(x_d - x) + (a_{13}a_{32} - a_{12}a_{33})(z_d - z)}{[a_{31}(x_d - x) + a_{32}(y_d - y) + a_{33}(z_d - z)]^2} \\ \partial p/\partial z &= -k_x f \frac{(a_{11}a_{33} - a_{13}a_{31})(x_d - x) + (a_{12}a_{33} - a_{13}a_{32})(y_d - y)}{[a_{31}(x_d - x) + a_{32}(y_d - y) + a_{33}(z_d - z)]^2} \\ \partial l/\partial x &= -k_y f \frac{(a_{22}a_{31} - a_{21}a_{32})(y_d - y) + (a_{23}a_{31} - a_{21}a_{33})(z_d - z)}{[a_{31}(x_d - x) + a_{32}(y_d - y) + a_{33}(z_d - z)]^2} \\ \partial l/\partial y &= -k_y f \frac{(a_{21}a_{32} - a_{22}a_{31})(x_d - x) + (a_{23}a_{32} - a_{22}a_{33})(z_d - z)}{[a_{31}(x_d - x) + a_{32}(y_d - y) + a_{33}(z_d - z)]^2} \end{aligned}$$

$$\partial l/\partial z = -k_y f \frac{(a_{21}a_{33} - a_{23}a_{31})(x_d - x) + (a_{22}a_{33} - a_{23}a_{32})(y_d - y)}{[a_{31}(x_d - x) + a_{32}(y_d - y) + a_{33}(z_d - z)]^2} \tag{8.33}$$

8.5 Navigation Beacon Planning Based on Geometric Observability Analysis

8.5.1 *Geometric Observability Analysis*

Geometric observability analysis only evaluates the influence of measurement equations on the performance of autonomous navigation system. Consider the measurement equation in the unit vector form. Equation (8.10) can be rewritten as

$$\tilde{\boldsymbol{b}}_i = \boldsymbol{C}_n^s \boldsymbol{n}_i^n + \boldsymbol{v}_i \tag{8.34}$$

where $\tilde{\boldsymbol{b}}_i = \tilde{\boldsymbol{n}}_i^s$. Then,

$$\begin{aligned} \boldsymbol{\mu}_{bi} &\triangleq \mathrm{E}\{\tilde{\boldsymbol{b}}_i\} = \boldsymbol{C}_n^s \boldsymbol{n}_i^n \\ cov(\tilde{\boldsymbol{b}}_i) &\triangleq \mathrm{E}\{(\tilde{\boldsymbol{b}}_i - \boldsymbol{\mu}_{b,i})(\tilde{\boldsymbol{b}}_i - \boldsymbol{\mu}_{b,i})^{\mathrm{T}}\} = \mathrm{E}\{\boldsymbol{v}_i \boldsymbol{v}_i^{\mathrm{T}}\} \end{aligned} \tag{8.35}$$

For the Gaussian noise approximation that satisfies Eq. (8.11), the probability density function of observations can be written as

$$p(\tilde{\boldsymbol{b}}_i; \boldsymbol{r}) = \prod_{i=1}^{n} \frac{1}{(2\pi)^{m/2}[\det(\sigma_i^2 \boldsymbol{I})]^{1/2}}$$

$$= \exp\{-\frac{1}{2}[\tilde{\boldsymbol{b}}_i - \boldsymbol{C}_n^s \boldsymbol{n}_i^n]^{\mathrm{T}}(\sigma_i^2 \boldsymbol{I})^{-1}[\tilde{\boldsymbol{b}}_i - \boldsymbol{C}_n^s \boldsymbol{n}_i^n]\} \tag{8.36}$$

where n is the observed number of navigation beacons. Taking the logarithm of Eq. (8.36) gives

$$\ln\big(p(\tilde{\boldsymbol{b}}_i; \boldsymbol{r})\big) = -\sum_{i=1}^{n} \frac{1}{2}\sigma_i^{-2} \|\tilde{\boldsymbol{b}}_i - \boldsymbol{C}_n^s \boldsymbol{n}_i^n]\|^2 + \text{const} \tag{8.37}$$

Let

$$J(\boldsymbol{r}) = \frac{1}{2}\sum_{i=1}^{n} \sigma_i^{-2} \|\tilde{\boldsymbol{b}}_i - \boldsymbol{C}_n^s \boldsymbol{n}_i^n]\|^2 \tag{8.38}$$

From Theorem 2.1, the Fisher information matrix can be derived as follows

$$\mathcal{I} = \mathrm{E}\left\{ \frac{\partial^2}{\partial \boldsymbol{r} \partial \boldsymbol{r}^{\mathrm{T}}} J(\boldsymbol{r}) \right\} \tag{8.39}$$

Denote $J_i = \sigma_i^{-2}(1 - \boldsymbol{C}_n^s \boldsymbol{n}_i^n)$ as the index function of a single measurement. Then

$$\begin{aligned} \boldsymbol{g}_i &\triangleq \frac{\partial J_i}{\partial \boldsymbol{r}} = \frac{\partial \boldsymbol{n}_i^n}{\partial \boldsymbol{r}} \frac{\partial J_i}{\partial \boldsymbol{n}_i^n} \\ &= \sigma_i^{-2} \frac{1}{r_{a,i}^2} [I - \boldsymbol{n}_i^n (\boldsymbol{n}_i^n)^{\mathrm{T}}] (\boldsymbol{C}_n^s)^{\mathrm{T}} \widetilde{\boldsymbol{b}}_i \end{aligned} \tag{8.40}$$

where $r_{a,i} = \left\| \boldsymbol{R}_{b,i}^n - \boldsymbol{r}^n \right\|$. Then

$$\begin{aligned} \mathrm{E}\{\frac{\partial \boldsymbol{g}_i}{\partial \boldsymbol{r}}\} &= -\mathrm{E}\{\frac{\partial \boldsymbol{g}_i}{\partial \boldsymbol{r}_{a,i}}\} \\ &= \sigma_i^{-2} \frac{1}{r_{a,i}^2} [\boldsymbol{I} - \boldsymbol{n}_i^n (\boldsymbol{n}_i^n)^{\mathrm{T}}] \end{aligned} \tag{8.41}$$

Substituting Eq. (8.41) into Eq. (8.39) gives

$$\begin{aligned} \mathcal{I} &= \sum_{i=1}^{n} \sigma_i^{-2} \frac{1}{r_{a,i}^2} [\boldsymbol{I} - \boldsymbol{n}_i^n (\boldsymbol{n}_i^n)^{\mathrm{T}}] \\ &= -\sum_{i=1}^{n} \sigma_i^{-2} \frac{1}{r_{a,i}^2} [\boldsymbol{n}_i^n \times]^2 \end{aligned} \tag{8.42}$$

According to the Cramer-Rao inequality (Theorem 2.1),

$$\begin{aligned} \boldsymbol{P} &\geq \mathcal{I}^{-1} \\ &= (-\sum_{i=1}^{n} \sigma_i^{-2} \frac{1}{r_{a,i}^2} [\boldsymbol{n}_i^n \times]^2)^{-1} \end{aligned} \tag{8.43}$$

where $\boldsymbol{P}$ is error covariance matrix. For autonomous navigation system, eigenvalues of the error covariance matrix can be used to measure its observability [28]. The autonomous navigation accuracy in the eigenvector direction corresponding to a large eigenvalue is low, while the navigation accuracy in the eigenvector direction corresponding to a small eigenvalue is high. Given two error covariances, the estimation accuracy of the error covariance matrix with the smaller trace is higher than that with the bigger trace. According to the above criteria, the following theorems can be obtained:

Theorem 8.1: For the optical autonomous navigation system based on LOS information, if only one LOS direction is observed, the component along this direction will not be observable. If two LOS directions are observed, the system will be always geometrically observable as long as the two LOSs are not collinear. Moreover, when the angle between the two LOSs is closer to 90°, the system has a larger degree of observability.

Proof First consider the case of one LOS observation, namely $i = 1$. Because the rank of $[\boldsymbol{n}_i^n\times]^2$ is 2, $\mathcal{I}$ is singular so that the system will not be observable. Let $p = 1/r_{a,i}^2$. The eigenvalues of $\mathcal{I}$ satisfy

$$|\lambda \boldsymbol{I} - \mathcal{I}| = |\lambda \boldsymbol{I} + \sigma_i^{-2} p [\boldsymbol{n}_i^n\times]^2| = 0 \tag{8.44}$$

Then $\lambda_1 = 0$, $\lambda_2 = \lambda_3 = \sigma_i^{-2} p$ where the eigenvector corresponding to the eigenvalue 0 is $\boldsymbol{n}_1^n$, indicating that the state along the LOS is not observable.

If two LOSs are used for autonomous navigation, then

$$\begin{aligned} \mathcal{I} &= -\sigma_1^{-2}\frac{1}{r_{a,1}^2}[\boldsymbol{n}_1^n\times]^2 - \sigma_2^{-2}\frac{1}{r_{a,2}^2}[\boldsymbol{n}_2^n\times]^2 \\ &= (a_1 + a_2)\boldsymbol{I} - [a_1\boldsymbol{n}_1^n(\boldsymbol{n}_1^n)^{\mathrm{T}} + a_2\boldsymbol{n}_2^n(\boldsymbol{n}_2^n)^{\mathrm{T}}] \end{aligned} \tag{8.45}$$

with $a_i = \sigma_i^{-2}/r_{a,i}^2$ $(i = 1, 2)$.

The three eigenvalues of $\mathcal{I}$ are

$$\begin{aligned} \lambda_1(\mathcal{I}) &= a_1 + a_2 \\ \lambda_2(\mathcal{I}) &= \frac{1}{2}\lambda_1(\mathcal{I})(1 + \gamma) \\ \lambda_3(\mathcal{I}) &= \frac{1}{2}\lambda_1(\mathcal{I})(1 - \gamma) \end{aligned} \tag{8.46}$$

where

$$\gamma = [1 - \frac{4a_1a_2|\boldsymbol{n}_1^n \times \boldsymbol{n}_2^n|^2}{\lambda_1(\mathcal{I})^2}]^{1/2} \tag{8.47}$$

Obviously, $\lambda_1(\mathcal{I}) > 0$, $\lambda_2(\mathcal{I}) > 0$, $\lambda_3(\mathcal{I}) \geq 0$. If and only if $\boldsymbol{n}_1^n \times \boldsymbol{n}_2^n = 0$ (i.e., $\boldsymbol{n}_1^n$ and $\boldsymbol{n}_2^n$ are collinear), $\lambda_3(\mathcal{I}) = 0$. Therefore, if the two LOSs are not collinear, the autonomous navigation system is geometrically observable.

The trace of the error covariance matrix can represent the degree of observability, and $\mathrm{Tr}[P] = \sum_{i=1}^{3}\lambda_i(P)$, $\lambda_i(P) = 1/\lambda_i(\mathcal{I})$, so

$$\mathrm{Tr}[P] = \frac{1}{\lambda_1(\mathcal{I})}(1 + \frac{4}{1 - \gamma^2})$$

$$= (\frac{1}{a_1} + \frac{1}{a_2})\frac{1}{|\boldsymbol{n}_1^n \times \boldsymbol{n}_2^n|^2} + \frac{1}{a_1 + a_2} \tag{8.48}$$

According to the above equation, if σ_i is given, Tr[P] will decrease with the distance from navigation beacon to the spacecraft. In addition, the $\boldsymbol{n}_1^n \times \boldsymbol{n}_2^n$ in the above equation reflects the angle between navigation beacons. The closer the angle is to 90°, the smaller Tr[P] will be. Therefore, the optical autonomous navigation system based on two LOS directions can improve the navigation accuracy by observing the navigation beacons near the spacecraft or two nearly orthogonal navigation beacons. This completes the proof.

Theorem 8.1 analyzes the observability in the cases of 1 LOS and 2 LOSs and gives a quantitative conclusion. For the observation along three lines of sight, the Fisher information matrix can be written as

$$\mathcal{I}_{123} = -a_1[\boldsymbol{n}_1^n \times]^2 - a_2[\boldsymbol{n}_2^n \times]^2 - a_3[\boldsymbol{n}_3^n \times]^2 \tag{8.49}$$

Firstly, the following theorem for a special case is presented:

Theorem 8.2 for the observation of three asteroids, if $a_1 = a_2 = a_3$, then the degree of observability will the greatest when $\boldsymbol{n}_1^n$, $\boldsymbol{n}_2^n$ and $\boldsymbol{n}_3^n$ are orthogonal to each other.

Proof: let $a = a_1 = a_2 = a_3$. Then Eq. (8.49) can be rewritten as

$$\begin{aligned}\mathcal{I}_{123} &= (a_1 + a_2 + a_3)\boldsymbol{I} - [a_1\boldsymbol{n}_1^n(\boldsymbol{n}_1^n)^{\mathrm{T}} + a_2\boldsymbol{n}_2^n(\boldsymbol{n}_2^n)^{\mathrm{T}} + a_3\boldsymbol{n}_3^n(\boldsymbol{n}_3^n)^{\mathrm{T}}] \\ &= 3a\boldsymbol{I} - a(\boldsymbol{n}_1^n(\boldsymbol{n}_1^n)^{\mathrm{T}} + \boldsymbol{n}_2^n(\boldsymbol{n}_2^n)^{\mathrm{T}} + \boldsymbol{n}_3^n(\boldsymbol{n}_3^n)^{\mathrm{T}})\end{aligned} \tag{8.50}$$

when $\boldsymbol{n}_1^n$, $\boldsymbol{n}_2^n$ and $\boldsymbol{n}_3^n$ are orthogonal to each other, it is obvious that

$$\lambda_1(\mathcal{I}) = \lambda_2(\mathcal{I}) = \lambda_3(\mathcal{I}) = 2a \tag{8.51}$$

Denote the error covariance in this case as $\boldsymbol{P}_\perp$. Then

$$Tr[\boldsymbol{P}_\perp] = \frac{3}{2a} \tag{8.52}$$

However, for $\forall \boldsymbol{n}_1, \boldsymbol{n}_2, \boldsymbol{n}_3$, it can be derived from Eq. (8.50) that

$$\begin{aligned}Tr[\mathcal{I}] &= \lambda_1(\mathcal{I}) + \lambda_2(\mathcal{I}) + \lambda_3(\mathcal{I}) \\ &= 2(a_1 + a_2 + a_3) = 6a\end{aligned} \tag{8.53}$$

By using the inequality

$$\frac{9}{x + y + z} \le \frac{1}{x} + \frac{1}{y} + \frac{1}{z}(x, y, z > 0), \tag{8.54}$$

the following result can be obtained

$$\begin{aligned}\mathrm{Tr}[\boldsymbol{P}] &= \frac{1}{\lambda_1(\mathcal{I})} + \frac{1}{\lambda_2(\mathcal{I})} + \frac{1}{\lambda_3(\mathcal{I})} \\ &\geq \frac{9}{\lambda_1(\mathcal{I}) + \lambda_2(\mathcal{I}) + \lambda_3(\mathcal{I})} = \frac{3}{2a} = \mathrm{Tr}[\boldsymbol{P}_\perp]\end{aligned} \tag{8.55}$$

thus Theorem 8.2 is proved. In the general case, it is difficult to get the analytical solution of the eigenvalue of information matrix.

8.5.2 Optimal Navigation Beacon Planning

As analyzed in the previous section, to ensure the observability of the optical autonomous navigation system based on LOS information, at least two non-collinear LOS vectors should be observed if the orbit dynamics are not considered. In the actual navigation system, the number of optional navigation beacons may be greater than 2, so the vectors in multiple LOS directions can be observed. In this case, the error covariance matrix of the system according to the optimal estimation in Eq. (8.43) can be written as

$$\boldsymbol{P} = \mathcal{I}^{-1} = (-\sum_{i=1}^{n} \sigma_i^{-2} \frac{1}{r_{a,i}^2} [\boldsymbol{n}_i^n \times]^2)^{-1} \tag{8.56}$$

As seen from Eq. (8.56), the factors affecting the variance matrix include σ_i, $r_{a,i}^2$ and $\boldsymbol{n}_i^n$. σ_i is the measurement accuracy of sensor. A smaller σ_i signifies a smaller variance matrix and better degree of observability. $r_{a,i}^2$ is the distance from navigation beacon to the spacecraft. A closer distance indicates a better degree of observability. $\boldsymbol{n}_i^n$ is the LOS direction of navigation beacon. It is difficult to make a quantitative conclusion about its effect on degree of observability when observing multiple landmarks. Therefore, the degree of observability of navigation system may be defined as

$$\kappa = \frac{1}{\mathrm{Tr}[\boldsymbol{P}]} = \frac{1}{\mathrm{Tr}\left[\left(-\sum_{i=1}^{n} \sigma_i^{-2} \frac{1}{r_{a,i}^2} [\boldsymbol{n}_i^n \times]^2\right)^{-1}\right]} \tag{8.57}$$

If the observation accuracy is constant, the degree of observability κ will be only related to the spatial distribution of navigation beacons. The spatial distribution contains two types of information: the distance from navigation beacon to the spacecraft and the LOS angle. A bigger κ signifies a smaller trace of error covariance matrix and a smaller navigation error. The analysis also shows that if the distances between navigation beacons and the spacecraft decrease by a factor of a, then $\kappa(m)$ will increase by a factor of a^2.

The optimal navigation beacon combination is selected in the following way. Suppose that in an autonomous navigation, n LOSs need to be observed, and $m(n \leq m)$ candidate beacons meet the selection criteria. Then, there will be $C(m, n)$[1] beacon combinations. Calculate the degree of observability of all the combinations, among which the combination with the greatest degree of observability will be the optimal.

8.6 Navigation Filtering Algorithm

Depending on the processing method of observation data, the navigation filter algorithm can be implemented through batch processing or recursive processing. Batch processing is a iteration algorithm to process a batch of observation data acquired in a period of time in order to obtain the optimal state estimate at a particular time in this period. Because the observation vector dimensions composed of observation data are much more than the dimensions of state variables, the least square method is usually used for solving. This method has a high accuracy and stable iterative operation process, but its estimation is a staged process that cannot produce the result in real time. Recursive processing updates the existing estimates through real-time data observation based on the determined initial orbit, so as to obtain new estimates. This is a sequential and recursive estimation algorithm suitable for the real-time processing of orbit attitude. It usually takes the forms of extended Kalman filtering and its variant algorithms. In this section, batch processing and Kalman navigation filtering algorithms will be presented, starting from the orbit dynamics model. The spacecraft orbit dynamics model used for navigation estimation is denoted as

$$\dot{x} = \boldsymbol{f}(\boldsymbol{x}, t) + \boldsymbol{w} \tag{8.58}$$

Detailed form of $\boldsymbol{f}(\boldsymbol{x}, t)$ are presented in Chapter 6.

8.6.1 Batch Filtering Algorithm

Denote $\tilde{z}_0, \tilde{z}_1, \cdots, \tilde{z}_N$ as the observations at $t_0, t_1, \cdots, t_N$. The batch least square (BLS) filtering algorithm processes these observation sequences in a centralized manner, and estimates $\boldsymbol{x}_0 \triangleq \boldsymbol{x}(t_0)$. The spacecraft state $\boldsymbol{x}(t)$ at any time can be written as a nonlinear algebraic function of $\boldsymbol{x}_0$, that is

$$\boldsymbol{x}(t) = \boldsymbol{\varphi}(\boldsymbol{x}_0, t_0, t) \tag{8.59}$$

[1] $C(m, n)$ is the number of the combinations each of which is composed of n elements taken from m elements.

which can be obtained by integrating differential Eq. (8.58). Denote the measurement equation as

$$\tilde{z} = h(x) + v \tag{8.60}$$

Then

$$\begin{aligned} \tilde{z}_0 &= h(x_0) + v_0 \\ \tilde{z}_1 &= h(x_1) + v_1 \\ &\vdots \\ \tilde{z}_N &= h(x_N) + v_N \end{aligned} \tag{8.61}$$

where $x_i \triangleq x(t = t_i)$. Substituting Eq. (8.59) into Eq. (8.61) gives

$$\begin{aligned} \tilde{z}_0 &= h(x_0) + v_0 \\ \tilde{z}_1 &= h(\varphi(x_0, t_0, t_1)) + v_1 \\ &\vdots \\ \tilde{z}_N &= h(\varphi(x_0, t_0, t_N)) + v_N \end{aligned} \tag{8.62}$$

Equation (8.62) is a nonlinear equation of x_0, and can be solved using iterative method. Given the prior estimate $\hat{x}_0^-$ of x_0 and its variance P_0^-, then the first-order Taylor expansion of $h(\varphi(x_0, t_0, t))$ at $\hat{x}_0^-$ is

$$h(\varphi(x_0, t_0, t)) = h(\varphi(\hat{x}_0^-, t_0, t)) + \frac{\partial h(\varphi(x_0, t_0, t))}{\partial x_0}\Big|_{x_0=\hat{x}_0^-}(x_0 - \hat{x}_0^-) \tag{8.63}$$

Substituting Eq. (8.63) into Eq. (8.62) gives

$$\begin{aligned} \Delta z_0 &= \tilde{H}_0 \Delta x_0 + v_0 \\ \Delta z_1 &= \tilde{H}_1 \Delta x_0 + v_1 \\ &\vdots \\ \Delta z_N &= \tilde{H}_N \Delta x_0 + v_N \end{aligned} \tag{8.64}$$

where

$$\begin{aligned} \Delta z_i &= \tilde{z}_i - h(\varphi(\hat{x}_0^-, t_0, t_i)) \\ \Delta x_0 &= x_0 - \hat{x}_0^- \\ \tilde{H}_i &= \frac{\partial h(\varphi(x_0, t_0, t_i))}{\partial x_0}\Big|_{x_0=\hat{x}_0^-} \end{aligned} \tag{8.65}$$

The calculation method of $\tilde{\boldsymbol{H}}_i$ will be addressed later. Equation (8.64) can be rewritten as

$$\Delta \boldsymbol{Z} = \mathcal{H}\Delta \boldsymbol{x}_0 + \boldsymbol{\nu} \tag{8.66}$$

where

$$\Delta \boldsymbol{Z} = \begin{bmatrix} \Delta z_0 \\ \Delta z_1 \\ \vdots \\ \Delta z_N \end{bmatrix}, \mathcal{H} = \begin{bmatrix} \tilde{\boldsymbol{H}}_0 \\ \tilde{\boldsymbol{H}}_1 \\ \vdots \\ \tilde{\boldsymbol{H}}_N \end{bmatrix}, \boldsymbol{\nu} = \begin{bmatrix} \boldsymbol{\nu}_0 \\ \boldsymbol{\nu}_1 \\ \vdots \\ \boldsymbol{\nu}_N \end{bmatrix} \tag{8.67}$$

According to Table 8.1, if ν meets the condition in Eq. (8.73) and $\mathcal{R} = \mathrm{E}\{\boldsymbol{\nu}\boldsymbol{\nu}^{\mathrm{T}}\}$, then the LMMSE estimates corresponding to Eq. (8.66) can be written as

$$\begin{aligned} \Delta\hat{\boldsymbol{x}}_0 &= (\boldsymbol{P}_0^{-1} + \mathcal{H}^{\mathrm{T}}\mathcal{R}^{-1}\mathcal{H})^{-1}\mathcal{H}^{\mathrm{T}}\mathcal{R}^{-1}\Delta \boldsymbol{Z} \\ \boldsymbol{P}_{\Delta\hat{\boldsymbol{x}}_0} &= (\boldsymbol{P}_0^{-1} + \mathcal{H}^{\mathrm{T}}\mathcal{R}^{-1}\mathcal{H})^{-1} \end{aligned} \tag{8.68}$$

Then, substitute $\hat{\boldsymbol{x}}_0^- + \Delta\hat{\boldsymbol{x}}_0$ as new $\hat{\boldsymbol{x}}_0^-$ and $\boldsymbol{P}_{\Delta\hat{\boldsymbol{x}}_0}$ as new $\boldsymbol{P}_0$ into Eq. (8.65), calculate new $\Delta\hat{\boldsymbol{x}}_0$ and $\boldsymbol{P}_{\Delta\hat{\boldsymbol{x}}_0}$ according to Eq. (8.68), and conduct continuous iterations until the termination criterion for iteration (e.g., $\|\Delta\hat{\boldsymbol{x}}_0\| < \varepsilon$ where ε is a very small positive number) is satisfied. The finally updated $\hat{\boldsymbol{x}}_0^-$ and $\boldsymbol{P}_0$ are regarded as the state estimate and error covariance at t_0. Once the state estimate and error covariance at

Table 8.1 LMMSE estimates under linear observation

Review of Chapter 2:

Consider the estimation algorithm in the case of linear measurement equation. $\boldsymbol{z}$ and $\boldsymbol{x}$ are given by the following linear measurement equation

$$\boldsymbol{z} = \boldsymbol{H}\boldsymbol{x} + \boldsymbol{v} \tag{8.72}$$

where $\boldsymbol{v}$ is measurement noise, while satisfying

$$\begin{aligned} &\mathrm{E}\{\boldsymbol{v}\} = \boldsymbol{0}, \mathrm{E}\{\boldsymbol{v}\boldsymbol{v}^{\mathrm{T}}\} = \boldsymbol{R}, \\ &\mathrm{E}\{\boldsymbol{x}\boldsymbol{v}^{\mathrm{T}}\} = \boldsymbol{0} \end{aligned} \tag{8.73}$$

Then, the LMMSE estimate $\hat{\boldsymbol{x}}^{\mathrm{LMMSE}}$ of Eq. (8.72) and the error covariance of $\tilde{\boldsymbol{x}}^{\mathrm{LMMSE}} \triangleq \hat{\boldsymbol{x}}^{\mathrm{LMMSE}} - \boldsymbol{x}$ can be written as

$$\begin{aligned} \hat{\boldsymbol{x}}^{\mathrm{LMMSE}} &= (\boldsymbol{P}_{xx}^{-1} + \boldsymbol{H}^{\mathrm{T}}\boldsymbol{R}^{-1}\boldsymbol{H})^{-1}(\boldsymbol{H}^{\mathrm{T}}\boldsymbol{R}^{-1}\boldsymbol{z} + \boldsymbol{P}_{xx}^{-1}\bar{\boldsymbol{x}}) \\ \boldsymbol{P}_{\tilde{\boldsymbol{x}}^{\mathrm{LMMSE}}} &= (\boldsymbol{P}_{xx}^{-1} + \boldsymbol{H}^{\mathrm{T}}\boldsymbol{R}^{-1}\boldsymbol{H})^{-1} \end{aligned} \tag{8.74}$$

where $\bar{\boldsymbol{x}} = \mathrm{E}\{\boldsymbol{x}\}$, $\boldsymbol{P}_{xx} = \mathrm{E}\{(\boldsymbol{x} - \bar{\boldsymbol{x}})(\boldsymbol{x} - \bar{\boldsymbol{x}})^{\mathrm{T}}\}$. If $\bar{\boldsymbol{x}}$ is unknown, then

$$\begin{aligned} \hat{\boldsymbol{x}}^{\mathrm{LMMSE}} &= (\boldsymbol{H}^{\mathrm{T}}\boldsymbol{R}^{-1}\boldsymbol{H})^{-1}\boldsymbol{H}^{\mathrm{T}}\boldsymbol{R}^{-1}\boldsymbol{z} \\ \boldsymbol{P}_{\tilde{\boldsymbol{x}}^{\mathrm{LMMSE}}} &= (\boldsymbol{H}^{\mathrm{T}}\boldsymbol{R}^{-1}\boldsymbol{H})^{-1} \end{aligned} \tag{8.75}$$

t_0 are obtained, the state estimates and covariances at other moments can be derived by propagating orbit dynamics.

As known from Eqs. (8.65) and (8.68), in order to implement the batch least squares filtering algorithm, $\tilde{\boldsymbol{H}}$ can be calculated with one of two typical methods:

Method 1: As $\tilde{\boldsymbol{H}}$ is the partial derivative of the observation at t to the state at t_0, small perturbations can be added to the component of $\hat{\boldsymbol{x}}_0^-$ in each direction at $\boldsymbol{x}_0$, and the numerical integration of orbital dynamics can be calculated. By combining the calculation result with measurement equation, the observation after perturbation can be obtained. Then, $\tilde{\boldsymbol{H}}$ is approximated [29] through difference operation to calculate the values of $\tilde{\boldsymbol{H}}$ at $t_1, t_2, \ldots, t_N$, respectively, namely $\tilde{\boldsymbol{H}}_1, \tilde{\boldsymbol{H}}_2, \ldots, \tilde{\boldsymbol{H}}_N$. From the measurement equation, the $\tilde{\boldsymbol{H}}$ at t_0 can be obtained directly as $\tilde{\boldsymbol{H}}_0 = \frac{\partial \boldsymbol{h}(\boldsymbol{x})}{\partial \boldsymbol{x}}|_{\boldsymbol{x}=\hat{\boldsymbol{x}}_0^-}$.

Method 2: $\tilde{\boldsymbol{H}}$ can be written as

$$\tilde{\boldsymbol{H}} = \frac{\partial \boldsymbol{h}(\boldsymbol{x})}{\partial \boldsymbol{x}} \frac{\partial \boldsymbol{x}}{\partial \boldsymbol{x}_0}|_{\boldsymbol{x}_0=\hat{\boldsymbol{x}}_0^-} \tag{8.69}$$

In this equation, $\frac{\partial \boldsymbol{h}(\boldsymbol{x})}{\partial \boldsymbol{x}}$ can be derived from the measurement equation, so the key is to calculate $\frac{\partial \boldsymbol{x}}{\partial \boldsymbol{x}_0}$ the state transition matrix of discrete dynamical system. Denote the state transition matrix from t_0 to t as $\boldsymbol{F}(t, t_0)$. Then,

$$\dot{\boldsymbol{F}}(t, t_0) = \boldsymbol{A}\boldsymbol{F}(t, t_0) \tag{8.70}$$

where $\boldsymbol{A} = \frac{\partial \boldsymbol{f}(\boldsymbol{x})}{\partial \boldsymbol{x}}$ and

$$\begin{gathered} \boldsymbol{F}(t, t_0) = \boldsymbol{I} \\ \boldsymbol{F}(t_2, t_0)\boldsymbol{F}(t_1, t_0) = \boldsymbol{F}(t_2, t_0) \end{gathered} \tag{8.71}$$

when $t - t_0$ is very small, the state transition matrix can be approximated by series expansion, as shown below

$$\boldsymbol{F}(t, t_0) = \boldsymbol{I} + \boldsymbol{A} \cdot (t - t_0) + \frac{\boldsymbol{A}^2 \cdot (t - t_0)^2}{2!} + \cdots$$

8.6.2 *Kalman Filtering Algorithm*

Since the orbit dynamics equation of spacecraft is nonlinear, the measurement equations are mostly nonlinear. If recursive filtering algorithm is used for estimation, a nonlinear Kalman filtering should be used. The fusion structure can be centralized Kalman filtering structure because a single optical imaging sensor is used for observation. In this section, extended Kalman filter (EKF) is taken as an example to illustrate the filtering implementation in optical autonomous navigation. Other nonlinear filtering algorithms can also be obtained in a similar way.

The EKF algorithm has been presented in Chap. 2. The EKF-based optical navigation is summarized as follows:

1. State prediction

$$\hat{\boldsymbol{x}}_{k,k-1} = \hat{\boldsymbol{x}}_{k-1} + \int_{t_{k-1}}^{t_k} \boldsymbol{f}(\boldsymbol{x})\mathrm{d}t \tag{8.76}$$

where $\hat{\boldsymbol{x}}_{k-1}$ and $\hat{\boldsymbol{x}}_{k,k-1}$ are state estimate and one-step state prediction, respectively; $\int_{t_{k-1}}^{t_k} \boldsymbol{f}(\boldsymbol{x})\mathrm{d}t$ is implemented through numerical integration.

2. Error covariance prediction

$$\boldsymbol{P}_{k,k-1} = \boldsymbol{F}_{k,k-1}\boldsymbol{P}_{k-1}\boldsymbol{F}_{k,k-1}^{\mathrm{T}} + \boldsymbol{Q}_{k-1} \tag{8.77}$$

where $\boldsymbol{P}_k$ and $\boldsymbol{P}_{k,k-1}$ are error covariance estimate and one-step error covariance prediction, respectively; $\boldsymbol{Q}_{k-1}$ is process noise variance matrix of orbital dynamics; $\boldsymbol{F}_{k,k-1}$ is the state transition matrix from t_{k-1} to t_k.

3. Kalman gain calculation

$$\boldsymbol{K}_k = \boldsymbol{P}_{k,k-1}\boldsymbol{H}_k^{\mathrm{T}}\left[\boldsymbol{H}_k\boldsymbol{P}_{k,k-1}\boldsymbol{H}_k^{\mathrm{T}} + \boldsymbol{R}_k\right]^{-1} \tag{8.78}$$

where $\boldsymbol{R}_k$ is the covariance of observation noise; $\boldsymbol{H}$ is the measurement sensitivity matrix, which is calculated by Eq. (8.9) for the apparent radius measurement or by Eq. (8.10) if for the LOS direction measurement.

4. Measurement update

$$\hat{\boldsymbol{x}}_k = \hat{\boldsymbol{x}}_{k,k-1} + \boldsymbol{K}_k\left[z_k - \hat{z}_k\right] \tag{8.79}$$

where $\hat{z}_k$ is the predicted observation, which is predicted by Eq. (8.8) for the apparent radius measurement or by Eq. (8.10) for LOS direction measurement

5. Error covariance update

$$\boldsymbol{P}_k = [\boldsymbol{I} - \boldsymbol{K}_k\boldsymbol{H}_k]\boldsymbol{P}_{k,k-1}[\boldsymbol{I} - \boldsymbol{K}_k\boldsymbol{H}_k]^{\mathrm{T}} + \boldsymbol{K}_k\boldsymbol{R}_k\boldsymbol{K}_k^{\mathrm{T}} \tag{8.80}$$

8.7 Numerical Simulation

8.7.1 Optical Navigation of the Transfer Phase

In this section, numerical simulation of the autonomous navigation based on asteroid observation used in the deep-space transfer is presented, which is divided into three parts. The first part demonstrates the navigation beacon planning algorithm. The second part compares and evaluates the performance of different filtering algorithms. The third part analyzes the impact of different error sources on navigation accuracy.

The epoch time of the starting point of transfer segment is August 5, 2007, namely JD2454317.5. Its initial position in the heliocentric J2000 ecliptic inertial frame is

$$\begin{bmatrix} 1.017978222619442 \times 10^8 \\ -1.123003524012581 \times 10^8 \\ -9.578654170367867 \times 10^4 \end{bmatrix} \mathrm{km}$$

and its initial velocity is

$$\begin{bmatrix} 21.40460801927876 \\ 25.08765627351607 \\ -2.007089018943564 \end{bmatrix} \mathrm{km/s}$$

The dynamic model that generates the true trajectory considers the gravitational perturbation of large planets in the solar system and the perturbation caused by solar radiation pressure. The orbital integrator is RKF78.

The optical imaging sensor has a focal length of 677 mm and an angular resolution of 13 μrad/pixel. As same with the observation strategy of "Deep Space 1", the navigation period is 7 days, and 12 asteroids are observed in every observation window.

8.7.1.1 Simulation of Navigation Beacon Planning

In this section, the navigation beacon planning algorithm is simulated and demonstrated. To facilitate comparison, EKF is used as the filtering algorithm. Other parameters are set as follows (all set as 1 σ, unless specially specified):

(1) Dynamic model in navigation filtering: perturbations of the planets and solar radiation pressure are considered. The orbital integrator is ODE45.
(2) The measurement equation is given by Eq. (8.31), and the measurement sensitivity matrix is given by Eq. (8.32).
(3) Initial navigation errors: each component of position error is 1000 km, and each component of velocity error is 1 m/s.
(4) Observation error of optical imaging sensor: 0.1 pixel.
(5) Attitude determination error: 1 urad.
(6) Asteroid ephemeris error: 100 km.

As shown in Table 8.2, two typical scenarios are considered. Scenario 1 observes 12 asteroids given by the planning algorithm, while scenario 2 randomly observes 12 asteroids. In each scenario, candidate navigation asteroids need to be selected from the asteroid database. Table 8.3 shows the selection criteria of asteroid apparent magnitude, distance, solar phase angle, apparent motion, and three-star probability. Considering these thresholds, the obtained screening results are shown in Table 8.4.

A single-simulation results are shown in Figs. 8.6, 8.7, 8.8, 8.9, 8.10, and 8.11. To comprehensively evaluate the navigation beacon planning algorithm, the Monte Carlo simulation technique is described in Chap. 4 is performed to generate n (n = 300) sets of orbital data and measurements, which are then substituted into EKF algorithm for state estimation to obtain n sets of estimates. The estimation error root mean square (RMS) of 3D position vector is calculated by the following method

Table 8.2 Beacon planning simulation scenarios

Scenario No.	Configuration
1	12 asteroids in each batch given by observation planning algorithm
2	12 randomly observed asteroids in each batch

Table 8.3 Selection criteria of asteroids

Criterion	Threshold
Apparent magnitude	≤ 12
Distance	$\leq 3 \times 10^8$ km
Solar phase angle	$\geq 135°$
Apparent motion	≤ 0.1 μrad/s
Three-star probability (magnitude $\leq$ 11)	$\geq 98\%$

Table 8.4 Number of candidate asteroids in each window

Window No.	1	2	3	4	5	6	7	8	9	10	11	12	13	14
Number of candidates	19	18	16	17	17	16	13	13	14	16	17	17	16	16
Window No.	15	16	17	18	19	20	21	22	23	24	25	26	27	28
Number of candidates	14	15	16	14	14	15	16	15	14	13	16	15	19	19
Window No.	29	30	31	32	33	34	35	36	37	38	39	40	41	42
Number of candidates	19	19	18	19	18	19	19	19	19	19	19	19	19	19

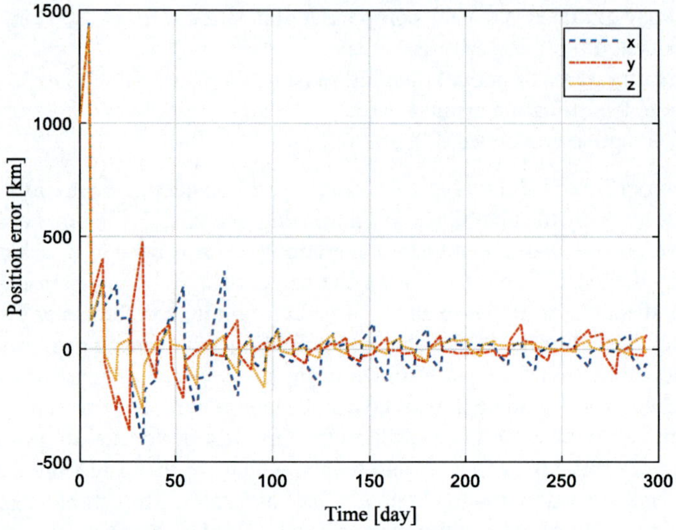

Fig. 8.6 Navigation position errors (scenario 1)

$$
\begin{aligned}
\sigma_{x,t} &= \sqrt{\frac{1}{n-1}\sum_{j=1}^{n}\left(r_{x,t}^{j} - \hat{r}_{x,t}^{j}\right)^{2}} \\
\sigma_{y,t} &= \sqrt{\frac{1}{n-1}\sum_{j=1}^{n}\left(r_{y,t}^{j} - \hat{r}_{y,t}^{j}\right)^{2}} \\
\sigma_{z,t} &= \sqrt{\frac{1}{n-1}\sum_{j=1}^{n}\left(r_{z,t}^{j} - \hat{r}_{z,t}^{j}\right)^{2}}
\end{aligned}
\tag{8.81}
$$

where $r_{x,t}^{j}$, $r_{y,t}^{j}$, $r_{z,t}^{j}$ and $\hat{r}_{x,t}^{j}$, $\hat{r}_{y,t}^{j}$, $\hat{r}_{z,t}^{j}$ represent, respectively, the truth value and estimated value of spacecraft position vector in the j-th set of simulation data. The average level of estimation error in a certain period is measured by calculating the average of triaxial positioning error MSEs $\sigma_{x,t}^{2}$, $\sigma_{y,t}^{2}$ and $\sigma_{z,t}^{2}$. The specific formula is

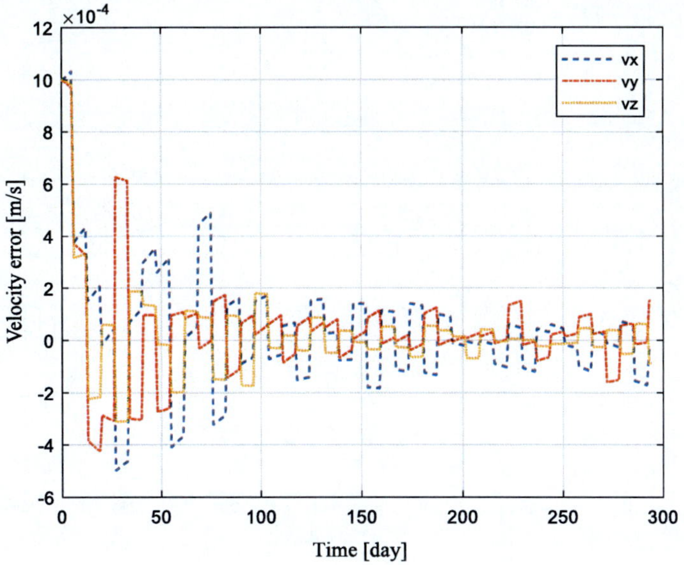

Fig. 8.7 Navigation velocity errors (scenario 1)

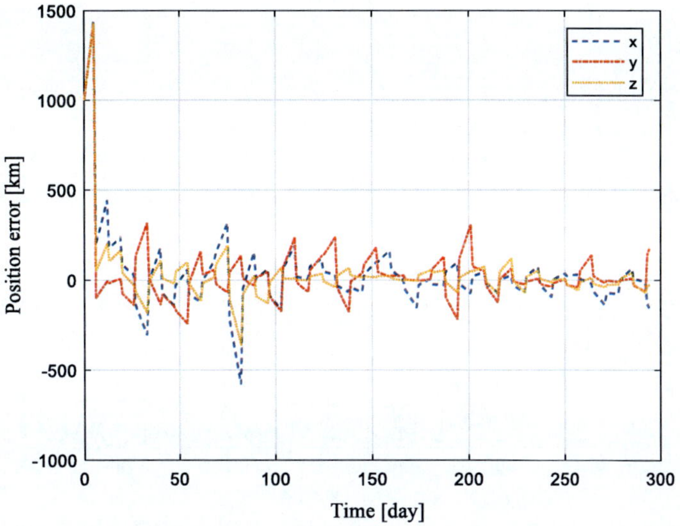

Fig. 8.8 Navigation position errors (scenario 2)

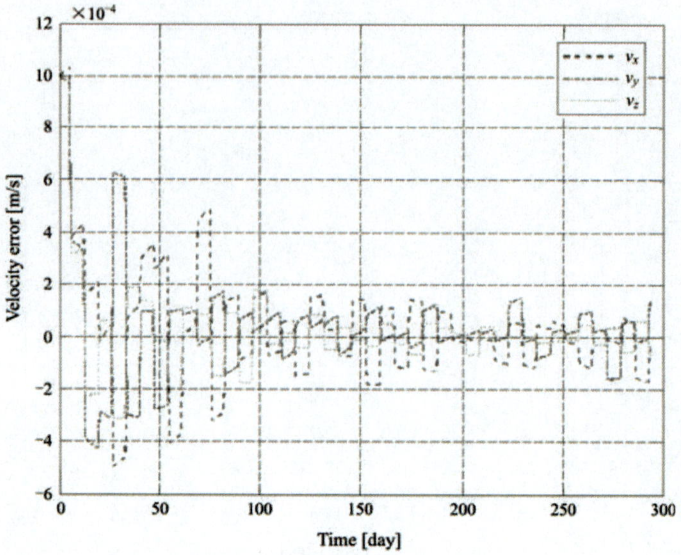

Fig. 8.9 Navigation velocity errors (scenario 2)

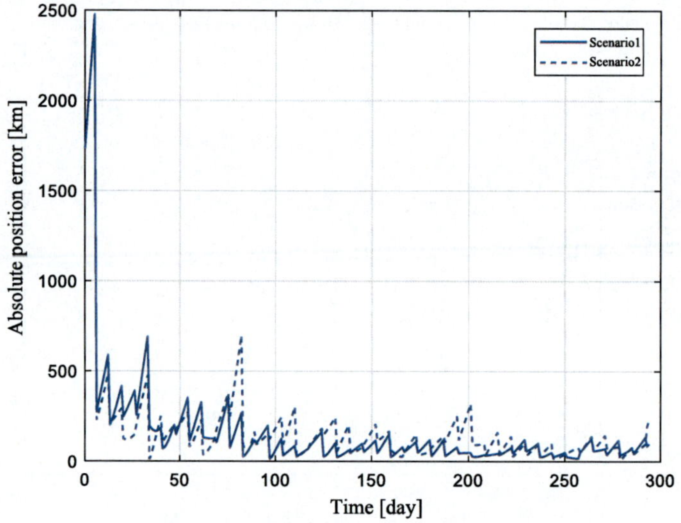

Fig. 8.10 Navigation position error comparisons (scenario 1 vs. scenario 2)

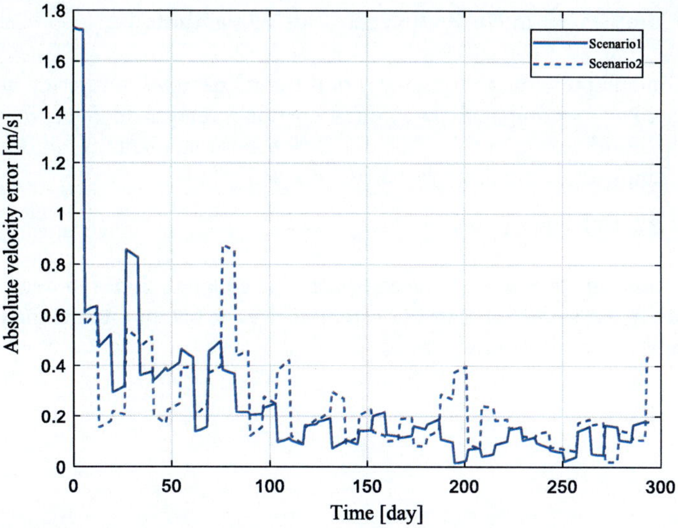

Fig. 8.11 Navigation velocity error comparisons (scenario 1 vs. scenario 2)

$$
\begin{aligned}
\sigma_y^2 &= \frac{1}{k}\sum_{i=1}^{k}\sigma_{y,t}^2 \\
\sigma_y^2 &= \frac{1}{k}\sum_{i=1}^{k}\sigma_{y,t}^2 \\
\sigma_z^2 &= \frac{1}{k}\sum_{i=1}^{k}\sigma_{z,t}^2
\end{aligned}
\tag{8.82}
$$

where k is the total number of estimation error data recorded during a steady-state period.

The Monte Carlo simulation results are shown in Table 8.5. Obviously, observing the asteroids given by the planning algorithm can effectively improve the estimation accuracy, thus verifying the effectiveness of navigation beacon planning algorithm.

Table 8.5 Comparison of simulation results

	Position RMS (km)			Velocity RMS (m/s)		
	σ_x	σ_y	σ_z	σ_{vx}	σ_{vy}	σ_{vz}
Scenario 1	85.8640	46.4867	40.5474	0.1555	0.0849	0.0757
Scenario 2	104.6719	93.7693	64.6903	0.1797	0.1531	0.1194

8.7.1.2 Comparison of Navigation Filtering Algorithms

In this section, simulation performances of different navigation filtering algorithms are compared. The compared parameters are the same as in Sect. 8.7.1.1. The observed asteroids are given by the navigation beacon planning algorithm. The adopted filtering algorithm is described as follows:

(1) Batch filtering algorithm

The batch filtering algorithm of "Deep Space I" is adopted, that is, to feed the latest four batches of measurement data (12 asteroids in each batch) into the filter for state estimation in one orbit determination.

(2) EKF

Refer to Sect. 8.6.2.

(3) UKF

Refer to Sect. 2.4.4. For some specific parameters, refer to Sect. 8.6.2.

Random sampling is performed for 300 times. The Monte Carlo simulation results are shown in Table 8.6. Table 8.7 shows a single-run time of three filtering algorithms.

The accuracy of the three filtering algorithms is basically the same. The time consumed by EKF is significantly less than that consumed by BLS and UKF, because the epoch conversion of measurement information in BLS and the unscented transformation in UKF require a large amount of computation. Given that in this example, the running time of EKF is the shortest under the same accuracy of state estimation, the following simulation will use only the EKF algorithm.

Table 8.6 Comparison of the state estimation accuracy of navigation filtering algorithms

Filter	Position RMS (km)			Velocity RMS (m/s)		
	σ_x	σ_y	σ_z	σ_{vx}	σ_{vy}	σ_{vz}
BLS	73.2591	51.2225	41.6800	0.1283	0.0912	0.0900
EKF	85.8640	46.4867	40.5474	0.1555	0.0849	0.0757
UKF	69.2337	50.0716	41.3984	0.1236	0.1399	0.0715

Table 8.7 Run time of navigation filtering algorithms

Algorithm	BLS	EKF	UKF
Time (s)	14.30	4.67	16.50

8.7.1.3 Error Analysis

The three major errors affecting the accuracy of autonomous navigation are attitude determination error, measurement error, and asteroid ephemeris error. In this section, the effects of the three errors on the performance of navigation system will be analyzed.

First, the effects of attitude determination error on the performance of navigation system is studied. The attitude determination error varies from 0 μrad to 5 μrad, and the ephemeris error and pixel error are both 0. For other parameters, refer to Sect. 8.7.1.1. In each case, random sampling is performed for 300 times. The statistical results are shown in Table 8.8. When the attitude determination error is 0 μrad, the position estimation error is $[6.2204, 6.3971, 1.0850]^{\mathrm{T}}$ km and the velocity estimation error is $[0.0215, 0.0227, 0.0025]^{\mathrm{T}}$ m/s. When the attitude determination error is 5 μrad, the position estimation error is $[295.2007, 227.5762, 173.4225]^{\mathrm{T}}$ km and the velocity estimation error is $[0.5104, 0.3940, 0.3212]^{\mathrm{T}}$ m/s.

Next, analyze the influence of measurement error on the performance of navigation system. The measurement error varies from 0 pixels to 0.5 pixels, and the ephemeris error and attitude determination error are both 0. For other parameters, refer to Sect. 8.7.1.1. In each case, random sampling is performed for 300 times. The simulation results are shown in Table 8.9. When the measurement error is 0 pixels, the position estimation error is $[6.2204, 6.3971, 1.0850]^{\mathrm{T}}$ km and the velocity estimation error is $[0.0215, 0.0227, 0.0025]^{\mathrm{T}}$ m/s. When the measurement error is 0.5 pixels, the position estimation error is $[88.8207, 99.0747, 67.9757]^{\mathrm{T}}$ km and the velocity estimation error is $[0.1527, 0.1796, 0.1199]^{\mathrm{T}}$ m/s.

Finally, analyze the influence of asteroid ephemeris error on the performance of navigation system. The asteroid ephemeris error varies from 0 km to 300 km,

Table 8.8 Different attitude determination errors and the corresponding state estimation accuracies

Attitude determination error (μrad)	Position RMS (km)			Velocity RMS (m/s)		
	σ_x	σ_y	σ_z	σ_{vx}	σ_{vy}	σ_{vz}
0	6.2204	6.3971	1.0850	0.0215	0.0227	0.0025
3	219.2010	152.7510	126.8422	0.3370	0.2615	0.2377
5	295.2007	227.5762	173.4225	0.5104	0.3940	0.3212

Table 8.9 Different measurement errors and the corresponding state estimation accuracies

Pixel error (pixel)	Position RMS (km)			Velocity RMS (m/s)		
	σ_x	σ_y	σ_z	σ_{vx}	σ_{vy}	σ_{vz}
0	6.2204	6.3971	1.0850	0.0215	0.0227	0.0025
0.1	18.7744	18.0593	10.6754	0.0393	0.0373	0.0205
0.3	55.5680	66.1118	44.5617	0.1026	0.1188	0.0794
0.5	88.8207	99.0747	67.9757	0.1527	0.1796	0.1199

Table 8.10 Different ephemeris errors and the corresponding state estimation accuracies

Asteroid ephemeris error (km)	Position RMS(km)			Velocity RMS (m/s)		
	σ_x	σ_y	σ_z	σ_{vx}	σ_{vy}	σ_{vz}
0	6.2204	6.3971	1.0850	0.0215	0.0227	0.0025
100	99.1258	95.4906	100.2497	0.0197	0.0215	0.0018
300	299.2996	295.7757	299.7761	0.0162	0.0193	0.0025

and the attitude determination error and measurement error are both 0. For other parameters, refer to Sect. 8.7.1.1. In each case, random sampling is performed for 300 times. The simulation results are shown in Table 8.10. When the ephemeris error is 0 km, the position estimation error is $[6.2204, 6.3971, 1.0850]^T$ km and the velocity estimation error is $[0.0215, 0.0227, 0.0025]^T$ m/s. When the ephemeris error is 300 km, the position estimation error is $[299.2996, 295.7757, 299.7761]^T$ km and the velocity estimation error is $[0.0162, 0.0193, 0.0025]^T$ m/s.

The comparison shows that the attitude determination error and asteroid ephemeris error have greater influences on the performance of navigation system, while the measurement error has a smaller influence. Measurement error is random noise, whose influence can be attenuated by filtering algorithm. Attitude determination error and ephemeris error are systematic errors, so they need to be corrected by systematic error correction algorithm.

8.7.2 Optical Navigation of the Approaching Phase

In this section, optical autonomous navigation for the Mars approach phase is simulated. The measurements used for navigation are LOS directions to Phobos, Deimos, and Mars center and the apparent radius of Mars. The initial orbital condition is shown in Table 8.11. The orbital model that generates flight trajectory considers solar gravitational perturbation, solar radiation pressure perturbation, and a 50 × 50 gravitational field model. The orbital integrator is ODE45.

Parameters of optical imaging sensor are as follows:

Focal length: 6 mm;

FOV: 60° × 60°;

Resolution: 1024 × 1024.

Other parameters are set as follows:

Table 8.11 Initial orbital condition of the approach phase

Epoch time	July 3, 1997 11:09:59.000 UTCG
Position (km)	[534059.29, 214496.39, 8822.61]
Velocity (km/s)	[− 4.94, −2.041, −0.085]

(1) Dynamics model in navigation filter: solar gravitational perturbation and a fourth-order gravitational field are considered.
(2) Initial navigation errors: 20 km position error in each axis and 2 m/s velocity error in each axis.
(3) Sensor noise: 0.1 pixels.
(4) Attitude determination error: 1 μrad.
(5) Measurement rate: 60 s.

As the number of Mars satellites is limited, navigation beacon planning is not needed, but the visibility of Mars satellites needs to be analyzed. The visibility analysis based on the positions of Mars, Mars satellites, and spacecraft shows that Deimos is more visible than Phobos and can be observed in most of the approaching segment. Since the observation variables include the LOS directions of Mars satellites and Mars center and the apparent radius of Mars, four typical navigation scenarios are considered, as shown in Table 8.12. Scenarios 1–3 are centralized-fusion simulation scenarios where the commonly used EKF algorithm is used for filtering. Scenario 4 is the distributed fusion of scenarios 2 and 3, where covariance intersection algorithm is used for fusion.

The influence of the ephemeris error of Martian satellites is not considered in the three scenarios. For scenario 1, when the attitude determination error is 0, the steady navigation position and velocity errors are $[2.0620\ \ 0.7395\ \ 0.2545]^{\mathrm{T}}$ km $[0.08700.04090.0220]^{\mathrm{T}}$ m/s, respectively, as shown in Figs. 8.12 and 8.13. When the attitude determination error is 6 urad, the steady navigation position and velocity errors are $[3.98761.05220.3881]^{\mathrm{T}}$ km $[0.15860.04920.0302]^{\mathrm{T}}$ m/s, respectively, as shown in Figs. 8.14 and 8.15. It is obvious that attitude determination error will adversely affect the navigation accuracy.

Simulation results of scenario 2 are shown in Figs. 8.16 and 8.17, where steady navigation position and velocity errors are $[2.58971.86692.3466]^{\mathrm{T}}$ km

Table 8.12 Simulation scenarios for approaching segment

Scenario No.	Configuration	Measurement equation	Measurement sensitivity matrix
1	Take the LOS directions of Phobos and Deimos as navigation observations	see Eq. (8.13)	see Eq. (8.14)
2	Take the angles between Martian satellites and Mars center as observations	see Eq. (8.19)	see Eq. (8.20)
3	Take the apparent radius of Mars and the LOS direction of Mars center as observations	see Eqs. (8.8)and (8.15)	see Eqs. (8.9) and (8.16)
4	Distributed fusion of scenarios 2 and 3	See the scenarios 2 and 3	

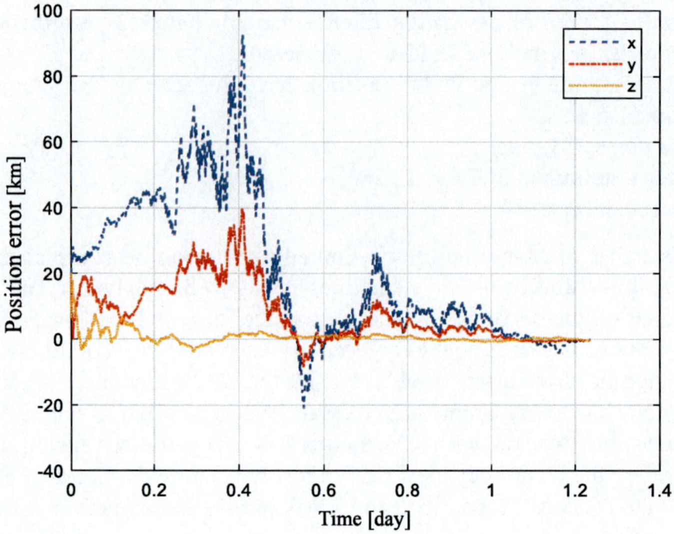

Fig. 8.12 Navigation position errors of scenario 1 (attitude determination error is 0)

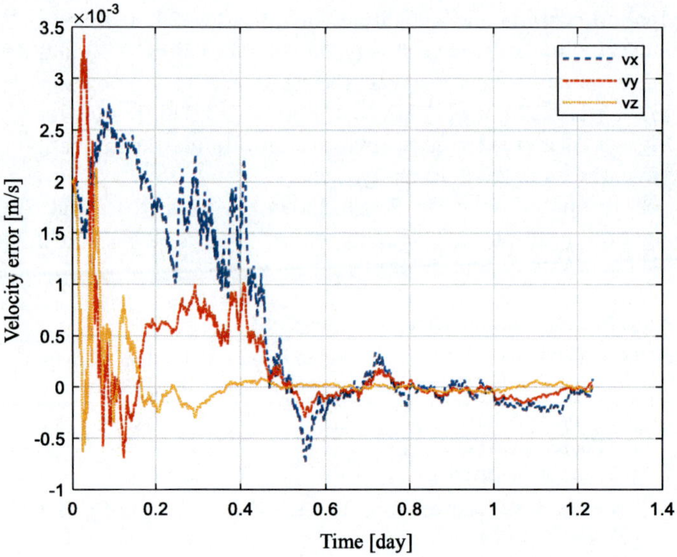

Fig. 8.13 Navigation velocity errors of scenario 1 (attitude determination error is 0)

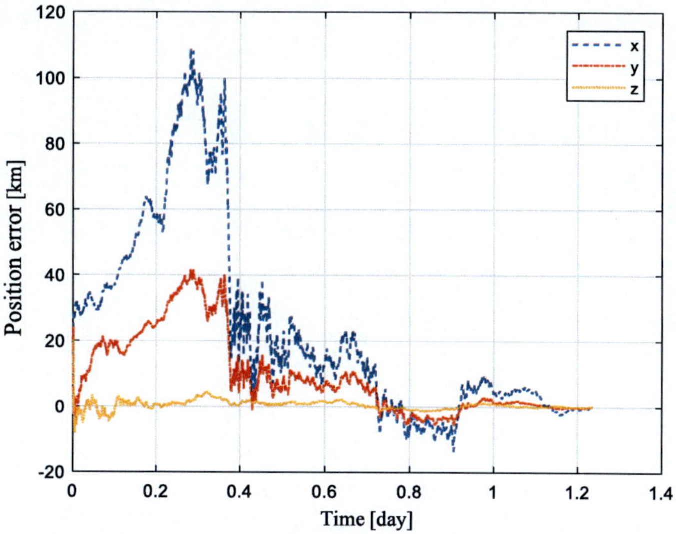

Fig. 8.14 Navigation position errors of scenario 1 (attitude determination error is 6 urad)

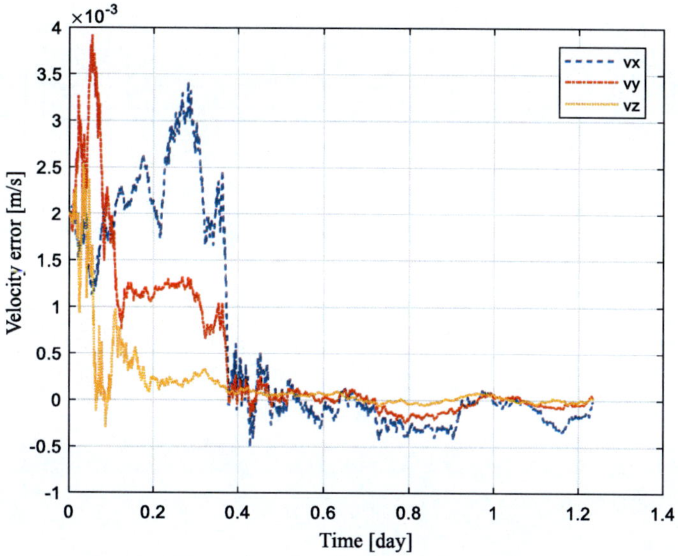

Fig. 8.15 Navigation velocity errors of scenario 1 (attitude determination error is 6 urad)

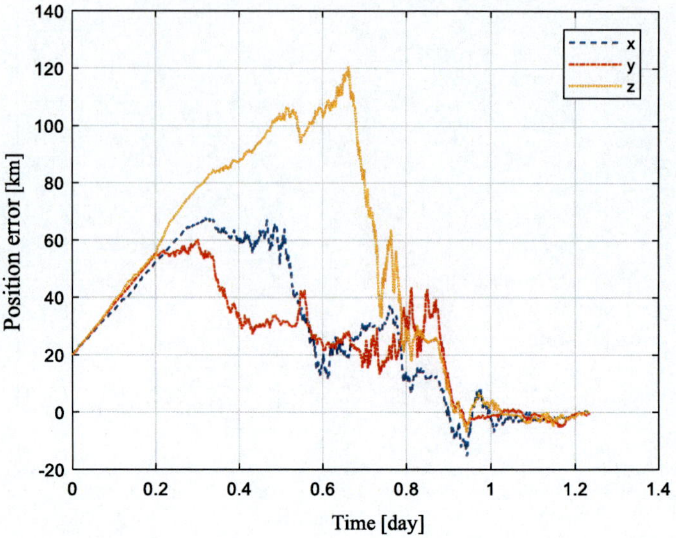

Fig. 8.16 Navigation position errors of scenario 2

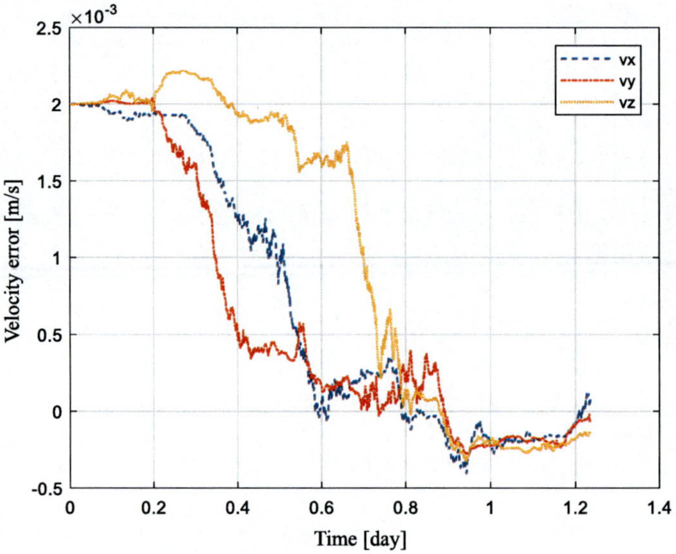

Fig. 8.17 Navigation velocity errors of scenario 2

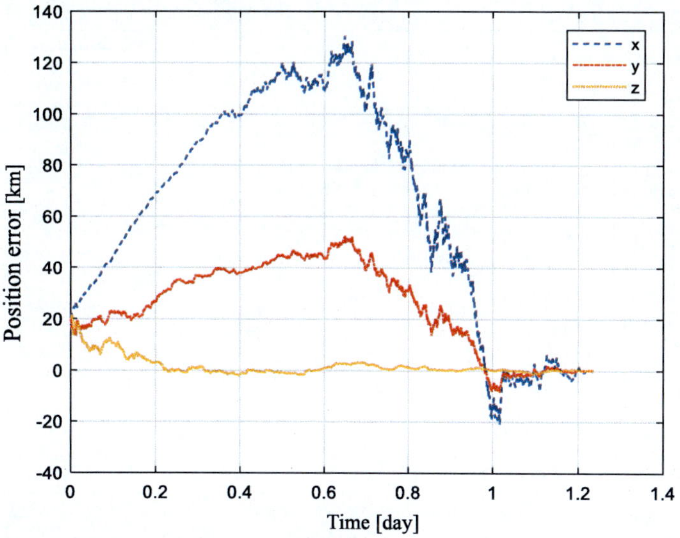

Fig. 8.18 Navigation position errors of scenario 3

[0.134640.123630.1243]^T m/s, respectively. The simulation accuracy in scenario 2 is slightly worse than that in the scenario 1 without attitude determination error, but better than that in the scenario 1 if attitude determination errors are presented.

Simulation results of scenario 3 (without attitude determination error) are shown in Figs. 8.18 and 8.19, where the steady navigation position error and velocity error are [6.14392.59190.47144]^T km and [0.276810.089860.11769]^T m/s, respectively, which are worse than the result of scenario 1 or 2.

The simulation results of scenario 4 are shown in Figs. 8.20 and 8.21, where the filtered steady position error and velocity error are [1.32580.296450.43873]^T km and [0.176150.0567250.030363]^T m/s, respectively. Obviously, the navigation performance after fusion is better than that in the scenarios 1, 2, and 3.

The filtering convergence speed in all the four navigation scenarios is slow. This is because at the early approaching phase, the change of spacecraft orbit relative to navigation beacon is not significant, thus leading to weak observability and slow convergence.

8.7.3 Optical Navigation of the Orbiting Phase

In this section, optical navigation for the Mars orbiting phase is simulated. The measurements are the LOS vectors to Phobos and Deimos. Initial orbital condition is shown in Table 8.13. The dynamic model that generates the true trajectory considers solar gravitational perturbation, solar radiation pressure perturbation, and

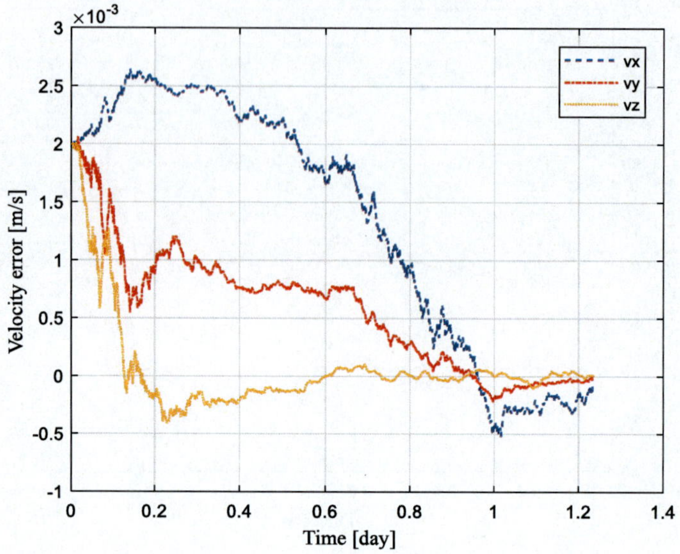

Fig. 8.19 Navigation velocity errors of scenario 3

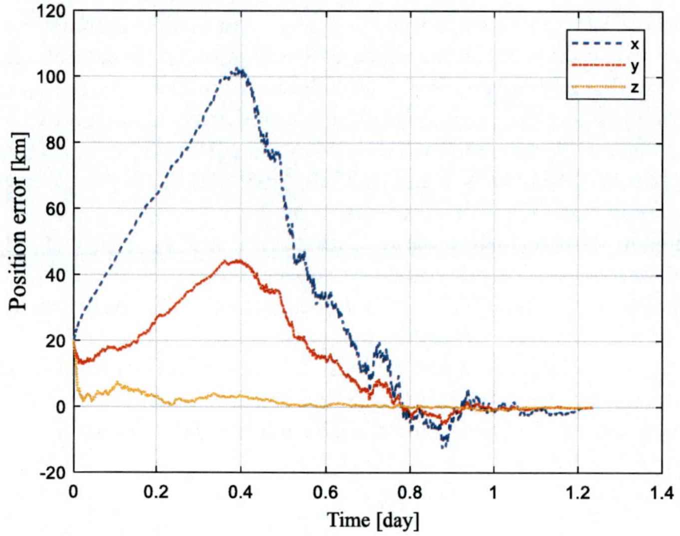

Fig. 8.20 Navigation position errors of scenario 4

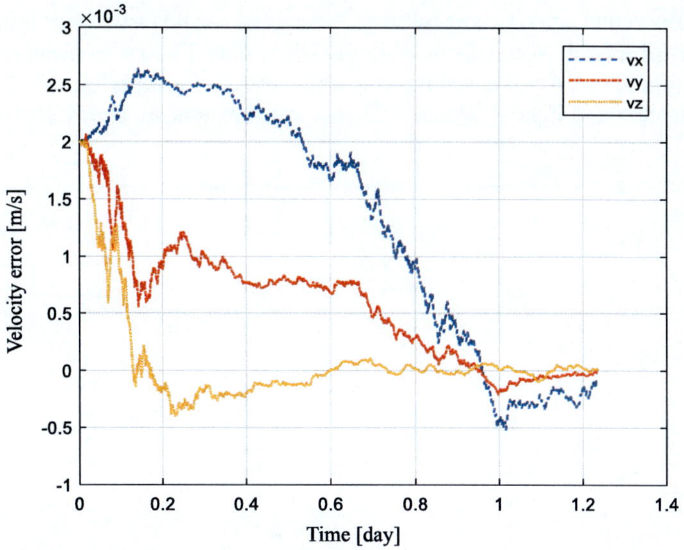

Fig. 8.21 Navigation velocity errors of scenario 4

Table 8.13 Orbital parameters of the starting point of orbiting phase

Epoch time	July 4, 1997 16:47:58.000 UTCG
Position (km)	[682.30, −4190.90, −252.308]
Velocity (km/s)	[− 3.15, −0.49, −0.01]

a 50th-order gravitational field model. The orbital integrator is a RKF78 numerical integrator.

Parameters of optical imaging sensor are as follows:

Focal length: 6 mm;

FOV: 60° × 60°;

Resolution: 1024 × 1024.

Other parameters are set as follows:

(1) dynamics model in navigation filter: solar gravitational perturbation and fifth-order gravitational field are considered.
(2) The measurement equation is given by Eq. (8.31), and the measurement sensitivity matrix is given by Eq. (8.32).
(3) Initial navigation errors: 3 km position error in each axis and 3 m/s velocity error in each axis.
(4) Sensor noise (1 σ): 0.1 pixels.
(5) Attitude determination error: 0 μrad.
(6) Measurement rate: 60 s.

The observation strategy is as follows. When Phobos and Deimos are both visible, Deimos is observed. When Deimos is invisible, only Phobos is observed. When neither Phobos nor Deimos is visible, only orbit propagation is performed. Simulation results are shown in Figs. 8.22 and 8.23, statistical results are shown in Table 8.14.

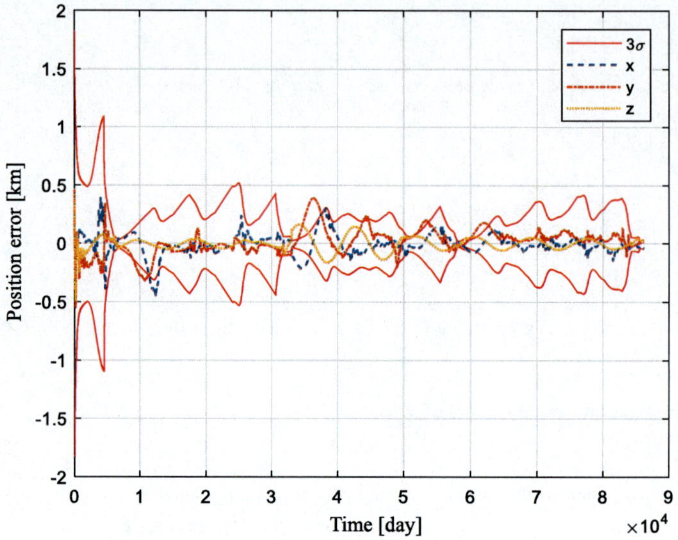

Fig. 8.22 Navigation position errors

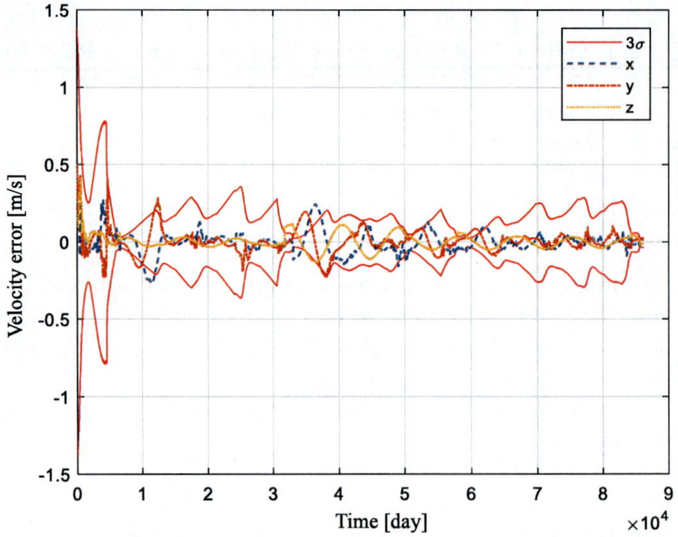

Fig. 8.23 Navigation velocity errors

Table 8.14 Autonomous navigation results during the orbiting phase

Position RMS (km)			Velocity RMS (m/s)		
σ_x	σ_y	σ_z	σ_{vx}	σ_{vy}	σ_{vz}
123.7710	166.3513	125.5043	0.0731	0.0788	0.0522

It can be seen that the navigation position error 3 σ is about 500 m, and the velocity error 3 σ is about 0.15 m/s.

8.8 Brief Summary

This chapter discusses the optical autonomous navigation technology from the aspects of autonomous navigation principle, navigation sensor, selection of candidate navigation beacon, optical measurement equation, navigation beacon planning, and filtering algorithm. Based on the measurement principle, this chapter tries to establish the relationship between optical autonomous navigations at different flight phases of a deep-space exploration mission. In terms of navigation filtering, this chapter focuses on the implementation of batch least squares and extended Kalman filtering algorithms, and simulates and demonstrates the optical autonomous navigation during the transfer, approaching, and orbiting phases. In terms of algorithm, the batch filtering, the extended Kalman filtering, and the unscented Kalman filtering algorithms as well as distributed covariance intersection algorithms are demonstrated here. In terms of observation source, the optical autonomous navigation techniques based on asteroid observation, planet observation, and planetary satellite observation are demonstrated in this chapter.

References

1. Riedel, J., et al. 2000. *Deep space 1 technology validation report: Autonomous optical navigation*. Pasadena: Jet Propulsion Laboratory.
2. Miso, T., T. Hashimoto, and K. Ninomiya. 1999. Optical guidance for autonomous landing of spacecraft. *IEEE Transactions on Aerospace and Electronic Systems* 35 (2): 459–473.
3. Rayman, M.D. 2002. *The successful conclusion of the deep space 1 mission: important results without a flashy title.*
4. Graf, J.E., et al. 2005. The Mars reconnaissance orbiter mission. *Acta Astronautica* 57 (2): 566–578.
5. Bhat, R.S., et al. 2004. *WILD2 approach maneuver strategy used for Stardust spacecraft*. Pasadena, CA: Jet Propulsion Laboratory, National Aeronautics and Space Administration.
6. Liu, Z., et al. 2015. High precision landing site mapping and rover localization for Chang'e-3 mission. *Science China Physics, Mechanics & Astronomy* 58 (1): 1–11.
7. Kubitschek, D.G., et al. 2006. *Deep impact autonomous navigation: the trials of targeting the unknown.*

8. Zhang, X.W., D.Y. Wang, and X.Y. Huang. 2009. Study on the selection of the beacon asteroids in autonomous optical navigation for interplanetary exploration. *Journal of Astronautics* 30 (3): 947–952.
9. Xu, W., et al. 2007. Selection and planning of asteroids for deep space autonomous optical navigation. *Acta Aeronautica Et Astronautica Sinica* 28 (4): 891–896.
10. Polle, B., et al. 2003. Autonomous on-board navigation for interplanetary missions. *Advances in the Astronautical Sciences* 113: 277–293.
11. Chausson, L. and S. Delavault. 2003. Optical navigation performance during interplanetary cruise. In *Proceedings of the 17th International Symposium on Space Flight Dynamics.*
12. Bhaskaran, S., et al. 1998. Orbit determination performance evaluation of the deep space 1 autonomous navigation system. In *AAS/AIAA Spaceflight Mechanics Meeting*, Monterrey, CA.
13. Wikipedia Contributors. *Apparent magnitude.* Wikipedia, The Free Encyclopedia 2017 January 15, 2020. Available from: https://en.wikipedia.org/w/index.php?title=Apparent_magnitude&oldid=935894559.
14. Bowell, E., et al. 1989. Application of photometric models to asteroids. In *Asteroids II*.
15. Romanishin, W., and S.C. Tegler. 2005. Accurate absolute magnitudes for Kuiper belt objects and Centaurs. *Icarus* 179 (2): 523–526.
16. Christian, J.A., and E.G. Lightsey. 2009. Review of options for autonomous cislunar navigation. *Journal of Spacecraft and Rockets* 46 (5): 1023–1036.
17. Shuster, M.D., and S.D. Oh. 1981. Three-axis attitude determination from vector observations. *Journal of Guidance, Control and Dynamics* 4 (1): 70–77.
18. Schlee, F., and N. Toda. 1967. Autonomous orbital navigation by optical tracking of unknown landmarks. *Journal of Spacecraft and Rockets* 4 (12): 1644–1648.
19. Levine, G.M. 1966. A method of orbital navigation using optical sightings to unknown landmarks. *AIAA Journal* 4 (11): 1928–1931.
20. Li, M., et al. 2013. Constrained estimation for autonomous navigation using unknown landmarks. In *Chinese Automation Congress (CAC)*, IEEE.
21. Keenan, R.V. and J.D. Regenhardt. 1962. *Star occultation measurements as an aid to navigation in cis-lunar space.* Massachusetts Institute of Technology.
22. Psiaki, M.L., and J.C. Hinks. 2007. Autonomous lunar orbit determination using star occultation measurements. In *AIAA Guidance, Navigation and Control Conference and Exhhibit.*
23. Landgraf, M., et al. 2006. Optical navigation for lunar exploration missions. In *57th International Astronautical Congress, I AC Paper IAC-06-C1.*
24. Ning, X., et al. 2013. Autonomous satellite navigation using starlight refraction angle measurements. *Advances in Space Research* 51 (9): 1761–1772.
25. Wang, X., J. Xie, and S. Ma. 2010. Starlight atmospheric refraction model for a continuous range of height. *Journal of Guidance, Control and Dynamics* 33 (2): 634.
26. Gounley, R., R. White, and E. Gai. 1984. Autonomous satellite navigation by stellar refraction. *Journal of Guidance* 7 (2): 129–134.
27. Fang, J., X. Ning, and Y. Tian. 2017. *Principle and method of autonomous celestial navigation system.* National Defense Industry Press.
28. Ham, F.M., and R.G. Brown. 1983. Observability, eigenvalues, and Kalman filtering. *IEEE Transactions on Aerospace and Electronic Systems* 2: 269–273.
29. Li, M., W. Jing, and X. Huang. 2012. Dual cone-scanning horizon sensor orbit and attitude corrections for Earth's oblateness. *Journal of Guidance, Control and Dynamics* 35 (1): 344–349.

Chapter 9
Optical/Pulsar Integrated Autonomous Navigation Technology

In this chapter, the fusion of optical autonomous navigation and pulsar-based autonomous navigation will be studied. The former can provide priori state information to the latter, while the latter can enhance the former's performance. Sections 9.1 and 9.2 briefly review measurement equations of optical autonomous navigation and pulsar-based autonomous navigation, respectively. Section 9.3 studies more deeply the navigation beacon planning algorithm. For the integrated navigation system fused on a single optical beacon and pulsar, the optimal beacons configuration is investigated based on geometric observability analysis. For a general optical/pulsar fused navigation system, the onboard planning algorithm of navigation beacons based on dynamic observability analysis is also studied. Section 9.4 presents optical/pulsar integrated navigation filtering. In Sect. 9.5, numerical simulation of the optical/pulsar integrated autonomous navigation technology for deep-space transfer, approaching, and surrounding phases are provided.

9.1 Review of Optical Autonomous Navigation

Optical autonomous navigation has been presented in Chap. 8. In this section, the measurements of the optical navigation are reviewed briefly. The measurements used for optical autonomous navigation can be classified into two categories: apparent radius and LOS direction. For the apparent radius measurement, the measurement equation can be written as

$$\rho = \arcsin \frac{\hat{R}_p}{r} + b + \nu \tag{9.1}$$

where ρ is apparent radius, $\hat{R}_p$ is the reference radius of observed celestial body, b is measurement systematic bias, and ν is measurement noise. If the filter state is composed of the position and velocity relative to the celestial body, i.e., $\boldsymbol{x} =$

D. Wang et al., *Spacecraft Autonomous Navigation Technologies Based on Multi-source Information Fusion*, Space Science and Technologies, https://doi.org/10.1007/978-981-15-4879-6_9

$[\boldsymbol{r}^{\mathrm{T}}, \boldsymbol{v}^{\mathrm{T}}]^{\mathrm{T}}$, then the measurement sensitivity matrix can be written as

$$\begin{aligned}\boldsymbol{H} &= \left[\, \frac{\partial \rho}{\partial \boldsymbol{r}} \; \frac{\partial \rho}{\partial \boldsymbol{v}} \,\right] \\ &= \left[-\frac{\hat{R}_p}{r^2\sqrt{r^2-\hat{R}_p^2}}\boldsymbol{r}^{\mathrm{T}} \; \boldsymbol{0}_{1\times 3} \right]\end{aligned} \tag{9.2}$$

For a LOS direction measurement, the measurement equation can be written as

$$\tilde{\boldsymbol{n}}^s = \boldsymbol{C}_n^s \frac{\boldsymbol{R}_b^n - \boldsymbol{r}^n}{\left\| \boldsymbol{R}_b^n - \boldsymbol{r}^n \right\|} + \boldsymbol{v} \tag{9.3}$$

where $\tilde{\boldsymbol{n}}$ is the LOS direction vector to the navigation beacon, $\boldsymbol{R}_b$ is the position vector of navigation beacon relative to the origin of coordinates, superscripts n and s indicate the expressions in navigation frame $\mathcal{F}_n$ and $\mathcal{F}_s$, respectively, and $\boldsymbol{v}$ is the measurement noise. For a small field of view (FOV) assumption, $\boldsymbol{v}$ satisfies

$$\begin{aligned}E\{\boldsymbol{v}\} &= \boldsymbol{0} \\ E\{\boldsymbol{v}\boldsymbol{v}^{\mathrm{T}}\} &= \sigma^2[\boldsymbol{I} - (\boldsymbol{C}_n^s\boldsymbol{n}^n)(\boldsymbol{C}_n^s\boldsymbol{n}^n)^{\mathrm{T}}]\end{aligned} \tag{9.4}$$

where $\boldsymbol{n}^n = \frac{\boldsymbol{R}_b^n - \boldsymbol{r}^n}{\left\| \boldsymbol{R}_b^n - \boldsymbol{r}^n \right\|}$. The measurement sensitivity matrix corresponding to Eq. (9.3) can be written as

$$\begin{aligned}\boldsymbol{H} &= \left[\, \frac{\partial \tilde{\boldsymbol{n}}^s}{\partial \boldsymbol{r}} \; \frac{\partial \tilde{\boldsymbol{n}}^s}{\partial \boldsymbol{v}} \,\right] \\ &= \left[\frac{1}{\left\| \boldsymbol{R}_b^n - \boldsymbol{r}^n \right\|}\boldsymbol{C}_n^s\left(\boldsymbol{I} - \boldsymbol{n}^n(\boldsymbol{n}^n)^{\mathrm{T}}\right) \; \boldsymbol{0}_{3\times 3} \right]\end{aligned} \tag{9.5}$$

Note that the attitude determination error affects navigation accuracy when using Eq. (9.3) as measurements. If the navigation sensor has a large FOV to observe multiple LOS directions at the same time, the angle between two LOS directions can be taken as measurement to eliminate the influence of attitude determination error. Then, the measurement equation can be written as

$$\begin{aligned}\theta_{12} &= \arccos(\boldsymbol{n}_1^n \cdot \boldsymbol{n}_2^n) + v \\ &= \arccos\left(\frac{\boldsymbol{R}_{b,1}^n - \boldsymbol{r}^n}{\left\| \boldsymbol{R}_{b,1}^n - \boldsymbol{r}^n \right\|} \cdot \frac{\boldsymbol{R}_{b,2}^n - \boldsymbol{r}^n}{\left\| \boldsymbol{R}_{b,2}^n - \boldsymbol{r}^n \right\|} \right) + v\end{aligned} \tag{9.6}$$

where θ_{12} is the angle between two LOS directions; $\boldsymbol{R}_{b,1}^n$ and $\boldsymbol{R}_{b,1}^n$ are the position vectors of two optical beacons in $\mathcal{F}_n$. The measurement sensitivity matrix corresponding to Eq. (9.6) can be written as

$$\boldsymbol{H} = \left[\, \frac{\partial \theta_{12}}{\partial \boldsymbol{r}} \; \frac{\partial \theta_{12}}{\partial \boldsymbol{v}} \,\right]$$

$$= \frac{1}{\sqrt{1-(\boldsymbol{n}_1^n \cdot \boldsymbol{n}_2^n)^2}} \left[\frac{1}{\|\boldsymbol{R}_{b,2}^n - \boldsymbol{r}^n\|} (\boldsymbol{n}_1^n)^{\mathrm{T}} \left(\boldsymbol{I} - \boldsymbol{n}_2^n (\boldsymbol{n}_2^n)^{\mathrm{T}}\right) \right.$$
$$\left. + \frac{1}{\|\boldsymbol{R}_{b,1}^n - \boldsymbol{r}^n\|} (\boldsymbol{n}_2^n)^{\mathrm{T}} \left(\boldsymbol{I} - \boldsymbol{n}_1^n (\boldsymbol{n}_1^n)^{\mathrm{T}}\right) \ \boldsymbol{0}_{1\times 3} \right] \tag{9.7}$$

where $\boldsymbol{n}_1^n = \frac{\boldsymbol{R}_{b,1}^n - \boldsymbol{r}^n}{\|\boldsymbol{R}_{b,1}^n - \boldsymbol{r}^n\|}$ and $\boldsymbol{n}_2^n = \frac{\boldsymbol{R}_{b,2}^n - \boldsymbol{r}^n}{\|\boldsymbol{R}_{b,2}^n - \boldsymbol{r}^n\|}$

If one of the optical beacons is the central celestial object (e.g., $\boldsymbol{R}_{b,2}^n = \boldsymbol{0}$), Eq. (9.6) can be written as

$$\theta_{12} = \arccos(\boldsymbol{n}_1^n \cdot \boldsymbol{n}_2^n) + v$$
$$= \frac{\boldsymbol{R}_{b,1}^n - \boldsymbol{r}^n}{\|\boldsymbol{R}_{b,1}^n - \boldsymbol{r}^n\|} \cdot \frac{-\boldsymbol{r}^n}{\|\boldsymbol{r}^n\|} + v \tag{9.8}$$

The corresponding measurement sensitivity matrix can be written as

$$\boldsymbol{H} = \left[\frac{\partial \theta_{12}}{\partial \boldsymbol{r}} \ \frac{\partial \theta_{12}}{\partial \boldsymbol{v}} \right]$$
$$= \frac{1}{\sqrt{1-(\boldsymbol{n}_1^n \cdot \boldsymbol{n}_2^n)^2}} \left[\frac{1}{\|\boldsymbol{r}^n\|} (\boldsymbol{n}_1^n)^{\mathrm{T}} \left(\boldsymbol{I} - \boldsymbol{n}_2^n (\boldsymbol{n}_2^n)^{\mathrm{T}}\right) \right.$$
$$\left. + \frac{1}{\|\boldsymbol{R}_{b,1}^n - \boldsymbol{r}^n\|} (\boldsymbol{n}_2^n)^{\mathrm{T}} \left(\boldsymbol{I} - \boldsymbol{n}_1^n (\boldsymbol{n}_1^n)^{\mathrm{T}}\right) \ \boldsymbol{0}_{1\times 3} \right] \tag{9.9}$$

9.2 Pulsar-Based Autonomous Navigation

9.2.1 *Basic Principle of Pulsar-Based Autonomous Navigation*

Pulsar is a neutron star with a high density, strong magnetic field, and high spinning speed. A pulsar typically has two magnetic poles that send pulse signals into space in a steady cycle. Due to the misalignment between magnetic pole axis and spin axis, when the pole beam cone generated by a spinning star sweeps the earth or the spacecraft, its pulse signals can be received by a detector. Pulsar-based autonomous navigation systems determine the position vector of the spacecraft relative to an inertial reference point (e.g., the solar system barycenter SSB) by measuring the arrival time difference of a pulse between the spacecraft and the inertial reference point.

Figure 9.1 shows the pulsar-based autonomous navigation geometry of a Martian probe, where c is light velocity, $\boldsymbol{n}$ is the unit LOS vector from the SSB to the pulsar, $\boldsymbol{r}_m$ and $\boldsymbol{r}_s$ represent the positions of Mars and spacecraft relative to the SSB, respectively, t_b and t_{sc} represent the arrival times at SSB and at the spacecraft for the same pulse

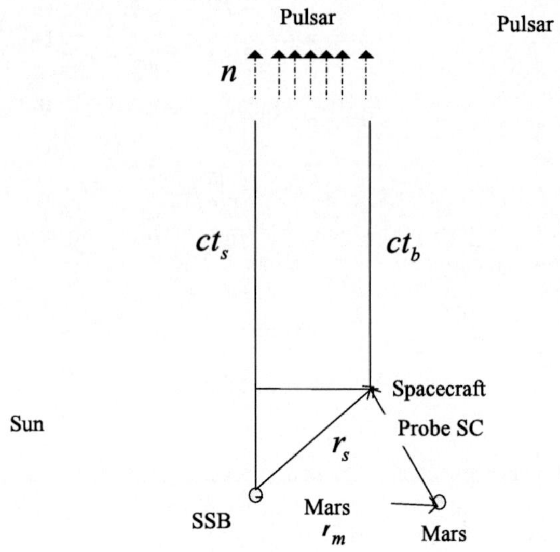

Fig. 9.1 Principle of spacecraft positioning based on pulsar observation

signal, respectively; t_{sc} is also called time of arrival (TOA), measured by a pulse detector, and t_b is accurately predicted by the SSB pulse-timing model, so the time difference of arrival (TDOA) $t_b - t_{sc}$ can be calculated. As seen from the figure, TDOA is related to the spacecraft position. Therefore, the autonomous navigation of a spacecraft can be achieved by observing multiple pulsars or combining the observation of a single pulsar with orbital dynamics.

The reason why pulse-timing model is established at the SSB is that the relative acceleration between the SSB and pulsars is small and the SSB and pulsars can be considered in the same inertial system. In addition, the pulse signal received at the SSB is a stable periodic signal so that only its cycle length and rate of change need to be determined in the modeling process. Conversely, if this model is established at a planet (such as the earth), the effects of planet revolution and relativity must be involved, thus complicating the model form [1].

In fact, pulsars can be classified into radio pulsars, optical pulsars, X-ray pulsars and γ-ray pulsars. Among those pulsars, X-ray pulsars are most suitable for navigation [2], because they are widely distributed, easier to identify, and can be detected by small-area detectors. Therefore, pulsar-based autonomous navigation usually refers to the X-ray pulsar-based navigation (XNAV), so does the pulsar-based autonomous navigation in this book, unless otherwise specially specified.

9.2.2 *Scheme and Process of Pulsar-Based Autonomous Navigation*

Pulsar-based autonomous navigation schemes can be divided into absolute navigation and delta-corrected navigation. In addition, some scholars have recently proposed the navigation algorithm based on phase tracking [3]. In this book, only absolute navigation and delta-corrected navigation will be discussed.

Similar to the GNSS system of an earth satellite, absolute navigation observes multiple pulsars simultaneously, and then compares the measured pulse profile with the standard pulse profile at the SSB to determine the position of the spacecraft relative to an inertial frame. The main advantage of absolute navigation is its independence of any external information, which makes it suitable for autonomous navigation after the fault recovery. However, this navigation method is difficult to implement, as it requires multiple X-ray detectors to observe multiple pulsars simultaneously. Depending on the size of spacecraft orbit, a single detector can sometimes observe different pulsars in a time-sharing manner to approximate the simultaneous observation of multiple pulsars [4]. In addition, due to the lack of priori state information, the measured pulse profile used in absolute navigation cannot be converted into the pulse profile at the SSB, thus ignoring the Doppler effect. If the orbital plane of the spacecraft is almost perpendicular to the LOS direction of a pulsar, the Doppler effect will be insignificant. On the contrary, if the spacecraft has a speed component along the LOS direction, the Doppler effect should be considerable. For example, multiple iterations can be conducted to reconstruct the pulse profile and determine the orbit.

In contrast, the delta-corrected navigation has a priori state information. It firstly observes one or more pulsars, and then uses the priori position information to convert the measured pulse into the pulse at the SSB in comparison with standard pulse. The phase difference (or the time difference) between the two pulses reflects the priori state error, which can thus be corrected by estimation algorithms. Figure 9.2 shows the implementation process of delta-corrected navigation, as specifically summarized below [5, 6]:

1. Measure proper times of X-ray photons radiated by pulsars.
2. Using spacecraft priori position and velocity information, transform the proper time of each photon into the TCB time and then transform the photon TOA at the spacecraft into that at the SSB.
3. Within the given observation time, fold the TCB time series of all the photons relative to SSB for synchronous averaging, and extract the measured pulse profile.
4. Compare the standard profile with measured pulse profile according to the pulse-timing model to obtain the TDOA caused by priori state error.
5. Correct position and velocity until the measured and standard pulse profile can be matched.

As stated above, it can be seen that absolute navigation is geometric orbit determination method, while delta-corrected navigation can be either geometric orbit determination or dynamic orbit determination.

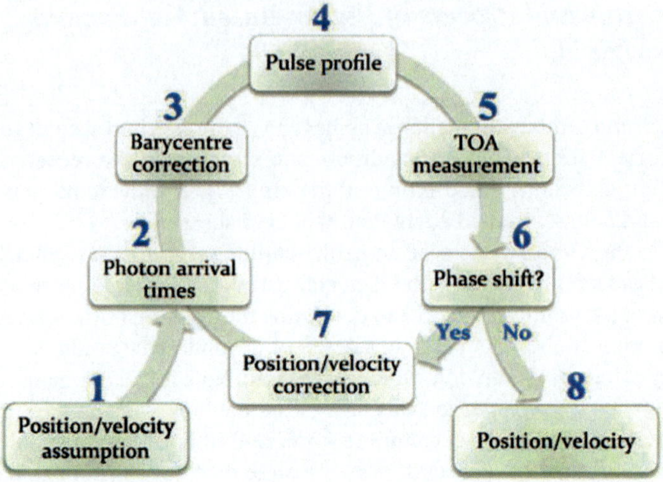

Fig. 9.2 Implementation process of delta-corrected navigation [5, 6]

It should be pointed out that in whatever navigation, the ambiguity of whole cycles needs to be considered. In other words, the difference in the integer number of cycles between measured and standard pulse profiles cannot be measured directly. The spacecraft can acquire an accurate initial navigation value through a ground station or other navigation techniques, unless it is "completely lost" in the solar system. For a millisecond pulsar, its shortest rotation period is about 1 ms and the corresponding light travel is about 300 km. The initial position estimation errors for the most of the existing navigation techniques are less than 300 km. In the pulsar observation period, the accuracy of spacecraft orbit dynamics propagation is also much better than 300 km [7]. Therefore, this book will not go deep into the issue of ambiguity of whole cycles.

9.2.3 Time-Scale Transformation of Photon TOA

The time transformation model of pulsars is obtained through two steps: the transformation from proper time to coordinate time, and the transformation from spacecraft position to the SSB. The former is the time-scale transformation of photon TOA, whereas the latter is the space scale transformation of photon TOA. This section will address the time-scale transformation, namely the transformation from proper time to coordinate time, while the next section will introduce the space scale transformation of photon TOA.

For a spacecraft orbiting major planets in the solar system, the transformation relationship between its proper time and coordinate time TCB is [2]

$$(t - t_0) = (\tau - \tau_0) + \frac{1}{c^2}(\boldsymbol{v}_{\mathrm{PB}} \cdot \dot{\boldsymbol{r}}_{\mathrm{SC/PB}})$$
$$+ \frac{1}{c^2}\int_{\tau_0}^{\tau}\left[U_{\mathrm{PB}}(\boldsymbol{r}_{\mathrm{SC/PB}}) + U_{\mathrm{exPB}}(\boldsymbol{r}_{\mathrm{SC}}) + \frac{1}{2}v_{\mathrm{PB}}^2 + \frac{1}{2}\dot{r}_{\mathrm{SC/PB}}^2\right]\mathrm{d}\tau \quad (9.10)$$

where t is the coordinate time; τ is proper time; $\boldsymbol{r}_{\mathrm{SC/PB}}$ is the position vector of the spacecraft relative to the barycenter of central celestial body in the BCRS coordinate system; $\boldsymbol{v}_{\mathrm{PB}}$ is the velocity vector of central celestial body in the BCRS coordinate system; $U_{\mathrm{PB}}(\boldsymbol{r}_{\mathrm{SC/PB}})$ is the gravitational potential of central celestial body in the spacecraft position; $U_{\mathrm{exPB}}(\boldsymbol{r}_{\mathrm{SC}})$ is the sum of gravitational potentials generated by other bodies in the solar system than central celestial body in the spacecraft position.

If the orbit is heliocentric, the formula of transformation between proper time and coordinate time is [8]

$$(t - t_0) = (\tau - \tau_0)\left(1 + \frac{3\mu_s}{2c^2 a}\right) + \frac{2}{c^2}(\boldsymbol{r} \cdot \boldsymbol{v} - \boldsymbol{r}_0 \cdot \boldsymbol{v}_0) \quad (9.11)$$

where μ_s denotes the gravitational constant of the sun, a denotes the semi-major axis of the heliocentric orbit, and $\boldsymbol{v}$ are the position and velocity vectors of the spacecraft in BCRS, respectively.

9.2.4 Space Scale Transformation of Photon TOA

Figure 9.3 shows the geometric relationships among spacecraft, pulsar, solar

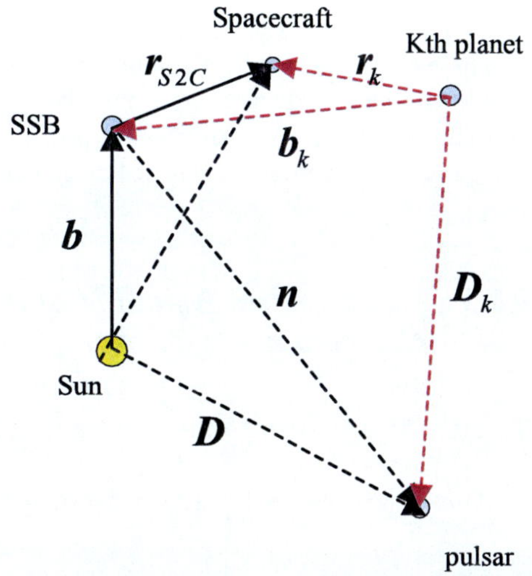

Fig. 9.3 Relationships between spacecraft and pulsar/other celestial bodies

barycenter, and the SSB in a heliocentric coordinate system. Neglecting higher-order terms such as the dispersion effect error, the gravitation of planets in the solar system and the pulsar motion action within the observation interval, the model of time transformation from spacecraft to the SSB can be written as [2]

$$\begin{aligned} t_{\mathrm{SSB}} &= t_{\mathrm{SC}} + \frac{\boldsymbol{n} \cdot \boldsymbol{r}_{\mathrm{S2C}}}{c} \\ &+ \frac{1}{2cD_0}\left[(\boldsymbol{n} \cdot \boldsymbol{r}_{\mathrm{S2C}})^2 - r_{\mathrm{S2C}}^2 - 2(\boldsymbol{b} \cdot \boldsymbol{r}_{\mathrm{S2C}}) + 2(\boldsymbol{n} \cdot \boldsymbol{b})(\boldsymbol{n} \cdot \boldsymbol{r}_{\mathrm{S2C}})\right] \\ &+ \frac{2\mu_s}{c^3} \ln\left|\frac{\boldsymbol{n} \cdot \boldsymbol{r}_{\mathrm{S2C}} + r_{\mathrm{S2C}}}{\boldsymbol{n} \cdot \boldsymbol{b} + b} + 1\right| + \delta t + v_t \end{aligned} \tag{9.12}$$

where t_{SSB} and t_{SC} are the photon TOAs at the SSB and spacecraft, respectively, in the TCB scale; $\boldsymbol{n}$ is the LOS direction vector of pulsar; $\boldsymbol{r}_{\mathrm{S2C}}$ is the position vector of the spacecraft relative to the SSB; D_0 represents the distance from the solar barycenter to pulsar; $\boldsymbol{b}$ is the position vector of SSB relative to the solar barycenter; δt is the satellite clock error; v_t is time measurement noise.

Equation (9.12) has a time transformation accuracy of about 10 μs and can be used as the time transformation equation of a deep-space exploration mission. It should be noted that, in the simulation of pulsar-based autonomous navigation system, the difference between TOA of the photon at spacecraft and TOA of the photon at the SSB, TDOA, is generally taken as observation variable, as shown below

$$\begin{aligned} t_{\mathrm{SSB}} - t_{\mathrm{SC}} &= \frac{\boldsymbol{n} \cdot \boldsymbol{r}_{\mathrm{S2C}}}{c} \\ &+ \frac{1}{2cD_0}\left[(\boldsymbol{n} \cdot \boldsymbol{r}_{\mathrm{S2C}})^2 - r_{\mathrm{S2C}}^2 - 2(\boldsymbol{b} \cdot \boldsymbol{r}_{\mathrm{S2C}}) + 2(\boldsymbol{n} \cdot \boldsymbol{b})(\boldsymbol{n} \cdot \boldsymbol{r}_{\mathrm{S2C}})\right] \\ &+ \frac{2\mu_s}{c^3} \ln\left|\frac{\boldsymbol{n} \cdot \boldsymbol{r}_{\mathrm{S2C}} + r_{\mathrm{S2C}}}{\boldsymbol{n} \cdot \boldsymbol{b} + b} + 1\right| + \delta t + v_t \end{aligned} \tag{9.13}$$

The establishment of pulse-photon TOA transformation model is to extract the profile of measured pulse and calculate its TOA, and then establish the cross-correlation between measured pulse profile and standard pulse profile to obtain the navigation observations of pulsar-based autonomous navigation system.

9.2.5 *Measurement Equation of Pulsar-Based Autonomous Navigation*

9.2.5.1 Measurement Equation for a Heliocentric Orbit

1. Direct measurement equation (without priori state information)

Equation (9.13) gives the pulsar measurement equation in the BCRS coordinate system. For a heliocentric orbit, the position vector of the spacecraft relative to the sun is interested and denoted as $\boldsymbol{r}$. Then,

$$\boldsymbol{r}_{\mathrm{S2C}} = \boldsymbol{r} - \boldsymbol{b} \tag{9.14}$$

Substituting Eq. (9.14) into Eq. (9.13) and multiplying it by c gives

$$\begin{aligned} ct_{\mathrm{SSB}} - ct_{\mathrm{SC}} &= \boldsymbol{n}\cdot(\boldsymbol{r}-\boldsymbol{b}) + \frac{1}{2D_0}\big[(\boldsymbol{n}\cdot(\boldsymbol{r}-\boldsymbol{b}))^2 - \|\boldsymbol{r}-\boldsymbol{b}\|^2 \\ &\quad -2(\boldsymbol{b}\cdot(\boldsymbol{r}-\boldsymbol{b})) + 2(\boldsymbol{n}\cdot\boldsymbol{b})(\boldsymbol{n}\cdot(\boldsymbol{r}-\boldsymbol{b}))\big] \\ &\quad + \frac{2\mu_s}{c^2}\ln\left|\frac{\boldsymbol{n}\cdot(\boldsymbol{r}-\boldsymbol{b}) + \|\boldsymbol{r}-\boldsymbol{b}\|}{\boldsymbol{n}\cdot\boldsymbol{b}+b} + 1\right| + c\delta t + \upsilon_l \end{aligned} \tag{9.15}$$

where $\upsilon_l = c\upsilon_t$. If written in the phase form, the left side of the above equation can be changed into $\lambda\varphi$, where φ is the difference between the pulse phase of arrival at spacecraft and that of arrival at the SSB, and λ is pulsar wavelength, whose relationship with the pulse period P is $\lambda = cP$. If only first-order terms are considered, Eq. (9.15) can be simplified as

$$ct_{\mathrm{SSB}} - ct_{\mathrm{SC}} = \lambda\varphi = \boldsymbol{n}\cdot(\boldsymbol{r}-\boldsymbol{b}) + c\delta t + \upsilon_l \tag{9.16}$$

The above equation is a linear equation that can be used for qualitative analysis of navigation system.

2. Measurement equation with priori state information

If the spacecraft has priori state information, the following equation can be derived from (9.7)

$$\begin{aligned} \hat{t}_{\mathrm{SSB}} &= \hat{t}_{\mathrm{SC}} + \frac{\boldsymbol{n}\cdot\hat{\boldsymbol{r}}_{\mathrm{S2C}}}{c} + \frac{1}{2cD_0}\big[(\boldsymbol{n}\cdot\hat{\boldsymbol{r}}_{\mathrm{S2C}})^2 - \hat{r}_{\mathrm{S2C}}^2 - 2(\boldsymbol{b}\cdot\hat{\boldsymbol{r}}_{\mathrm{S2C}}) + 2(\boldsymbol{n}\cdot\boldsymbol{b})(\boldsymbol{n}\cdot\hat{\boldsymbol{r}}_{\mathrm{S2C}})\big] \\ &\quad + \frac{2\mu_s}{c^3}\ln\left|\frac{\boldsymbol{n}\cdot\hat{\boldsymbol{r}}_{\mathrm{S2C}} + \hat{r}_{\mathrm{S2C}}}{\boldsymbol{n}\cdot\boldsymbol{b}+b} + 1\right| \end{aligned} \tag{9.17}$$

where the superscript ^ represents the physical quantity corresponding to priori information, and $\hat{\boldsymbol{r}}_{\mathrm{S2C}} = \hat{\boldsymbol{r}} + \boldsymbol{b}$. Denote

$$\begin{aligned} \delta t_{\mathrm{SSB}} &= t_{\mathrm{SSB}} - \hat{t}_{\mathrm{SSB}} \\ \delta t_{\mathrm{SC}} &= t_{\mathrm{SC}} - \hat{t}_{\mathrm{SC}} \\ \delta\boldsymbol{r} &= \boldsymbol{r}_{\mathrm{S2C}} - \hat{\boldsymbol{r}}_{\mathrm{S2C}} = \boldsymbol{r} - \hat{\boldsymbol{r}} \end{aligned} \tag{9.18}$$

Subtracting Eq. (9.17) from (9.13) and neglecting higher-order terms, the following equation can be obtained

$$\begin{aligned}\delta t_{\mathrm{SSB}} - \delta t_{\mathrm{SC}} = & \frac{\boldsymbol{n}\cdot\delta\boldsymbol{r}}{c} \\ & + \frac{1}{cD_0}\left[(\boldsymbol{n}\cdot\hat{\boldsymbol{r}}_{\mathrm{S2C}})(\boldsymbol{n}\cdot\delta\boldsymbol{r}) - \hat{\boldsymbol{r}}_{\mathrm{S2C}}\cdot\delta\boldsymbol{r} - (\boldsymbol{b}\cdot\delta\boldsymbol{r}) + (\boldsymbol{n}\cdot\boldsymbol{b})(\boldsymbol{n}\cdot\delta\boldsymbol{r})\right] \\ & + \frac{2\mu_s}{c^3}\frac{\boldsymbol{n}\cdot\delta\boldsymbol{r} + (\hat{\boldsymbol{r}}_{\mathrm{S2C}}\cdot\delta\boldsymbol{r})/\hat{r}_{\mathrm{S2C}}}{\boldsymbol{n}\cdot\hat{\boldsymbol{r}}_{\mathrm{S2C}} + \boldsymbol{n}\cdot\boldsymbol{b} + \hat{r}_{\mathrm{S2C}} + b} + \delta t + v_t\end{aligned} \tag{9.19}$$

Equation (9.19) can be written in the distance form, as shown below

$$\begin{aligned}c\delta t_{\mathrm{SSB}} - c\delta t_{\mathrm{SC}} = & \, \boldsymbol{n}\cdot\delta\boldsymbol{r} \\ & + \frac{1}{D_0}\left[(\boldsymbol{n}\cdot\hat{\boldsymbol{r}}_{\mathrm{S2C}})(\boldsymbol{n}\cdot\delta\boldsymbol{r}) - \hat{\boldsymbol{r}}_{\mathrm{S2C}}\cdot\delta\boldsymbol{r} - (\boldsymbol{b}\cdot\delta\boldsymbol{r}) + (\boldsymbol{n}\cdot\boldsymbol{b})(\boldsymbol{n}\cdot\delta\boldsymbol{r})\right] \\ & + \frac{2\mu_s}{c^2}\frac{\boldsymbol{n}\cdot\delta\boldsymbol{r} + (\hat{\boldsymbol{r}}_{\mathrm{S2C}}\cdot\delta\boldsymbol{r})/\hat{r}_{\mathrm{S2C}}}{\boldsymbol{n}\cdot\hat{\boldsymbol{r}}_{\mathrm{S2C}} + \boldsymbol{n}\cdot\boldsymbol{b} + \hat{r}_{\mathrm{S2C}} + b} + c\delta t + v_l\end{aligned} \tag{9.20}$$

If written in the phase form, then

$$\begin{aligned}\lambda\delta\varphi = & \, \boldsymbol{n}\cdot\delta\boldsymbol{r} \\ & + \frac{1}{D_0}\left[(\boldsymbol{n}\cdot\hat{\boldsymbol{r}}_{\mathrm{S2C}})(\boldsymbol{n}\cdot\delta\boldsymbol{r}) - \hat{\boldsymbol{r}}_{\mathrm{S2C}}\cdot\delta\boldsymbol{r} - (\boldsymbol{b}\cdot\delta\boldsymbol{r}) + (\boldsymbol{n}\cdot\boldsymbol{b})(\boldsymbol{n}\cdot\delta\boldsymbol{r})\right] \\ & + \frac{2\mu_s}{c^2}\frac{\boldsymbol{n}\cdot\delta\boldsymbol{r} + (\hat{\boldsymbol{r}}_{\mathrm{S2C}}\cdot\delta\boldsymbol{r})/\hat{r}_{\mathrm{S2C}}}{\boldsymbol{n}\cdot\hat{\boldsymbol{r}}_{\mathrm{S2C}} + \boldsymbol{n}\cdot\boldsymbol{b} + \hat{r}_{\mathrm{S2C}} + b} + c\delta t + v_l\end{aligned} \tag{9.21}$$

with $\delta\varphi = \varphi - \hat{\varphi}$.

Equations (9.20) and (9.21) can be written in the following standard form

$$\delta z = \boldsymbol{H}\delta\boldsymbol{x} + \nu \tag{9.22}$$

where $\delta z = c\delta t_{\mathrm{SSB}} - c\delta t_{\mathrm{SC}} = \lambda\delta\varphi$. If $\delta\boldsymbol{x} = [\delta\boldsymbol{r}^{\mathrm{T}}, \delta\boldsymbol{v}^{\mathrm{T}}]^{\mathrm{T}}$, then

$$\begin{aligned}\boldsymbol{H} = & \left[\boldsymbol{n}^{T} + \frac{1}{D_0}[(\boldsymbol{n}\cdot\hat{\boldsymbol{r}}_{\mathrm{S2C}})\boldsymbol{n}^{\mathrm{T}} - \hat{\boldsymbol{r}}_{\mathrm{S2C}}^{\mathrm{T}} - \boldsymbol{b}^{\mathrm{T}} + (\boldsymbol{n}\cdot\boldsymbol{b})\boldsymbol{n}^{\mathrm{T}}]\right. \\ & \left. + \frac{2\mu_s}{c^2}\frac{\boldsymbol{n}^{\mathrm{T}} + \hat{\boldsymbol{r}}_{\mathrm{S2C}}^{\mathrm{T}}/\hat{r}_{\mathrm{S2C}}}{\boldsymbol{n}\cdot\hat{\boldsymbol{r}}_{\mathrm{S2C}} + \boldsymbol{n}\cdot\boldsymbol{b} + \hat{r}_{\mathrm{S2C}} + b}, \boldsymbol{0}_{1\times 3}\right] \\ \nu = & \, c\delta t + v_l\end{aligned} \tag{9.23}$$

If $\delta\boldsymbol{x} = [\delta\boldsymbol{r}^{\mathrm{T}}, \delta\boldsymbol{v}^{\mathrm{T}}, \delta t]^{\mathrm{T}}$, then

$$\begin{aligned}\boldsymbol{H} = & \left[\boldsymbol{n}^{\mathrm{T}} + \frac{1}{D_0}[(\boldsymbol{n}\cdot\hat{\boldsymbol{r}}_{\mathrm{S2C}})\boldsymbol{n}^{\mathrm{T}} - \hat{\boldsymbol{r}}_{\mathrm{S2C}}^{\mathrm{T}} - \boldsymbol{b}^{\mathrm{T}} + (\boldsymbol{n}\cdot\boldsymbol{b})\boldsymbol{n}^{\mathrm{T}}]\right. \\ & \left. + \frac{2\mu_s}{c^2}\frac{\boldsymbol{n}^{\mathrm{T}} + \hat{\boldsymbol{r}}_{\mathrm{S2C}}^{\mathrm{T}}/\hat{r}_{\mathrm{S2C}}}{\boldsymbol{n}\cdot\hat{\boldsymbol{r}}_{\mathrm{S2C}} + \boldsymbol{n}\cdot\boldsymbol{b} + \hat{r}_{\mathrm{S2C}} + b}, \boldsymbol{0}_{1\times 3}, c\right]\end{aligned}$$

$$\nu = \upsilon_l \tag{9.24}$$

9.2.5.2 Measurement Equations for Flight Phases Around the Target Object

1. Direct measurement equation (without priori state information)

For the flight phases around the target object, such as approaching and surrounding phases, the measurement equation shall be written as a function of the spacecraft position relative to the target. Take Martian probe as an example. Figure 9.4 shows the geometric relationships among spacecraft, pulsar, Mars, the sun, and the SSB at the approaching and surrounding phases. In this case, the time when the pulse arrives at the Mars, denoted as t_{Mars}, can be accurately predicted through A pulse-timing model. From the time transformation equation, the following equation can be derived

$$\begin{aligned} t_{\text{SSB}} - t_{\text{Mars}} = & \frac{\boldsymbol{n} \cdot \boldsymbol{r}_{\text{S2M}}}{c} \\ & + \frac{1}{2cD_0}\left[(\boldsymbol{n} \cdot \boldsymbol{r}_{\text{S2M}})^2 - r_{\text{S2M}}^2 - 2(\boldsymbol{b} \cdot \boldsymbol{r}_{\text{S2M}}) + 2(\boldsymbol{n} \cdot \boldsymbol{b})(\boldsymbol{n} \cdot \boldsymbol{r}_{\text{S2M}})\right] \end{aligned}$$

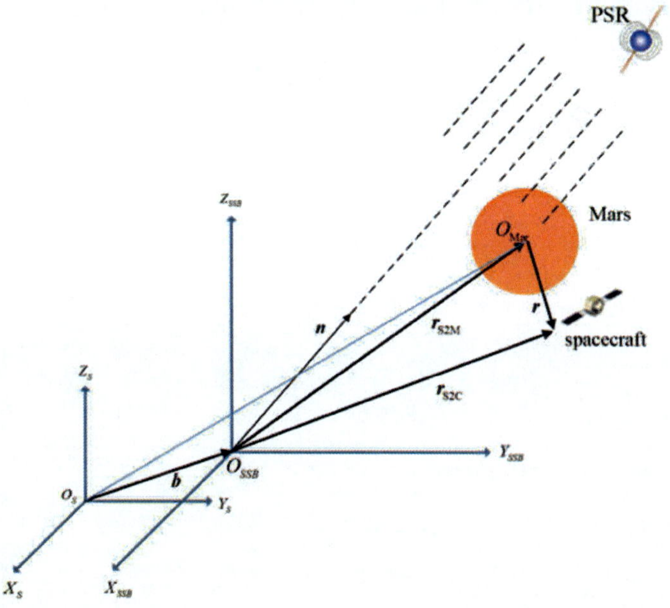

Fig. 9.4 Geometric relationships among spacecraft, pulsar, mars barycenter, and solar barycenter in the SSB coordinate system

$$+\frac{2\mu_s}{c^3}\ln\left|\frac{\boldsymbol{n}\cdot\boldsymbol{r}_{\mathrm{S2M}}+r_{\mathrm{S2M}}}{\boldsymbol{n}\cdot\boldsymbol{b}+b}+1\right| \tag{9.25}$$

where $\boldsymbol{r}_{\mathrm{S2M}}$ is the position vector of Mars relative to SSB, satisfying $\boldsymbol{r}_{\mathrm{S2M}}=\boldsymbol{r}_{\mathrm{S2C}}-\boldsymbol{r}$. The subtraction of Eq. (9.25) from Eq. (9.13) gives

$$\begin{aligned} t_{\mathrm{Mars}}-t_{\mathrm{SC}}=&\frac{\boldsymbol{n}\cdot\boldsymbol{r}}{c}+\frac{1}{cD_0}[(\boldsymbol{n}\cdot\boldsymbol{r}_{\mathrm{S2M}})(\boldsymbol{n}\cdot\boldsymbol{r})-\boldsymbol{r}_{\mathrm{S2M}}\cdot\boldsymbol{r}-(\boldsymbol{b}\cdot\boldsymbol{r})+(\boldsymbol{n}\cdot\boldsymbol{b})(\boldsymbol{n}\cdot\boldsymbol{r})]\\ &+\frac{2\mu_s}{c^3}\ln\left|\frac{\boldsymbol{n}\cdot\boldsymbol{r}+(\boldsymbol{r}_{\mathrm{S2M}}\cdot\boldsymbol{r})/r_{\mathrm{S2M}}}{\boldsymbol{n}\cdot\boldsymbol{r}_{\mathrm{S2M}}+\boldsymbol{n}\cdot\boldsymbol{b}+\hat{r}_{\mathrm{S2M}}+b}+1\right|+\delta t+v_t \end{aligned} \tag{9.26}$$

The order of magnitude of the second and third terms on the right side of the above equation is as small as nanosecond [9], so the two terms can be neglected. Therefore, Eq. (9.26) can be simplified as

$$t_{\mathrm{Mars}}-t_{\mathrm{SC}}=\frac{\boldsymbol{n}\cdot\boldsymbol{r}}{c}+\delta t+v_t \tag{9.27}$$

Equation (9.27) can be used as the pulsar-based autonomous navigation measurement equation for flight phases around the target object.

2. Measurement equation without priori state information

If the priori information on spacecraft state is available, the following equation can be derived from the pulsar time transformation equation

$$\hat{t}_{\mathrm{Mars}}=\hat{t}_{\mathrm{SC}}+\frac{\boldsymbol{n}\cdot\hat{\boldsymbol{r}}}{c} \tag{9.28}$$

By subtracting Eq. (9.27) from Eq. (9.28) and multiplying it by the velocity of light, the following equation can be obtained

$$\delta z=\boldsymbol{H}\delta\boldsymbol{x}+v \tag{9.29}$$

where $\delta z=c\delta t_{\mathrm{SSB}}-c\delta t_{\mathrm{SC}}=\lambda\delta\varphi$. If $\delta\boldsymbol{x}=[\delta\boldsymbol{r}^{\mathrm{T}},\delta\boldsymbol{v}^{\mathrm{T}}]^{\mathrm{T}}$, then

$$\begin{aligned} \boldsymbol{H}&=\left[\boldsymbol{n}^{\mathrm{T}},\boldsymbol{0}_{1\times3}\right]\\ v&=c\delta t+v_l \end{aligned} \tag{9.30}$$

If $\delta\boldsymbol{x}=[\delta\boldsymbol{r}^{\mathrm{T}},\delta\boldsymbol{v}^{\mathrm{T}},\delta t]^{\mathrm{T}}$, then

$$\begin{aligned} \boldsymbol{H}&=\left[\boldsymbol{n}^{\mathrm{T}},\boldsymbol{0}_{1\times3},c\right]\\ v&=v_l \end{aligned} \tag{9.31}$$

9.3 Navigation Beacon Planning Algorithm

9.3.1 Navigation Beacon Planning Algorithm Based on Geometric Observability Analysis

This section studies navigation beacon planning algorithm based on geometric observability analysis. Here, we only consider the case where a single LOS direction and a single pulsar TOA are fused. According to Eq. (8.43), the information matrix of an optical autonomous navigation system based on a single LOS direction can be written as

$$\mathcal{I}_1 = -\kappa_o[\boldsymbol{n}_o\times]^2 \tag{9.32}$$

where $\boldsymbol{n}_o$ is the LOS direction of the optical beacon expressed in $\mathcal{F}_n$, $\kappa_o > 0$ (see Sect. 8.5.1), and subscript o indicates a quantity in the optical navigation system. For geometric orbital determination, the state along the LOS direction is unobservable.

For XNAV system based on a single pulsar measurement, if higher-order terms of measurement equation are neglected, the information matrix is given by

$$\mathcal{I}_2 = \kappa_p\boldsymbol{n}_p\boldsymbol{n}_p^{\mathrm{T}} \tag{9.33}$$

where $\boldsymbol{n}_p$ is the LOS vector of the pulsar, $\kappa_p > 0$, and subscript p indicates a quantity in the XNAV system.

Assuming that the measurement noises between the XNAV and optical navigation systems are independent with each other, the information matrix of a fusion-based autonomous navigation system can be written as

$$\mathcal{I} = \mathcal{I}_1 + \mathcal{I}_2 = -\kappa_o[\boldsymbol{n}_o\times]^2 + \kappa_p\boldsymbol{n}_p\boldsymbol{n}_p^{\mathrm{T}} \tag{9.34}$$

The key of geometric observability analysis is to analyze eigenvalues of $\mathcal{I}$. If $\mathcal{I}$ has the eigenvalue 0, the system is unobservable along the eigenvector direction corresponding to the eigenvalue 0. Depending on the geometry relationship between $\boldsymbol{n}_o$ and $\boldsymbol{n}_p$, three cases can be considered.

(1) Case 1: $\boldsymbol{n}_o = \boldsymbol{n}_p$

With $[\boldsymbol{n}_o\times]^2 = -\boldsymbol{I}_3 + \boldsymbol{n}_o\boldsymbol{n}_o^{\mathrm{T}}$, the following equation can be obtained

$$\mathcal{I} = \kappa_o\boldsymbol{I}_3 - (\kappa_o - \kappa_p)\boldsymbol{n}_p\boldsymbol{n}_p^{\mathrm{T}} \tag{9.35}$$

Obviously,

$$\mathcal{I}\boldsymbol{n}_p = \kappa_p\boldsymbol{n}_p \tag{9.36}$$

This means that one eigenvalue of $\mathcal{I}$ is κ_p, and its corresponding eigenvector is $\boldsymbol{n}_p$.

Denote $\boldsymbol{n}_p, \boldsymbol{n}_{p\perp 1}, \boldsymbol{n}_{p\perp 2}$ as a set of vectors that are orthogonal with each other. Then

$$\begin{aligned} \mathcal{I}\boldsymbol{n}_{p\perp 1} &= \kappa_o \boldsymbol{n}_{p\perp 1} \\ \mathcal{I}\boldsymbol{n}_{p\perp 2} &= \kappa_o \boldsymbol{n}_{p\perp 2} \end{aligned} \tag{9.37}$$

This means that the other two eigenvalues are κ_o and their corresponding eigenvectors are $\boldsymbol{n}_{p\perp 1}, \boldsymbol{n}_{p\perp 2}$.

Because the three eigenvalues of $\mathcal{I}$, namely κ_p, κ_o, and κ_o, are all larger than 0,the system is observable in the case of $\boldsymbol{n}_o = \boldsymbol{n}_p$. In this case, the trace of error covariance matrix is

$$\begin{aligned} \mathrm{Tr}[\boldsymbol{P}] &= \sum \lambda_i(\boldsymbol{P}) \\ &= \sum 1/\lambda_i(\mathcal{I}) \\ &= \frac{2}{\kappa_o} + \frac{1}{\kappa_p} \end{aligned} \tag{9.38}$$

(2) Case 2: $\boldsymbol{n}_o \perp \boldsymbol{n}_p$

With $[\boldsymbol{n}_o\times]^2 = -\boldsymbol{I}_3 + \boldsymbol{n}_o\boldsymbol{n}_o^{\mathrm{T}}$, the following equation can be obtained

$$\mathcal{I} = \kappa_o \boldsymbol{I}_3 - \kappa_o \boldsymbol{n}_o \boldsymbol{n}_o^{\mathrm{T}} + \kappa_p \boldsymbol{n}_p \boldsymbol{n}_p^{\mathrm{T}} \tag{9.39}$$

Obviously,

$$\begin{aligned} \mathcal{I}(\boldsymbol{n}_o \times \boldsymbol{n}_p) &= \kappa_o(\boldsymbol{n}_o \times \boldsymbol{n}_p) \\ \mathcal{I}\boldsymbol{n}_o &= 0\boldsymbol{n}_o \\ \mathcal{I}\boldsymbol{n}_p &= (\kappa_o + \kappa_p)\boldsymbol{n}_p \end{aligned} \tag{9.40}$$

This means that the three eigenvalues of $\mathcal{I}$ are κ_o, 0, and $(\kappa_o + \kappa_p)$, respectively, whose corresponding eigenvectors are $\boldsymbol{n}_o \times \boldsymbol{n}_p$, $\boldsymbol{n}_o$, and $\boldsymbol{n}_p$. In this case, the system is unobservable along the LOS direction of the optical beacon.

(3) $\boldsymbol{n}_o \neq \boldsymbol{n}_p$, and $\boldsymbol{n}_o, \boldsymbol{n}_p$ are not orthogonal to each other

With $[\boldsymbol{n}_o\times]^2 = -\boldsymbol{I}_3 + \boldsymbol{n}_o\boldsymbol{n}_o^{\mathrm{T}}$, the following equation can be obtained

$$\mathcal{I} = \kappa_o \boldsymbol{I}_3 + \mathcal{M} \tag{9.41}$$

where

$$\mathcal{M} = -\kappa_o \boldsymbol{n}_o \boldsymbol{n}_o^{\mathrm{T}} + \kappa_p \boldsymbol{n}_p \boldsymbol{n}_p^{\mathrm{T}} \tag{9.42}$$

Denote the eigenvalue of $\mathcal{M}$ as $\lambda_{M,i}$ and its eigenvector as $\boldsymbol{w}_i$. Then, the eigenvalue of $\mathcal{I}$ is $\kappa_o + \lambda_{M,i}$ and its eigenvector will be $\boldsymbol{w}_i$. Next, resolve the eigenvalue of $\mathcal{M}$. Because

$$\begin{aligned} \boldsymbol{n}_o^{\mathrm{T}} \cdot (\boldsymbol{n}_o \times \boldsymbol{n}_p) = 0 \\ \boldsymbol{n}_p^{\mathrm{T}} \cdot (\boldsymbol{n}_o \times \boldsymbol{n}_p) = 0 \end{aligned} \tag{9.43}$$

Therefore,

$$\mathcal{M}(\boldsymbol{n}_o \times \boldsymbol{n}_p) = \boldsymbol{0} \tag{9.44}$$

This means that one eigenvalue of $\mathcal{M}$, denoted as $\lambda_{M,1}$, is 0, and its corresponding eigenvector $\boldsymbol{w}_i$ is $\boldsymbol{n}_o \times \boldsymbol{n}_p$. Because $\mathcal{M}$ is a 3×3 matrix, $\det(\lambda_M \boldsymbol{I} - \mathcal{M})$ can be written as

$$\lambda_M^3 - a\lambda_M^2 + b\lambda_M + c = 0 \tag{9.45}$$

Because the eigenvalue $\lambda_{M,1}$ of $\mathcal{M}$ is 0, therefore $c = 0$. The above equation can be simplified as

$$\lambda_M^2 - a\lambda_M + b = 0 \tag{9.46}$$

According to the properties of the roots of a polynomial equation, the following equations can be obtained

$$\begin{aligned} \lambda_{M,2} + \lambda_{M,3} &= a \\ \lambda_{M,2}\lambda_{M,3} &= b \end{aligned} \tag{9.47}$$

Because

$$\begin{aligned} \mathrm{Tr}[\mathcal{M}] &= \sum_{i=1}^{3} \lambda_{M,i} \\ &= \lambda_{M,2} + \lambda_{M,3} \\ &= a \end{aligned} \tag{9.48}$$

and according to Eq. (9.42), there is

$$\mathrm{Tr}[\mathcal{M}] = -\kappa_o + \kappa_p \tag{9.49}$$

Comparing Eq. (9.48) with Eq. (9.49) gives

$$a = -\kappa_o + \kappa_p \tag{9.50}$$

For Eq. (9.46), the other two eigenvalues can be derived only by resolving b. The expansion of $\det(\lambda_M \boldsymbol{I} - \mathcal{M})$ is

$$\begin{aligned}\det(\lambda_M \boldsymbol{I} - \mathcal{M}) &= \det\begin{bmatrix} \lambda_M - m_{11} & m_{12} & m_{13} \\ m_{12} & \lambda_M - m_{22} & m_{23} \\ m_{13} & m_{23} & \lambda_M - m_{33} \end{bmatrix} \\ &= (\lambda_M - m_{11})\det(\boldsymbol{A}_{11}) - m_{12}\det(\boldsymbol{A}_{12}) + m_{13}\det(\boldsymbol{A}_{13}) \end{aligned} \tag{9.51}$$

where m_{ij} is the element in the ith row and jth column of M.

$$\begin{aligned} \boldsymbol{A}_{11} &= \begin{bmatrix} \lambda_M - m_{22} & m_{23} \\ m_{23} & \lambda_M - m_{33} \end{bmatrix} \\ \boldsymbol{A}_{12} &= \begin{bmatrix} m_{12} & m_{13} \\ m_{23} & \lambda_M - m_{33} \end{bmatrix} \\ \boldsymbol{A}_{12} &= \begin{bmatrix} m_{12} & m_{13} \\ \lambda_M - m_{32} & m_{23} \end{bmatrix} \end{aligned} \tag{9.52}$$

As known from Eq. (9.45), b is the coefficient of the term λ_M in the expansion of $\det(\lambda_M \boldsymbol{I} - \mathcal{M})$. Considering the Eq. (9.51), there will be

$$\begin{aligned} b &= m_{22}m_{33} + m_{11}m_{22} + m_{11}m_{33} - m_{23}^2 - m_{12}^2 - m_{13}^2 \\ &= -\kappa_o\kappa_p \left\| \boldsymbol{n}_o \times \boldsymbol{n}_p \right\|^2 \end{aligned} \tag{9.53}$$

So the two roots of Eq. (9.46) are

$$\begin{aligned} \lambda_{M,2,3} &= \frac{1}{2}(a \pm \sqrt{a^2 - 4b}) \\ &= \frac{1}{2}\left(-\kappa_o + \kappa_p \pm \sqrt{(-\kappa_o + \kappa_p)^2 + 4\kappa_o\kappa_p \left\| \boldsymbol{n}_o \times \boldsymbol{n}_p \right\|^2} \right) \end{aligned} \tag{9.54}$$

Next, resolve the eigenvector $\boldsymbol{w}_{2,3}$ corresponding to $\lambda_{M,2,3}$. Because the three eigenvectors of M are orthogonal with each other and $\boldsymbol{w}_1$ is $\boldsymbol{n}_o \times \boldsymbol{n}_p$, the other two eigenvectors lie in the plane formed by $\boldsymbol{n}_o$, $\boldsymbol{n}_p$. Since $\boldsymbol{n}_o$ is clearly not an eigenvector of M, let

$$\boldsymbol{w}_{2,3} = \boldsymbol{n}_p + c_n \boldsymbol{n}_o \tag{9.55}$$

where c_n is a coefficient to be determined. So

$$\begin{aligned} \lambda_{M,2,3}(\boldsymbol{n}_p + c_n\boldsymbol{n}_o) &= \mathcal{M}(\boldsymbol{n}_p + c_n\boldsymbol{n}_o) \\ &= (-\kappa_o \boldsymbol{n}_o \boldsymbol{n}_o^T + \kappa_p \boldsymbol{n}_p \boldsymbol{n}_p^{\mathrm{T}})(\boldsymbol{n}_p + c_n\boldsymbol{n}_o) \\ &= -\kappa_o \boldsymbol{n}_o^T \boldsymbol{n}_p \boldsymbol{n}_o + \kappa_p \boldsymbol{n}_p - \kappa_o c_n \boldsymbol{n}_o + \kappa_p c_n \boldsymbol{n}_p^{\mathrm{T}} \boldsymbol{n}_o \boldsymbol{n}_p \end{aligned}$$

$$= \kappa_p(1 + c_n \boldsymbol{n}_p^{\mathrm{T}} \boldsymbol{n}_o)\boldsymbol{n}_p - \kappa_o(\boldsymbol{n}_o^{\mathrm{T}} \boldsymbol{n}_p + c_n)\boldsymbol{n}_o \tag{9.56}$$

Let the coefficients of $\boldsymbol{n}_p$ and $\boldsymbol{n}_o$ on both sides of the above equation be equal. Then,

$$\begin{aligned} \lambda_{M,2,3} &= \kappa_p(1 + c_n \boldsymbol{n}_p^{\mathrm{T}} \boldsymbol{n}_o) \\ \lambda_{M,2,3} c_n &= -\kappa_o(\boldsymbol{n}_o^{\mathrm{T}} \boldsymbol{n}_p + c_n) \end{aligned} \tag{9.57}$$

So that the solution of c_n can be obtained as

$$c_n = \frac{-(\kappa_p + \kappa_o) \pm \sqrt{(\kappa_p + \kappa_o)^2 - 4\kappa_p\kappa_o(\boldsymbol{n}_p^{\mathrm{T}} \boldsymbol{n}_o)^2}}{2\kappa_p \boldsymbol{n}_p^{\mathrm{T}} \boldsymbol{n}_o} \tag{9.58}$$

So far, all the eigenvalues and eigenvectors of M have been obtained and summarized as follows:

$$\begin{aligned} \lambda_{M,1} &= 0 \\ \boldsymbol{w}_1 &= \boldsymbol{n}_o \times \boldsymbol{n}_p \\ \lambda_{M,2,3} &= \frac{1}{2}\left(-\kappa_o + \kappa_p \pm \sqrt{(-\kappa_o + \kappa_p)^2 + 4\kappa_o\kappa_p \left\|\boldsymbol{n}_o \times \boldsymbol{n}_p\right\|^2}\right) \\ \boldsymbol{w}_{2,3} &= \boldsymbol{n}_p + c_n \boldsymbol{n}_o \end{aligned} \tag{9.59}$$

where c_n is given by Eq. (9.58). Therefore, the eigenvalues and eigenvectors of $\mathcal{I}$ are as follows

$$\begin{aligned} \lambda_{I,1} &= \kappa_o \\ \boldsymbol{w}_1 &= \boldsymbol{n}_o \times \boldsymbol{n}_p \\ \lambda_{I,2,3} &= \frac{1}{2}\left(\kappa_o + \kappa_p \pm \sqrt{(-\kappa_o + \kappa_p)^2 + 4\kappa_o\kappa_p \left\|\boldsymbol{n}_o \times \boldsymbol{n}_p\right\|^2}\right) \\ \boldsymbol{w}_{2,3} &= \boldsymbol{n}_p + c_n \boldsymbol{n}_o \end{aligned} \tag{9.60}$$

where c_n is given by Eq. (9.58). Here, all the eigenvalues of $\mathcal{I}$ are >0, which means that the system is observable. The trace of error covariance matrix is given by

$$\begin{aligned} \mathrm{Tr}[\boldsymbol{P}] &= \sum \lambda_i(\boldsymbol{P}) \\ &= \sum 1/\lambda_i(\mathcal{I}) \\ &= \frac{1}{\kappa_o} + \frac{\kappa_o + \kappa_p}{\kappa_o\kappa_p\left(1 - \left\|\boldsymbol{n}_o \times \boldsymbol{n}_p\right\|^2\right)} \end{aligned} \tag{9.61}$$

When $\boldsymbol{n}_o \neq \boldsymbol{n}_p$ and $\boldsymbol{n}_o, \boldsymbol{n}_p$ are not orthogonal to each other, there will be $0 < 1 - \left\|\boldsymbol{n}_o \times \boldsymbol{n}_p\right\|^2 < 1$ and the degree of observability will be lower than that in the case of $\boldsymbol{n}_o = \boldsymbol{n}_p$. Therefore, it can be concluded that in the case of $\boldsymbol{n}_o = \boldsymbol{n}_p$, the system has the largest degree of observability.

9.3.2 Navigation Beacon Planning Algorithm Based on Dynamic Observability Analysis

The navigation beacon planning algorithm based on geometric observability analysis considers only the effect of spatial distribution of beacons on navigation performance. In practice, autonomous navigation can be based on the orbital dynamics and observation model of a spacecraft. Orbital dynamics is used to analyze the observability (called dynamic observability in this book), and then plan the navigation beacons. Therefore, this section will study the navigation beacon planning algorithm based on dynamic observability analysis. First, the following lemma is given.

Lemma 9.1 *Consider the following system*

$$\begin{aligned} \boldsymbol{x}_k &= \boldsymbol{F}_{k,k-1}\boldsymbol{x}_{k-1} + \boldsymbol{w}_k \\ \boldsymbol{z}_k &= \boldsymbol{H}_k\boldsymbol{x}_k + \boldsymbol{v}_k \end{aligned} \tag{9.62}$$

where $\boldsymbol{w}$ *and* $\boldsymbol{v}$ *are process noise and measurement noise, respectively. For an optimal estimator, denote the error covariance matrix at* t_k *without observations as* $\boldsymbol{P}_{k|k-1}$, *and the error covariance matrix with the observation of an additional beacon as* $\boldsymbol{P}_k$. *Then*

$$\mathrm{Tr}[\boldsymbol{P}_{k|k-1}] \geq \mathrm{Tr}[\boldsymbol{P}_k] \tag{9.63}$$

Proof first, it can be derived from Cramer-Rao inequality that

$$\begin{aligned} \boldsymbol{P}_{k|k-1} &= \mathcal{I}_{k|k-1}^{-1} \\ \boldsymbol{P}_k &= \mathcal{I}_k^{-1} \end{aligned} \tag{9.64}$$

where the relation between $\mathcal{I}_{k|k-1}$ and $\mathcal{I}_k$ is given by [1]

$$\mathcal{I}_k = \mathcal{I}_{k|k-1} + \boldsymbol{H}_k^{\mathrm{T}}\boldsymbol{R}_k^{-1}\boldsymbol{H}_k \tag{9.65}$$

Noting that $\mathcal{I}_{k|k-1} > \boldsymbol{0}$ and $\boldsymbol{H}^{\mathrm{T}}\boldsymbol{R}^{-1}\boldsymbol{H} \geq \boldsymbol{0}$, there exists a non-singular matrix $\boldsymbol{T}$ such that $\mathcal{I}_{k|k-1} = \boldsymbol{T}\boldsymbol{T}^{\mathrm{T}}$ and $\boldsymbol{H}^{\mathrm{T}}\boldsymbol{R}^{-1}\boldsymbol{H} = \boldsymbol{T}\boldsymbol{\Upsilon}\boldsymbol{T}^{\mathrm{T}}$ [10]. From Eq. (9.64), it can be derived that

$$\boldsymbol{P}_{k|k-1} = \mathcal{I}_{k|k-1}^{-1} = \boldsymbol{T}^{-\mathrm{T}}\boldsymbol{T}^{-1}$$

$$P_k = \mathcal{I}_k^{-1} = T^{-\mathrm{T}}(I + \Upsilon)^{-1}T^{-1} \tag{9.66}$$

with $\Upsilon \geq \mathbf{0}$, so

$$\mathrm{Tr}[P_{k|k-1}] - \mathrm{Tr}[P_k] = \mathrm{Tr}\big[T^{-\mathrm{T}}(I - (I + \Upsilon)^{-1})T^{-1}\big] \geq 0 \tag{9.67}$$

Q.E.D.

Remark: Lemma 9.1 indicates that the system's observability can be improved by taking more measurements.

The next issue to be studied is how to select navigation beacon at t_k to improve the observability of navigation system as much as possible. As known from Eq. (9.67), the larger the right-hand side of the equation is, the more the degree of observability will be improved. Because $(I + \Upsilon)$ is a diagonal matrix, it can be derived that

$$I - (I + \Upsilon)^{-1} = \mathrm{diag}[\mu_1/(1 + \mu_1), \mu_2/(1 + \mu_2), \ldots, \mu_6/(1 + \mu_6)] \tag{9.68}$$

where μ_i is the diagonal element of Υ.

Note that $\mathrm{Tr}[P_{k|k-1}] = \sum_i^3 g_i$ and $\mathrm{Tr}[P_k] = \sum_i^6 g_i \frac{1}{1+\mu_i}$, where g_i is the ith diagonal element of $G^{\mathrm{T}}G$ and $G = T^{-1}$. A relative degree of observability can be defined as

$$D_r = \frac{\mathrm{Tr}[P_k]}{\mathrm{Tr}[P_{k|k-1}]} = \frac{\sum_1^6 g_i \frac{1}{1+\mu_i}}{\sum_1^6 g_i}. \tag{9.69}$$

A smaller D_r signifies higher improvement of degree of observability. It can be seen from the above equation that, the improvement of degree of observability depends on both g_i and μ_i. Denote the minimum of μ_i as $\mu_{\mathrm{i,min}}$. Then, $\mu_i \geq \mu_{\mathrm{i,min}} > 0$, and

$$\mathrm{Tr}[P_k] \leq \frac{1}{1 + \mu_{\mathrm{i,min}}}\mathrm{Tr}[P_{k|k-1}] < \mathrm{Tr}[P_{k|k-1}] \tag{9.70}$$

Through the above approximation relation, the effect of g_i and μ_i on degree of observability can be decoupled. From the above equation, $D_r \leq \frac{1}{1+\mu_{i,\mathrm{min}}} < 1$ can be derived. $\frac{1}{1+\mu_{i,\mathrm{min}}}$ can be used to quantitatively evaluate the degree to which the observation of additional navigation beacon improves the degree of observability of autonomous navigation system. A smaller $\frac{1}{1+\mu_{i,\mathrm{min}}}$ indicates a higher improvement degree; and vice versa. Therefore, from the perspective of degree of observability, the navigation beacon minimizing $\frac{1}{1+\mu_{i,\mathrm{min}}}$ should be selected for observation when planning navigation beacons.

From the above analysis, a new planning algorithm can be obtained as below:

1. At t_1, $\mathcal{I}_1 = P_1^{-1}$ due to the lack of observations

2. $k = 2 \ldots N$

   (a) The predicted information matrix at t_k is

$$\mathcal{I}_{k|k-1} = \boldsymbol{Q}_k^{-1} - \boldsymbol{Q}_k^{-1}\boldsymbol{F}_{k,k-1}[\mathcal{I}_{k-1} + (\boldsymbol{F}_{k,k-1}^{\mathrm{T}}\boldsymbol{Q}_k^{-1}\boldsymbol{F}_{k,k-1})]^{-1}\boldsymbol{F}_{k,k-1}\boldsymbol{Q}_k^{-1} \tag{9.71}$$

   (b) The singular value decomposition (SVD) of $\mathcal{I}_{k|k-1}$ is given by

$$\mathcal{I}_{k|k-1} = \boldsymbol{U}\boldsymbol{Q}^2\boldsymbol{U}^{\mathrm{T}} \tag{9.72}$$

   (c) Calculate $\boldsymbol{H}^{\mathrm{T}}\boldsymbol{R}^{-1}\boldsymbol{H}$ for all the candidate navigation beacons

$$\Delta\mathcal{I}_j = \boldsymbol{H}^{\mathrm{T}}\boldsymbol{R}^{-1}\boldsymbol{H} \tag{9.73}$$

   with $j = 1, 2, \ldots, m$.

   (d) Construct the matrix

$$\boldsymbol{C}_j = \boldsymbol{Q}^{-1}\boldsymbol{U}^{\mathrm{T}}\Delta\mathcal{I}_j\boldsymbol{U}\boldsymbol{Q}^{-1} \tag{9.74}$$

   Perform the SVD of $\boldsymbol{C}_j$

$$\boldsymbol{C}_j = \boldsymbol{V}\boldsymbol{\Upsilon}_j\boldsymbol{V}^{\mathrm{T}} \tag{9.75}$$

   (e) Denote the diagonal vector of Υ_j as $\boldsymbol{\mu}_j$ and the smallest element of $\boldsymbol{\mu}_j$ as $\mu_{j,\min}$. Choose the navigation beacon with the maximum $\mu_{j,\min}$ to be observed.

   (f) $k = k + 1$.

3. Accomplishment of planning.

Figure 9.5 shows the flowchart of navigation beacon planning based on dynamic observability analysis. Compared with traditional planning algorithm, this algorithm is a recursive planning algorithm with a small calculation load. The number of its searches in each operation is $C(m, 2) + (m - 2) + (m - 3) + \cdots + (m - n + 1) = C(m, 2) + (n - 2)(2m - n - 1)/2$. In the case of $n = 2$, the number of its searches is equal to $C(m, n)$. However, when n is close to $m/2$, the number of its searches will be smaller than $C(m, n)$ (for example, given $m = 14, n = 7$, if the method of exhaustion is adopted, the traces of $C(14, 7) = 3432$ combinations need to be calculated in each observation, while the number of searches in this method is only $C(14, 2) + (7 - 2)(28 - 7 - 1)/2 = 141$).

In addition, when m is very large, the computational advantages of this algorithm will also loom large. Figures 9.6 and 9.7 show the search times of this new algorithm and of traditional algorithm, respectively, when 10 of m candidate navigation beacons are selected for observation. It can be seen from the figure that, the search times of the two algorithms will increase with m but at very different orders of magnitude. In

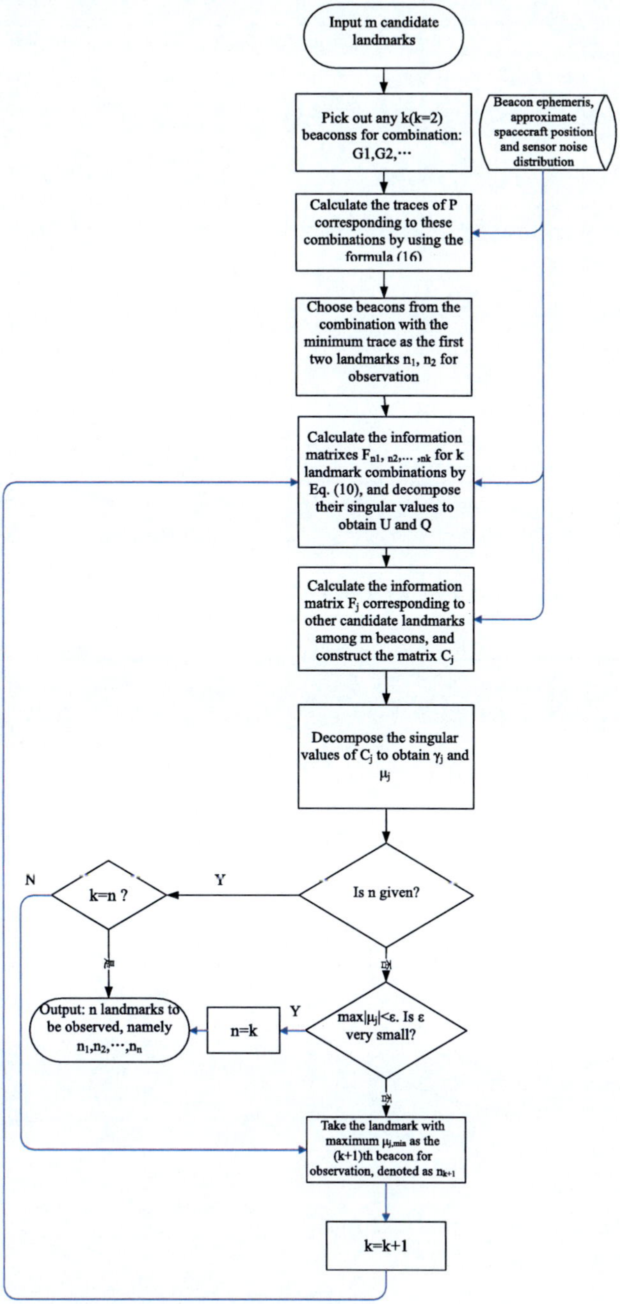

Fig. 9.5 Flowchart of the navigation beacon planning

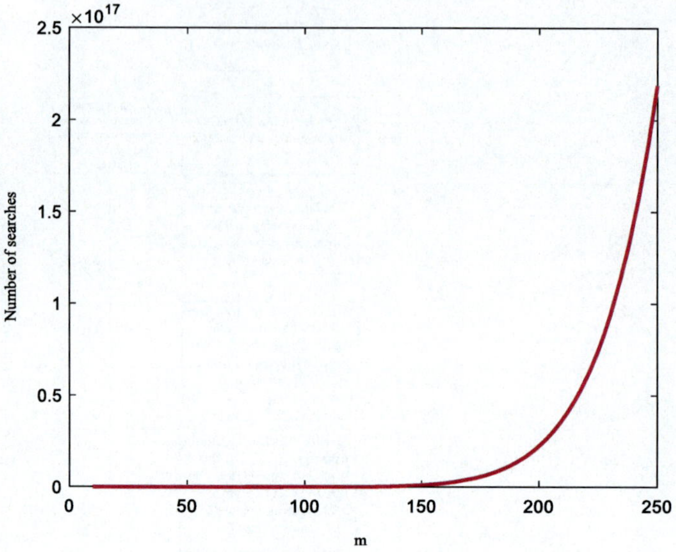

Fig. 9.6 Number of searches by traditional algorithm

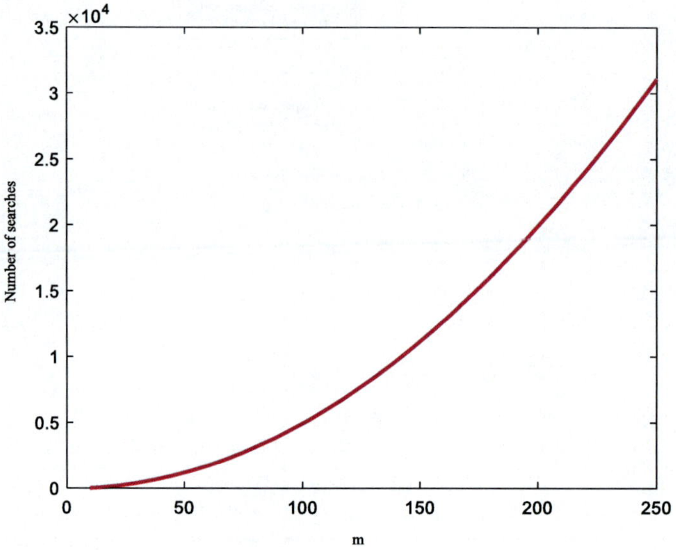

Fig. 9.7 Number of searches by new algorithm

the case of $m = 250$, the search times of traditional algorithm are up to 10^{17} orders of magnitude, while the search times of the new algorithm are much smaller, only 10^4 orders of magnitude.

By using the method of induction, the search times of the new algorithm can prove to be fewer than or equal to those of traditional algorithm and can quantitatively evaluate the influence of the number of navigation landmarks on navigation performance. When

$$|\boldsymbol{\mu}_j| \to 0 \tag{9.76}$$

navigation performance will not be greatly improved even if the number of observed beacons is increased, so additional observation is not needed.

9.4 Integrated Autonomous Navigation Filtering

This section presents different navigation filters for the XNAV/optical integrated navigation system. The orbit dynamics equation of the spacecraft as

$$\dot{\boldsymbol{x}} = \boldsymbol{f}(\boldsymbol{x}, t) + \boldsymbol{w}_d \tag{9.77}$$

where $\boldsymbol{x} = [\boldsymbol{r}^{\mathrm{T}}, \boldsymbol{v}^{\mathrm{T}}]^{\mathrm{T}}$.

Accordingly, the measurement equations of optical autonomous navigation and pulsar-based autonomous navigation are denoted as

$$\begin{aligned} \tilde{z}_o &= \boldsymbol{h}_o(\boldsymbol{x}) + \boldsymbol{v}_o \\ \tilde{z}_p &= \boldsymbol{h}_p(\boldsymbol{x}) + \boldsymbol{v}_p \end{aligned} \tag{9.78}$$

For the centralized fusion architecture, noting that it takes some time to establish the pulse profile for the XNAV, a sequential filtering can be carried out to process optical and pulsar measurements successively. In addition, since the matching of pulsar pulse profile needs priori spacecraft position information, optical autonomous navigation is usually adopted initially. Figure 9.8 shows the corresponding implementation process, of which the process of navigation filtering algorithm is shown in Chap. 8 and the steps of navigation beacon planning algorithm are given in Sect. 9.3.2.

When updating the measurements of optical autonomous navigation, the measurement prediction and measurement sensitivity equations are given as below

$$\begin{aligned} \hat{z}_o &= \boldsymbol{h}_o(\hat{\boldsymbol{x}}) \\ \boldsymbol{H}_o &= \frac{\partial \tilde{z}_o}{\partial \boldsymbol{x}} \end{aligned} \tag{9.79}$$

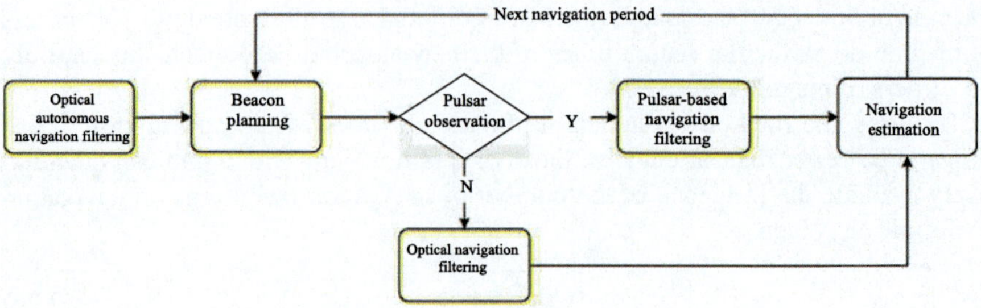

Fig. 9.8 Centralized fusion for XNAV/optical integrated fusion

When updating the measurements of XNAV, measurement prediction and measurement sensitivity equations are given as below

$$\hat{z}_p = \boldsymbol{h}_p(\hat{\boldsymbol{x}})$$
$$\boldsymbol{H}_p = \frac{\partial \tilde{z}_p}{\partial \boldsymbol{x}} \tag{9.80}$$

If the filter is a federated Kalman filter structure, it can be divided into the local filter for optical autonomous navigation and the local filter for XNAV. According to Chap. 3, the specific steps of federated Kalman filtering algorithm are as follows for the XNAV/optical integrated navigation system:

1. Global initialization of $\hat{\boldsymbol{x}}_{g,k}$, $\boldsymbol{P}_{g,0}$, $\boldsymbol{Q}_{g,0}$
2. Information allocation

$$(\boldsymbol{P}_{i,0})^{-1} = \beta_i^{-1}\boldsymbol{P}_{g,0}^{-1}$$
$$\boldsymbol{Q}_{i,k-1} = \beta_i^{-1}\boldsymbol{Q}_{g,k-1}$$
$$\hat{\boldsymbol{x}}_{i,k-1} = \hat{\boldsymbol{x}}_{g,k-1} \tag{9.81}$$

where subscript $i = o, p$.

3. Conduct Kalman filtering for local filters and obtain $\hat{\boldsymbol{x}}_{o,k}$, $\hat{\boldsymbol{x}}_{p,k}$ and $\boldsymbol{P}_{o,k}$, $\boldsymbol{P}_{p,k}$
4. Global estimation

$$\boldsymbol{P}_{g,k} = \left[(\boldsymbol{P}_{o,k})^{-1} + (\boldsymbol{P}_{p,k})^{-1}\right]^{-1}$$
$$\hat{\boldsymbol{x}}_{g,k} = \boldsymbol{P}_{g,k}\left[(\boldsymbol{P}_{o,k})^{-1}\hat{\boldsymbol{x}}_{o,k} + (\boldsymbol{P}_{p,k})^{-1}\hat{\boldsymbol{x}}_{p,k}\right] \tag{9.82}$$

5. Local estimation resetting

$$
\begin{aligned}
(\boldsymbol{P}_{i,k})^{-1} &= \beta_i^{-1} \boldsymbol{P}_{g,k}^{-1} \\
\hat{\boldsymbol{x}}_{i,k} &= \hat{\boldsymbol{x}}_{g,k}
\end{aligned} \tag{9.83}
$$

If covariance intersection fusion algorithm is adopted, the main steps without feedback are as follows:

1. Initialization of $\hat{\boldsymbol{x}}_{g,k}$, $\boldsymbol{P}_{g,0}$

$$
\begin{aligned}
(\boldsymbol{P}_0^{(i)})^{-1} &= \boldsymbol{P}_{g,0}^{-1} \\
\hat{\boldsymbol{x}}_k^{(i)} &= \hat{\boldsymbol{x}}_{g,k}
\end{aligned} \tag{9.84}
$$

2. Conduct Kalman filtering for local filters and obtain $\hat{\boldsymbol{x}}_{o,k}$, $\hat{\boldsymbol{x}}_{p,k}$ and $\boldsymbol{P}_{o,k}$, $\boldsymbol{P}_{p,k}$
3. Fusion by the fusion center

$$
\begin{aligned}
(\hat{\boldsymbol{P}}_k)^{-1} &= \omega_o(\hat{\boldsymbol{P}}_{o,k})^{-1} + \omega_p(\hat{\boldsymbol{P}}_{p,k})^{-1} \\
(\hat{\boldsymbol{P}}_k)^{-1}\hat{\boldsymbol{x}}_k &= \omega_o(\hat{\boldsymbol{P}}_{o,k})^{-1}\hat{\boldsymbol{x}}_{o,k} + \omega_p(\hat{\boldsymbol{P}}_{p,k})^{-1}\hat{\boldsymbol{x}}_{p,k}
\end{aligned} \tag{9.85}
$$

where $\omega_o, \omega_p \geq 0$ and $\omega_o + \omega_p = 1$. The optimal values of ω_o and ω_p are obtained by minimizing the trace of $\hat{\boldsymbol{P}}_k$.

If feedback is added, local filters need to be reset after each fusion

$$
\begin{aligned}
\hat{\boldsymbol{x}}_{o,k} &= \hat{\boldsymbol{x}}_k \\
\hat{\boldsymbol{P}}_{p,k} &= \hat{\boldsymbol{P}}_k
\end{aligned} \tag{9.86}
$$

9.5 Numerical Simulations

9.5.1 Autonomous Navigation Based on Fused Asteroid and Pulsar Observations at the Deep-Space Transfer Phase

In this section, numerical simulations for the integrated navigation based on optical and pulsar measurements are presented, consisting of three parts. The first and second parts evaluate the navigation beacon planning algorithms based on geometric observability analysis and dynamic observability analysis, respectively. The third part compares different fusion algorithms.

The initial time is August 5, 2007, 00:00:00 UTCG. The initial position and velocity in heliocentric ecliptic J2000.0 frame are $[1.017978222619442 \times 10^8, -1.123003524012581 \times 10^8, -9.578654170367867 \times 10^4]$ km, and $[21.40460801927876, 25.08765627351607, -2.007089018943564]$ km/s, respectively. For the perturbations of true orbit dynamics, the gravitational perturbation of large planets in the solar system and the perturbation caused by solar radiation pressure are considered. The true mission orbit is obtained using RKF78 integrator.

The optical imaging sensor has a focal length of 677 mm and an angular resolution of 13 μrad/pixel.

9.5.1.1 Navigation Beacon Planning Simulation Based on Geometric Observability Analysis

This section demonstrates the navigation beacon planning simulation based on geometric observability analysis. According to the selection criteria presented in Chap. 8, ten candidate navigation asteroids are selected, the label of which are 2063, 3800, 2368, 2100, 4954, 4464,3920, 843, 2329, 1065. For XNAV, three candidate pulsars are selected: B0531+21, B1821-24, B1937+21.

The parameters related to optical autonomous navigation are as follows:

1. Optical sensor noise: 0.1 pixel
2. Attitude knowledge error: 1 μrad
3. Asteroid ephemeris error: 0.

Noises and error terms of XNAV are shown in Table 9.1.

The initial position and velocity estimation errors are 1000 km and 1 m/s in each axis, respectively. Every two days, an asteroid and a pulsar information are

Table 9.1 Noise and error terms for XNAV

Name	Value
Timing resolution in meters	600 m
Time-scale transformation error of TOA (1σ)	0
Space scale transformation error of TOA (1σ)	3 km
TOA measurement error (1σ)	See Table 9.2
Ephemeris error	0

Table 9.2 TOA measurement errors of candidate pulsars

Number	Name of pulsar	TOA measurement error (1σ, m)
1	B0531+21	99.4891
2	B1821-24	297.4315
3	B1937+21	314.5242

Table 9.3 Navigation simulation scenarios

Scenario	Configuration
Scenario 1	The angle between the observed asteroid and the pulsar is the smallest
Scenario 2	The angle between the observed asteroid and the pulsar is closest to 90°

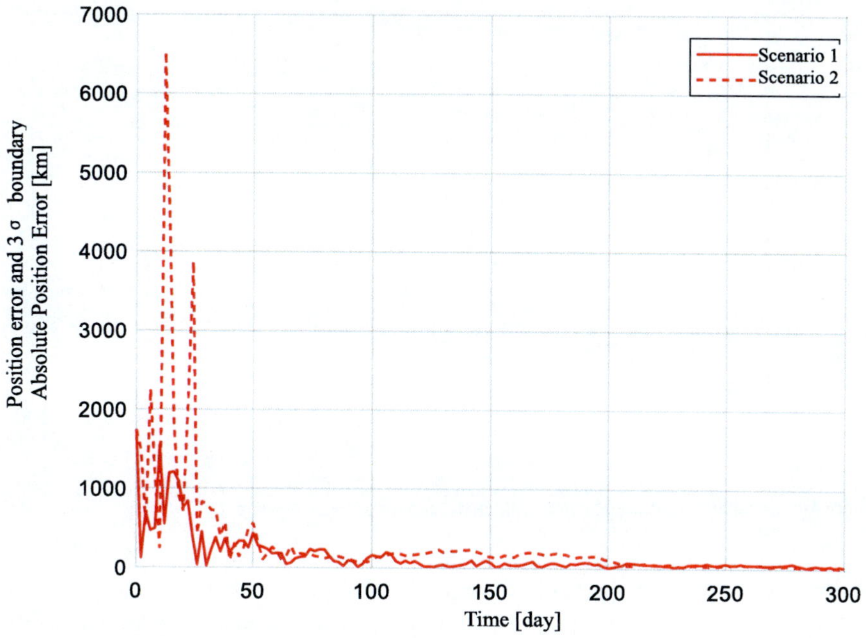

Fig. 9.9 Comparison of the absolute position errors of scenarios 1 and 2

measured. In order to demonstrate the navigation beacon planning algorithm based on geometric observability analysis, two typical scenarios shown in Table 9.3 are considered. Simulation results are shown in Figs. 9.9 and 9.10. It can be seen that filters in both scenarios are convergent. Because the beacons configuration in scenario 1 has a better geometric observability, the navigation performance of the scenario from 0 to 200 days is better than that of the scenario 2. After 200 days, the effect of the geometric observability on the navigation accuracy is not significant, and both scenarios have similar navigation performance.

9.5.1.2 Navigation Beacon Planning Simulation Based on Dynamic Observability Analysis

The main advantages of the navigation beacon planning algorithm based on dynamic observability analysis are high planning efficiency and the avoidance of unnecessary

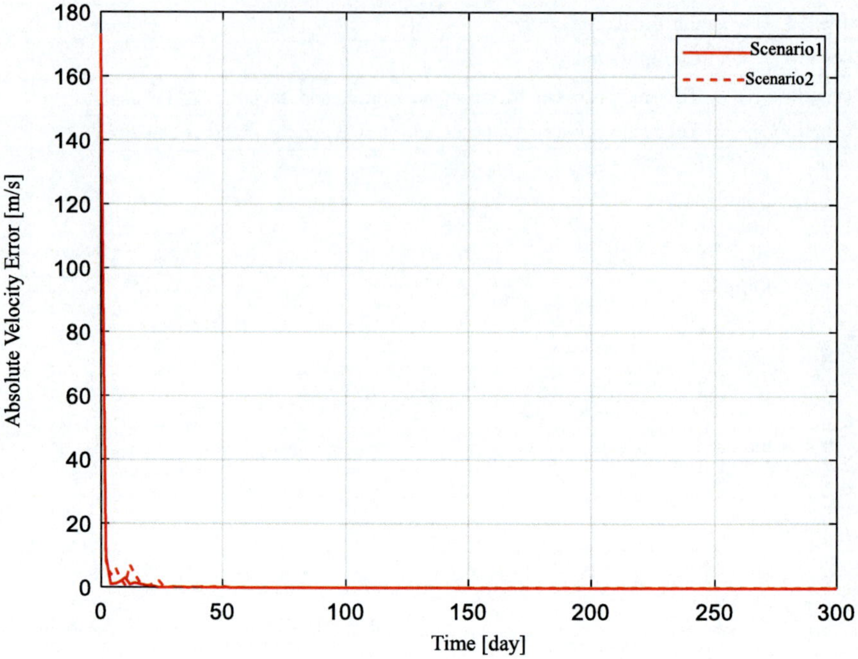

Fig. 9.10 Comparison of the absolute velocity errors of scenarios 1 and 2

observation so that it is particularly suitable for autonomous navigation based on asteroid observation at the deep-space transfer phase. Therefore, this section evaluates the navigation beacon planning algorithm based on dynamic observability analysis for deep-space transfer optical navigation.

The centralized sequential Kalman filtering is used, and $\boldsymbol{n}$ asteroids are observed every two days. Other related parameters are the same with Sect. 9.5.1.1.

Intuitively, the observation of more navigation asteroids would improve navigation performance, but puts a strain on the sensors. The beacon planning algorithm presented in Sect. 9.3.2 can avoid unnecessary observation while satisfying the performance of navigation system. According to Fig. 9.5, the best optical beacon can be selected at the every observation time, and then evaluate the $|\mu_j|$ of Eq. (9.75). If $\max\{|\mu_j|\} < \varepsilon$ (e.g., $\varepsilon = 0.1$), then the additional observation would not improve the navigation performance.

Figure 9.11 shows the navigation comparisons for two scenarios. In one scenario, without consider the threshold, ten asteroids are observed under any circumstances. The other scenario considers the threshold and only those asteroids satisfying $\max\{|\mu_j|\} \geq \varepsilon$ are used for navigation. It can be seen from the figure that the navigation simulation accuracy is similar in each of the two scenarios. Figure 9.12

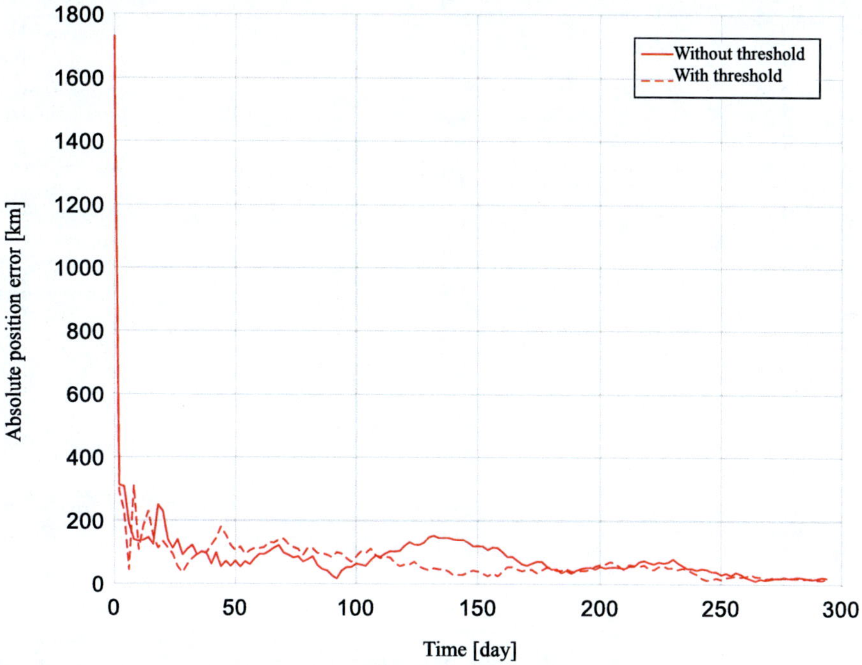

Fig. 9.11 Comparison of navigation results with and without threshold

shows the number of observed asteroids given by the planning algorithm. Obviously, the planning algorithm avoids unnecessary observation, thus demonstrating its effectiveness.

9.5.1.3 Simulation Comparison of Different Fusion Filtering Algorithms

This section simulates the integrated navigation based on optical and pulsar measurement information during the deep-space transfer phase and compares the performance of centralized Kalman filtering algorithm, federated filtering algorithm, and CI algorithm. Simulation results are shown in Figs. 9.13 and 9.14, and statistical results are shown in Table 9.4. It can be seen that the performance of the three algorithms is similar, but the performance of centralized Kalman filtering is slightly better than the other two algorithms.

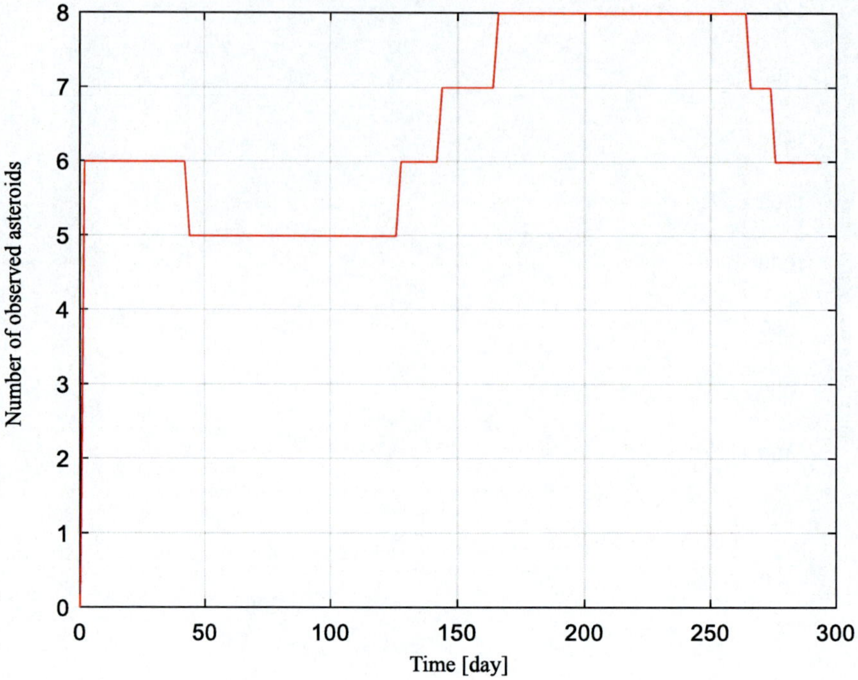

Fig. 9.12 Number of observed asteroids given by the planning algorithm

9.5.2 XNAV/Optical Integrated Navigation During the Approach Phase

In this section, the XNAV/optical integrated navigation during the Mars approach phase is considered. The optical measurements are the LOS directions to Phobos, Deimos, and Mars center. Table 9.5 shows the initial position and velocity of the spacecraft in the Mars J2000 frame. For the true orbit dynamics, solar gravitational perturbation, solar radiation pressure perturbation, and 50th-order gravitational field model are considered. The true mission orbit is obtained using RKF78 integrator.

The parameters of optical imaging sensor are given by

Focal length: 6 mm
FOV: 60° × 60°
Resolution: 1024 × 1024
Measurement period: 60 s.

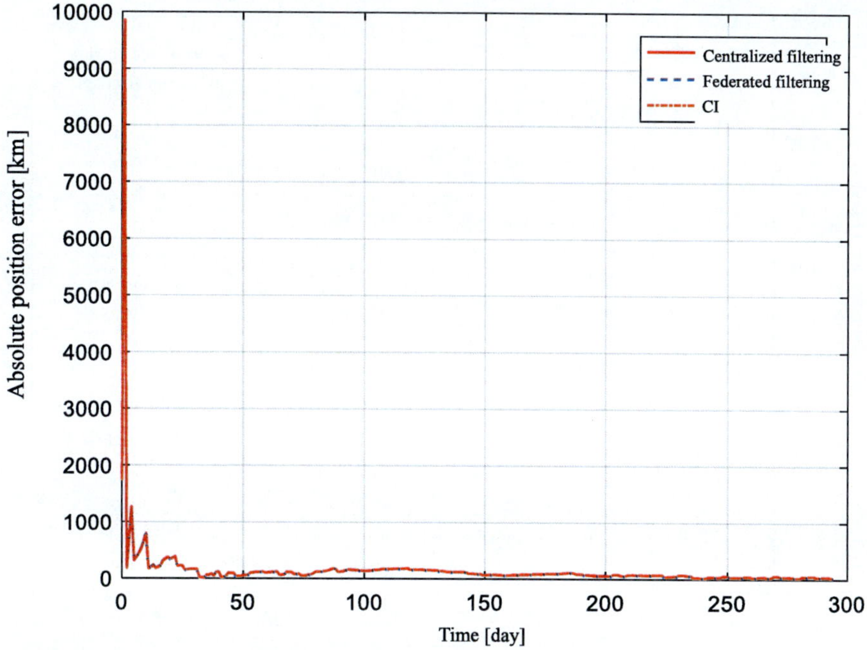

Fig. 9.13 Navigation position errors of different algorithms

The filter update period is set as 60 s. The pulsar detector observes three pulsars (B0531+21, B1821-24, B1937+21) in a time-sharing manner, and outputs the measurements once every 600 s. Parameters of the XNAV are shown in Table 9.1.

First, the performance of optical navigation, XNAV, and XNAV/optical integrated navigation is compared using the centralized sequential Kalman filtering algorithm. Simulation results are shown in Figs. 9.15 and 9.16. Navigation errors of different navigation modes in steady state are shown in Table 9.6. It can be seen that the fusion navigation has the best performance and converges much faster than optical navigation during filtering.

Next, simulation results of different fusion algorithms are compared. Simulation results are shown in Figs. 9.17 and 9.18, and their navigation errors in steady state are shown in Table 9.7. It can be seen that the performance of the three filters is similar, but the accuracy of federated filtering algorithm is the worst. This is because the federated filtering assumes that the measurements are independent of each other, but the measurements here in optical autonomous navigation are dependent upon each other, which violates the assumption and degrades the filter performance.

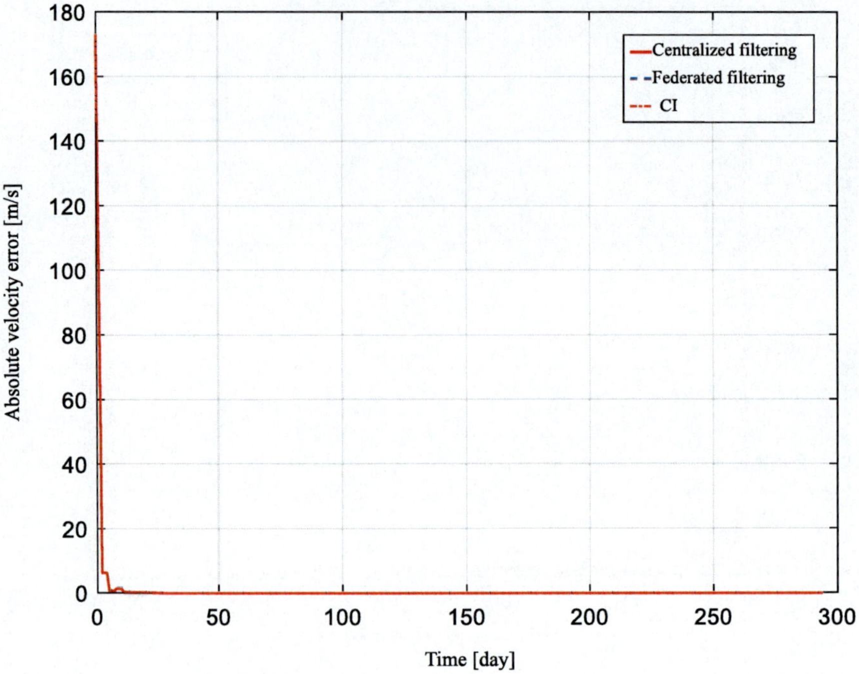

Fig. 9.14 Navigation velocity errors of different algorithms

Table 9.4 Comparison of three algorithms

	Statistical index	Position estimation error (km)			Velocity estimation error (m/s)		
Centralized Kalman	Mean	24.97	9.97	29.86	0.052	0.012	−0.044
	1σ	28.86	26.91	66.71	0.065	0.046	0.12
Federated Kalman	Mean	25.56	10.22	30.23	0.055	0.017	−0.044
	1σ	28.87	26.91	66.76	0.067	0.050	0.14
CI	Mean	25.56	10.28	30.29	0.055	0.017	−0.045
	1σ	28.97	27.03	66.87	0.067	0.056	0.18

Table 9.5 Initial orbital parameters of the approaching phase

Epoch time	July 3, 1997 11:09:59.000 UTCG
Position [km]	[534,059.29, 214,496.39, 8822.61]
Velocity [km/s]	[−4.94, −2.041, −0.085]

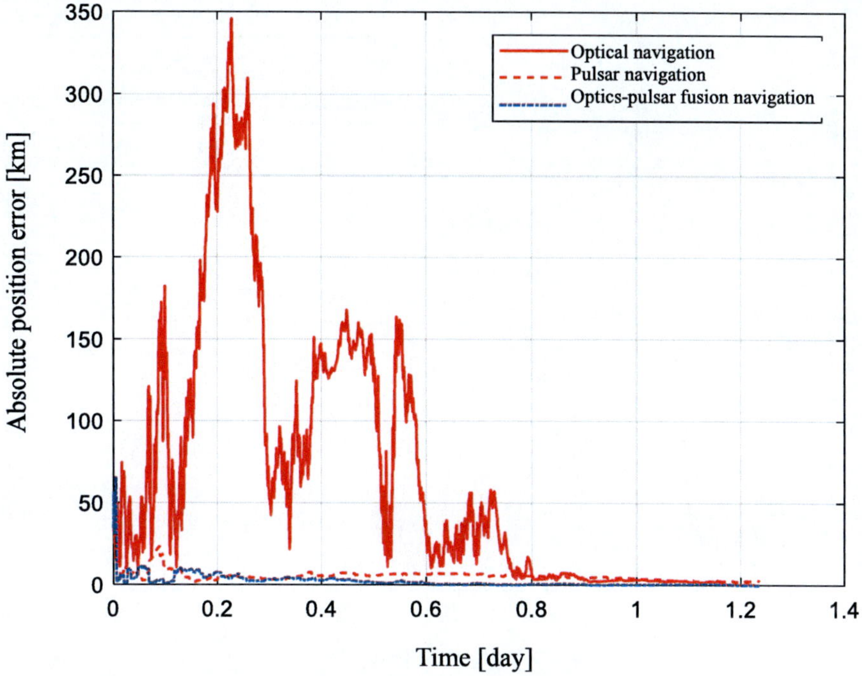

Fig. 9.15 Navigation position errors in different navigation modes

9.5.3 XNAV/Optical Integrated Navigation During the Surrounding Phase

In this section, the XNAV/optical integrated navigation during the Mars surrounding phase is considered, where LOS vectors to Phobos and Deimos, and TOA from three pulsars B0531+21, B1821-24, and B1937+21 are observed.

Table 9.8 shows the initial position and velocity of the spacecraft in the Mars J2000 frame. The dynamic orbit is generated where solar gravitational perturbation, solar radiation pressure perturbation, and 50th-order gravitational field model are considered. The orbital integrator is a *RKF78* numerical integrator.

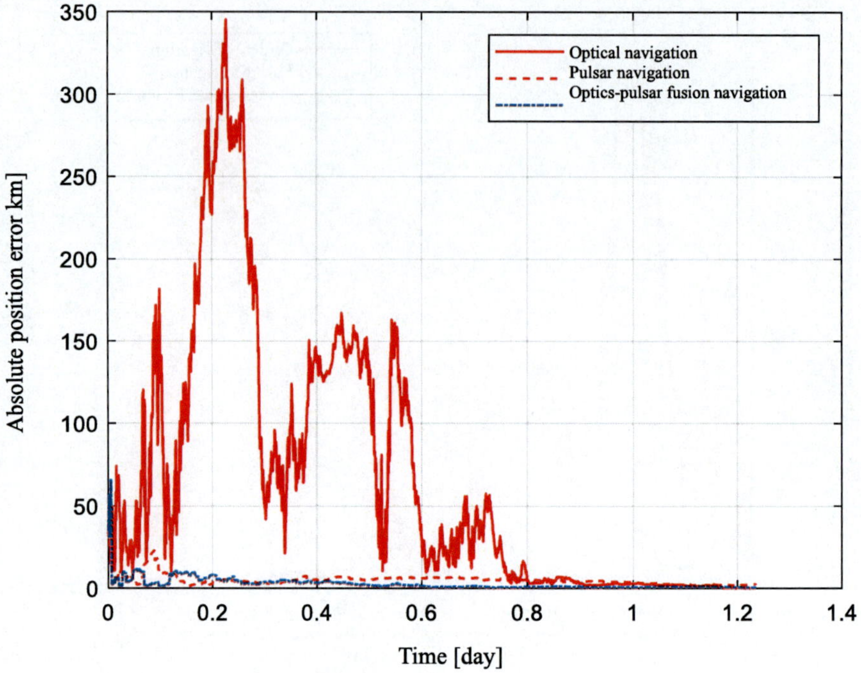

Fig. 9.16 Navigation velocity errors in different navigation modes

Table 9.6 Comparison of simulation results in different navigation modes

Navigation mode	Position estimation error RMS (km)			Velocity estimation error RMS (m/s)		
	σ_{rx}	σ_{ry}	σ_{rz}	σ_{vx}	σ_{vy}	σ_{vz}
Optical navigation	0.8608	0.5796	1.7526	0.0240	0.0198	0.0765
XNAV	2.7212	0.1613	0.4971	0.1097	0.1047	0.0132
XNAV/optical integrated navigation	0.2302	0.1001	0.4350	0.008	0.003	0.015

In this section, centralized filtering, federated filtering, and CI algorithm are simulated. The main parameters are set as follows:

1. The parameters of optical imaging sensor:

   Focal length: 6 mm
   FOV: 60° × 60°
   Resolution: 1024 × 1024
   Measurement noise: 0.1 pixels
   Attitude knowledge error: 1 μrad
   Observation period: 60 s.

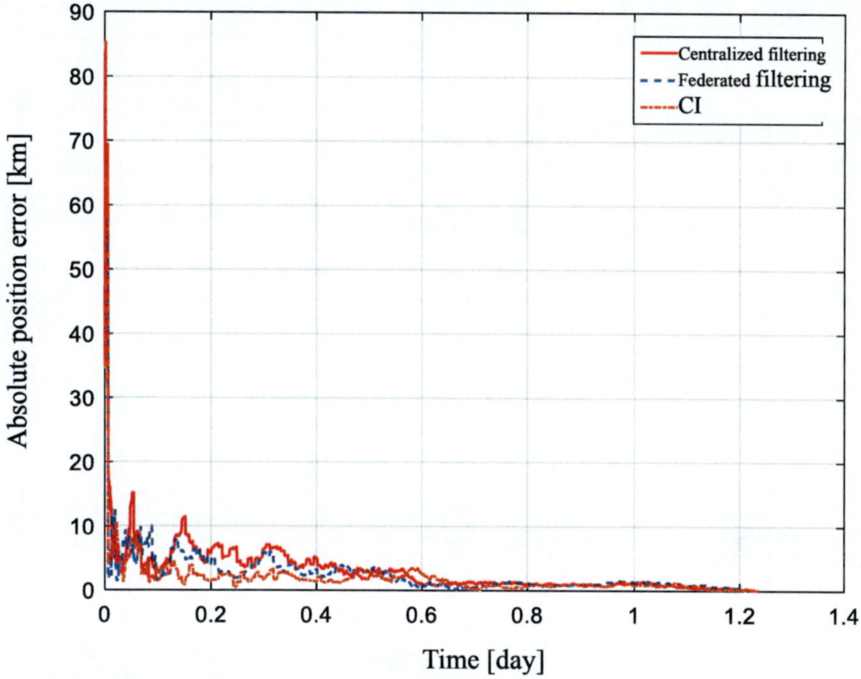

Fig. 9.17 Position estimation errors of different fusion algorithms

2. For the pulsar navigation, the error settings are shown in Table 9.1, and the observation period is 600 s.
3. The navigation filtering period is 10 s.
4. Initial navigation errors: each component of position error is 3 km and each component of velocity error is 3 m/s.

Simulation results are shown in Figs. 9.19 and 9.20, and steady navigation errors are shown in Table 9.9. The steady navigation position errors of the three filtering algorithms are all smaller than 100 m. According to the simulation results in Chaps. 7 and 8, the position accuracy of optical navigation alone is about 150 m and that of pulsar navigation is about 1 km, so the integrated navigation accuracy is better than that of a single optical navigation or a XNAV system.

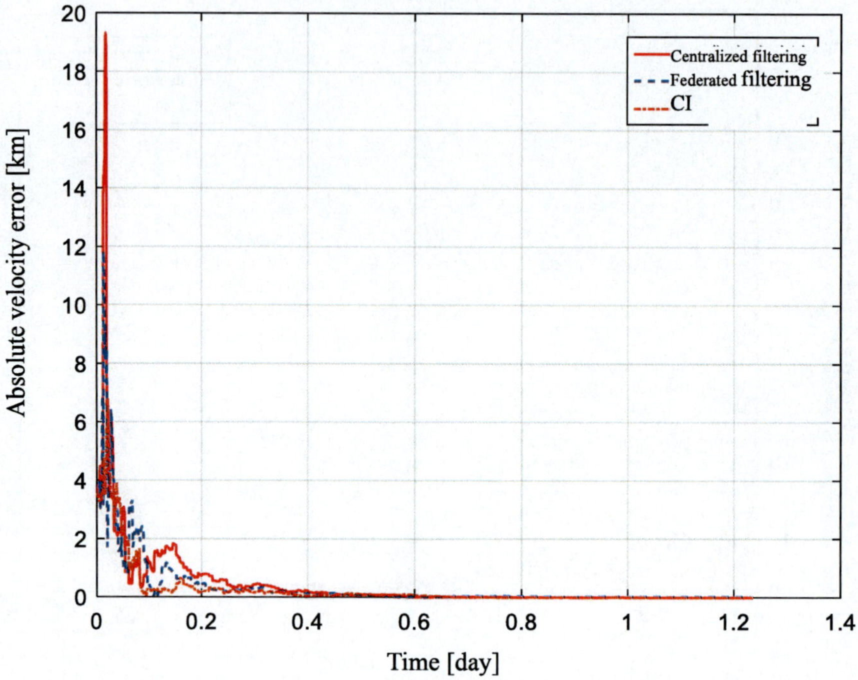

Fig. 9.18 Velocity estimations errors of different fusion algorithms

Table 9.7 Comparison of simulation results of different navigation algorithms

Navigation mode	Position estimation error RMS (m)			Velocity estimation error RMS (m/s)		
	σ_{rx}	σ_{ry}	σ_{rz}	σ_{vx}	σ_{vy}	σ_{vz}
Centralized	478.5504	188.8362	473.3856	0.0099	0.0055	0.0056
Federated	273.9184	298.6835	855.2817	0.0094	0.0080	0.0123
CI	334.7286	132.1582	597.1518	0.0076	0.0050	0.0075

Table 9.8 Initial orbital parameters of surrounding phase

Epoch time	July 4, 1997 16:47:58.000 UTCG
Position [km]	[682.30, −4190.90, −252.308]
Velocity [km/s]	[−3.15, −0.49, −0.01]

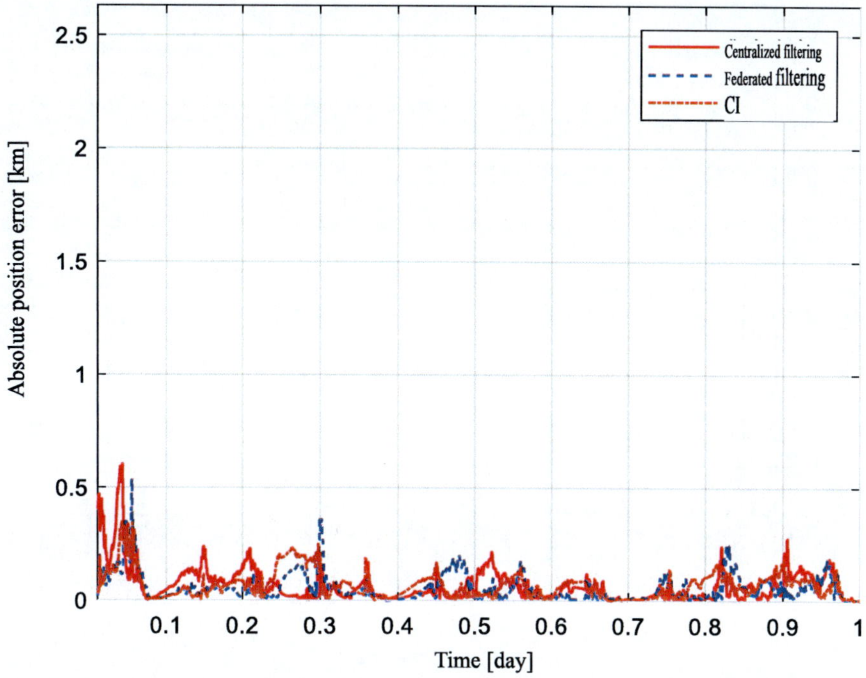

Fig. 9.19 Position estimation errors of fusion navigation (surrounding phase)

9.6 Brief Summary

This chapter has studied the integrated navigation technology based on the fusion of optical beacons and pulsars, with the main contribution of proposing a new navigation beacon planning algorithm. Compared with traditional algorithms, this algorithm is easy to calculate and is suitable for on-orbit operation. In terms of fusion navigation filtering, this chapter also presented the fusion process of centralized Kalman filtering, federated Kalman filtering, and CI algorithm through the combination with systematic error correction technique. Finally, the effectiveness of the navigation beacon planning and fusion algorithms is demonstrated through numerical simulation.

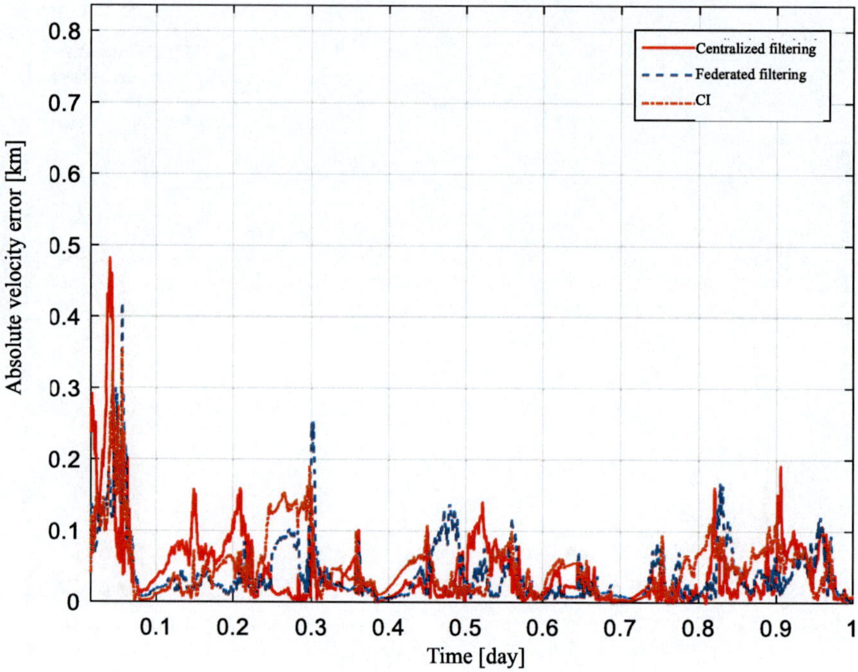

Fig. 9.20 Velocity estimation errors of fusion navigation (surrounding phase)

Table 9.9 Comparison of simulation results of different navigation algorithms

Navigation mode	Position estimation error RMS (m)			Velocity estimation error RMS (m/s)		
	σ_{rx}	σ_{ry}	σ_{rz}	σ_{vx}	σ_{vy}	σ_{vz}
Centralized	45.8818	54.6471	5.7518	0.0341	0.0350	0.0035
Federated	51.1386	53.0162	8.2649	0.0335	0.0369	0.0058
CI	59.7070	57.9090	5.4928	0.0389	0.0434	0.0041

References

1. Xiong, K. 2008. *Research on the spacecraft autonomous navigation based on pulsars*. Beijing: China Academy of Space Technology.
2. Sheikh, S.I. 2005. *The use of variable celestial X-ray sources for spacecraft navigation*, Ph.D. thesis. Maryland: University of Maryland.
3. Anderson, K.D., D.J. Pines, and S.I. Sheikh. 2015. Validation of pulsar phase tracking for spacecraft navigation. *Journal of Guidance, Control and Dynamics* 38 (10): 1885–1897.
4. Deng, X., et al. 2013. Interplanetary spacecraft navigation using pulsars. *Advances in Space Research* 52 (9): 1602–1621.
5. Becker, W., M.G. Bernhardt, and A. Jessner. 2013. Autonomous spacecraft navigation with pulsars. *Acta Futura* 7: 11–28.

6. Shuai, P., et al. 2009. *Principle and method of X ray pulsar navigation system*. Beijing: China Astronautic Publishing House (in Chinese).
7. Wang, Y.D. 2016. *X-ray pulsar-based navigation: Signal processing and positioning algorithms*. Chang Sha, China: National University of Defense Technology.
8. Bradford, P.W., J. Spilker, and P. Enge. 1996. *Global positioning system: Theory and applications*, 109. Washington DC: AIAA.
9. Ma, P., et al. 2017. Autonomous navigation of Mars probes by single X-ray pulsar measurement and optical data of viewing Martian moons. *The Journal of Navigation* 70 (1): 18–32.
10. Bellman, R. 1997. *Introduction to matrix analysis*. PA, USA: Society for Industrial and Applied Mathematics.

Chapter 10 Altimeter and Velocimeter-/Optical-Aided Inertial Navigation Technology

Inertial navigation has been successfully applied in soft landing missions. The inertial navigation error is gradually increased caused by initial navigation error and inertial measurement unit (IMU) measurement error. In order to improve the navigation accuracy of the lander relative to the celestial body surface, direct measurements of the lander relative to the celestial body surface are usually used to correct inertial navigation results. Candidate sensors include altimeter, velocimeter, and optical imaging sensor. This chapter focuses on the altimeter-/velocimeter-aided inertial navigation and optical-aided inertial navigation for soft landing missions.

10.1 Soft Landing Flight Process

In this section, the flight process of Chang'e-3 lunar soft landing and a general Mars soft landing missions are briefly introduced. Since moon is an airless body, the velocity reduction is achieved by a main braking engine. As shown in Fig. 10.1, the landing process of the Chang'e-3 includes main braking, rapid adjustment, approaching, hovering, obstacle avoidance, and slow descent phases.

(1) Flight procedure of main braking phase
The main braking phase is responsible for soft landing braking, that is, through deceleration by a thrust engine (referred to as main engine here), to slow down the lander's velocity relative to the moon from about 1.7 km/s a specific value and to reduce its altitude relative to the moon to a specific altitude. According to the mission requirement of eliminating high initial horizontal velocity at the powered descent stage, propellant consumption optimization is the main design goal of guidance law at the main braking stage. The design requirements for autonomy, real-time performance, and connection with subsequent task states should also be considered. According to the mission requirements, the terminal constraints for the guidance law in this phase are the altitude and

D. Wang et al., *Spacecraft Autonomous Navigation Technologies Based on Multi-source Information Fusion*, Space Science and Technologies, https://doi.org/10.1007/978-981-15-4879-6_10

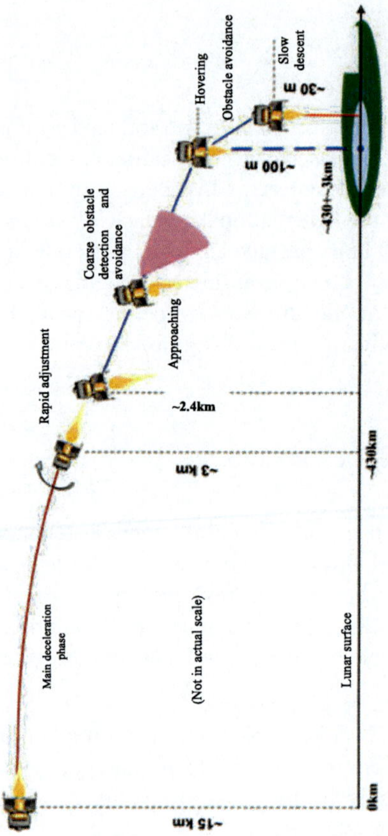

Fig. 10.1 Powered descent process of the Chang'e-3

three-dimensional velocity constraints. The altitude reduction of this phase is about 15 km.

(2) Flight procedure of rapid adjustment phase
At the end of the main braking stage, the lander attitude is still nearly horizontal, and the main engine still works with the maximum thrust. However, at the subsequent approaching stage, the lander attitude is required to be nearly vertical, the main engine is required to work under a low thrust, and the altitude, velocity, and acceleration are required to satisfy a certain constraint. Therefore, the main function of rapid adjustment is to connect main braking with the subsequent approaching phase. The lander in rapid adjustment should adjust to the entrance attitude of approaching phase through rapid attitude maneuvers. In this phase, the engine thrust should be gradually reduced to a low level, so as to ensure fast and smooth transition from the end of main braking phase to the entrance of approaching phase, with an altitude reduction of about 600 m.

(3) Flight procedure of approaching phase
The main task of approaching phase is to observe the predetermined landing site and carry out coarse obstacle avoidance. According to the requirements for coarse obstacle avoidance, the flight path should ensure that: (1) The imaging sensor can continuously observe the predetermined landing site; (2) the lander can arrive at the landing site without maneuvering. That is, the flight path of approaching phase meets the requirements for a specific attitude and descent trajectory and approaches the target landing site.
Note that the invisible area brought about by the plume of the 7500 N main engine is a cone with a semi-cone angle of about 25° and that the FOV of imaging sensor is 30°, and the LOS of imaging sensor is offset by 40° to avoid the influence of the plume of the main engine on the imaging sensor and shorten the sensor's LOS distance as much as possible. To ensure that the landing site can be observed in the FOV of imaging sensor at the approaching stage, a straight descent mode where the angle between descent trajectory and horizontal plane is close to 45° is adopted to approach the landing site step by step.

(4) Flight procedure of hovering phase
The main task of hovering phase is to take high-precision three-dimensional images of lunar surface, accurately detect obstacles in the landing area, and determine a safe landing site. Therefore, the lander needs to maintain a hovering state: Its velocity and angle velocity are close to 0, such that the lander's attitude and position remain unchanged. According to the working range of the laser 3D imaging sensor and the observation requirements of a sufficiently large landing site, hovering control is carried out at the attitude of 90–110 m. In the hovering state, the laser 3D imaging sensor accurately observes the landing area, processes the 3D image data, determines a safe landing site, and helps the lander switch to the obstacle avoidance phase. The desired states used for guidance of this phase are as follows: The lander altitude is the navigation altitude at the entrance of hovering phase, the velocities in three directions are 0 m/s, and the acceleration in the altitude direction is 0 m/s^2.

(5) Flight procedure of obstacle avoidance phase

The main task of obstacle avoidance phase is precise obstacle avoidance and descent. To ensure the landing accuracy and to save propellant, precise obstacle avoidance and descent are carried out at the same time. In light of the determined safe landing site, the lander descends from the altitude of about 100–30 m above the landing site. Its velocity relative to the lunar surface drops to -1.5 m/s, and its horizontal velocity is close to zero.

(6) Flight procedure of slow descent phase
To ensure the control accuracy of lunar landing velocity/attitude, the slow descent phase is vertical descent with a preset small constant speed in order to eliminate horizontal velocity and acceleration. The main engine will not be shut down until a measurement signal from the shutdown sensor is received or the acceleration measurement is bigger than the preset value. When descending vertically from an attitude of about 30 m to the attitude above the landing site, the descent velocity is chosen as -2 m/s considering the factors of propellant consumption and navigation position drift.

For a Mars soft landing mission, the flight process can be divided into three phases [1, 2]:

(1) Atmospheric entry phase. The lander enters the Mars atmosphere and uses its aerodynamic configuration to slow down. To withstand a high aerodynamic load and heating, the lander is protected by a heat shield.
(2) Parachute descent phase. The parachute is deployed to slow the lander when meeting the deployment condition, and then heat shield is separated.
(3) Terminal landing phase. After reaching the predetermined states, the parachute is separated from the lander, and the orbit control engine and attitude control thruster start to work to carry out active deceleration and obstacle avoidance. This phase is similar to the soft landing on a celestial body without atmosphere.

The principle of autonomous navigation in the soft landing process is basically the same regardless of whether there is an atmosphere on the surface of targeted celestial body or not. However, the working environments of IMU, range/velocity measurement sensors, and optical imaging sensors are different in the two cases (with/without atmosphere). These differences need to be considered in the design and development of navigation sensors.

10.2 Autonomous Navigation System for Soft Landing

10.2.1 System Composition and Workflow

The soft landing autonomous navigation system mainly includes IMU (accelerometer and gyro), altimeter, velocimeter, optical imaging sensor, star tracker, and a GNC computer. The system composition is shown in Fig. 10.2. The IMU measures the angle velocity and non-gravitational acceleration of a lander. The altimeter measures

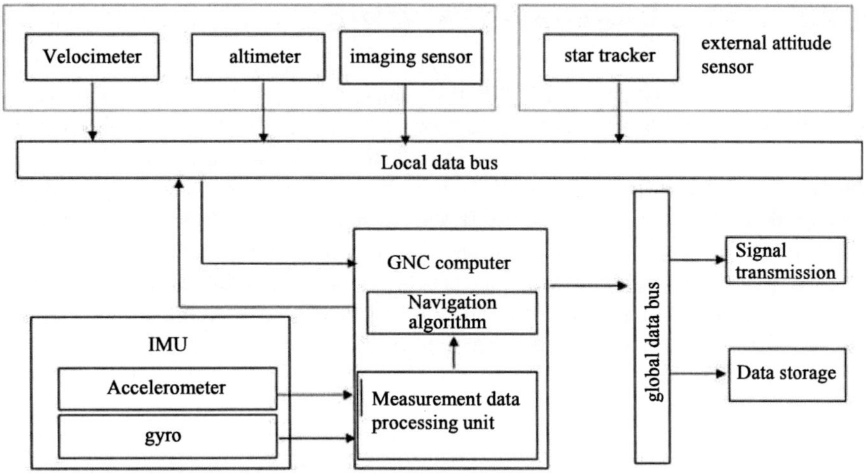

Fig. 10.2 Block diagram of composition of soft landing autonomous navigation system

the slant range to celestial body surface, and the velocimeter measures the velocity of lander relative to celestial body surface. The optical imaging sensor captures images of the celestial body surface. The star tracker provides an initial inertial attitude reference. The GNC computer is responsible for receiving and processing the measurement information from various sensors and performing navigation solutions.

The general workflow of soft landing autonomous navigation system is as follows: The initial inertial attitude is determined by GNC computer using the measurement data from star tracker and gyro, and the initial orbit by ground transmission or onboard propagation. The GNC computer collects measurement information from gyro and accelerometer, and performs data processing and inertial navigation calculation to estimate the attitude, position, and velocity of lander. When using the measured range/velocity information for navigation correction, the GNC computer collects the measurement data from altimeter and velocimeter, and conducts data processing and navigation correction to update the position and velocity of lander. When using image information for navigation correction, the GNC computer collects the surface images of targeted celestial body from optical imaging sensor and carries out image processing and navigation correction to update the position and velocity of lander. During the obstacle recognition, the GNC computer collects the optical surface images or 3D terrain images of the moon from optical imaging sensor, then processes those images, identifies obstacles, and selects a safe landing site.

10.2.2 Inertial Measurement Unit (IMU)

The measurement principle of IMU has been addressed in Chap. 7. The gyroscope package used in soft landing missions generally includes six single-degree-of-freedom (SDOF) gyroscopes, which are used to measure the angle velocities of lander relative to inertial space. Any three of the six gyros can constitute a working mode, independently completing the task of attitude angle velocity measurement. To ensure the redundancy configuration of acceleration measurement information, the IMU is usually equipped with four accelerometers, including three upright and one angled, to ensure that any three accelerometers can constitute a working mode to independently complete the measurement of triaxial gravitational accelerations.

10.2.3 Integrated Altimeter/Velocimeter Sensor

Integrated altimeter/velocimeter sensor (shorten for altimeter/velocimeter) is one of the key navigation sensors in the soft landing process. It can directly measure the lander's LOS velocity and distance relative to celestial body surface. Its measurement data can be used to correct the inertial navigation results and eliminate the influence of initial orbit determination error and IMU measurement error, so as to improve the accuracy of autonomous navigation and ensure the landing safety.

Note that the measurement ranges for the lander's distance and velocity relative to celestial body surface are relatively large during the soft landing process, and the requirements for the measurement ranges and accuracies of altimeter/velocimeter are usually demanding. Among them, the range measurement is generally required to be 0–30 km, with the highest accuracy of 0.1 m, while the velocity measurement required to be 0 m/s–1 km/s, with the highest accuracy of 0.2 m/s.

The attitude variation range is very large during soft landing. To ensure that the measuring beams of altimeter/velocimeter can hit the celestial body surface, an appropriate way of beam installation shall be determined for soft landing. Suppose that the lander is equipped with four measuring beams, denoted as L_1–L_4, each of which can work independently. To ensure that at least three beams can hit the celestial body surface, one usable way to install them has been developed, as shown below. L_4 is along the X-axis (the longitudinal axis of lander), and L_1 along the Z-axis of lander. The projections of L_2 and L_3 in the XOZ plane are 45° to X-axis and Z-axis, the projection of L_2 in the YOZ plane is 45° to Y-axis and Z-axis, and the projection of L_3 in the YOZ plane is 45° to Y-axis and Z-axis (Figs. 10.3 and 10.4).

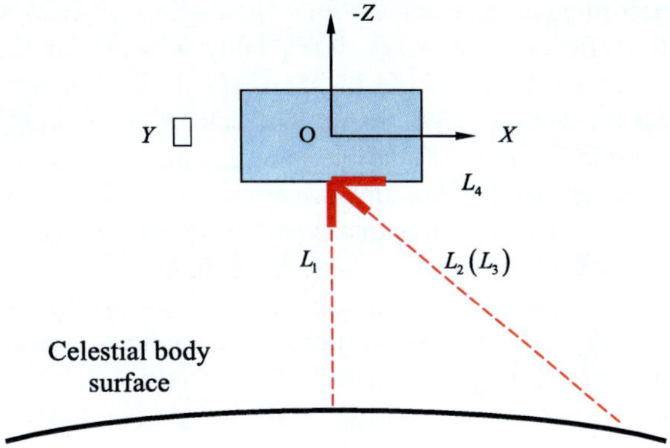

Fig. 10.3 Beam pointing at the beginning of landing

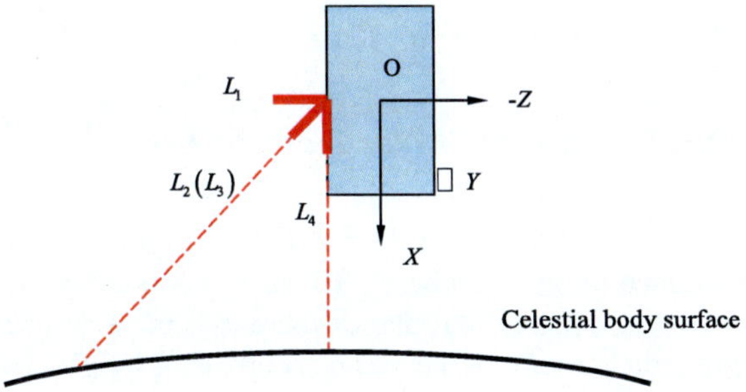

Fig. 10.4 Beam pointing during vertical descent

10.2.4 Optical Imaging Sensor

Optical imaging sensor is used to image the landing areas. The obtained images can be used for terrain-relative navigation to detect and track feature points and for safe landing site selection to avoid obstacle. Optical imaging sensors can be divided into 3D active imaging sensors and optical cameras. 3D active imaging sensors can work in two ways, e.g., laser 3D imaging and microwave 3D imaging.

Laser 3D imaging sensors are working under two different mechanisms: scanning and non-scanning. The scanning sensor first acquires the altitude information of each point in the FOV through point-by-point measurement and then splices and processes the acquired data. The non-scanning sensor illuminates the whole target once with laser and measures the intensity and distance information of each pixel on the target once. Compared with the scanning mechanism, the non-scanning mechanism has a faster imaging speed and higher measurement accuracy.

Microwave 3D imaging sensors are also divided into scanning type and non-scanning type in terms of working mechanism. Scanning-type microwave 3D imaging is similar to scanning-type laser 3D imaging. The non-scanning type is to obtain the average altitude and coarseness of celestial body surface in each imaging unit according to the power–time distribution of echo signals and then acquire the topographic features of celestial body surface from these two parameters.

Typical technical specifications of a 3D active imaging sensor are: Range measurement accuracy is less than 0.1 m, data sample period is less than 0.1 s, the FOV is large than 30° × 30°, the operating distance range is 30–500 m, and planar resolution is better than 0.2 m × 0.2 m.

The advantage of optical cameras is their fast imaging speed, which can fall in two types: CCD and CMOS APS. According to the lander's altitude relative to celestial body and the sunlight condition, the optical camera can recover the shade and shape of optical images to detect obstacles. Typical technical specifications of an optical camera are as follows: The FOV is greater than 30° × 30°, the planar resolution is better than 0.1 m × 0.1 m, and the imaging period is less than 10 ms.

10.3 Measurement Equation

10.3.1 Ranging Measurement Equation

Define a landing frame $\mathcal{F}_l$, where the origin o_l is the center of the celestial body (assumed to be moon here). The axis $o_l x_l$ points to the powered descent point, and $o_l - x_l y_l$ is the orbital plane where $o_l y_l$ is perpendicular to $o_l x_l$ and points to the direction of motion. The axis $o_l z_l$ completes a right-handed frame. The measurement data of the altimeter is the LOS distance of the measuring beam pointing to the lunar surface, which needs to be converted into the lander's altitude relative to the lunar surface before for navigation correction. The measurement equation can be written as

$$\tilde{z}_l = \|\boldsymbol{r}\| - R_m + \nu_l \tag{10.1}$$

where $\boldsymbol{r}$ is the position vector of the lander, R_m is the reference radius of the moon, and ν_l is ranging noise.

Define the filter state $\boldsymbol{x}$ as $\boldsymbol{x} = [\boldsymbol{r}^T, \boldsymbol{v}^T]^T$. Then, the measurement sensitivity matrix corresponding to Eq. (10.1) can be written as

$$\begin{aligned} \boldsymbol{H}_l &= \begin{bmatrix} \frac{\partial \tilde{z}_l}{\partial \boldsymbol{r}} & \frac{\partial \tilde{z}_l}{\partial \boldsymbol{v}} \end{bmatrix} \\ &= \begin{bmatrix} \frac{\boldsymbol{r}^T}{r} & \boldsymbol{0}_{1\times 3} \end{bmatrix} \end{aligned} \tag{10.2}$$

Equations (10.1) and (10.2) can be rewritten as

$$\begin{aligned} \tilde{z}_l &= \sqrt{x^2 + y^2 + z^2} - R_m + \nu_l \\ \boldsymbol{H}_l &= \begin{bmatrix} x/r & y/r & z/r & \boldsymbol{0}_{1\times 3} \end{bmatrix} \end{aligned} \tag{10.3}$$

where x, y, z are the three elements of $\boldsymbol{r}$ expressed in $\mathcal{F}_{\mathrm{l}}$.

10.3.2 Velocity Measurement Equation

The velocimeter can measure the slant rate along each beam. Denote $\tilde{v}_{p,(i)}$ as the slant of the ith beam. Then, the measurement equation can be written as

$$\tilde{v}_{p,(i)} = \left(\boldsymbol{p}_{(i)}^{\mathrm{b}}\right)^T \boldsymbol{v}_{\mathrm{gb}}^{\mathrm{b}} + v_v \tag{10.4}$$

where $\boldsymbol{p}^{\mathrm{b}}$ is beam's direction vector, subscript (i) indicates a quantity along the ith beam, and $\boldsymbol{v}_{\mathrm{gb}}^{\mathrm{b}} = \boldsymbol{C}_{\mathrm{l}}^{\mathrm{b}}[\dot{x}, \dot{y}, \dot{z}]^T$ is the velocity related to the moon.

If a number of available beams are greater than three, then the velocity related to the moon can be obtained from Eq. (10.4) and denoted as $\tilde{\boldsymbol{v}}_{\mathrm{gb}}^{\mathrm{b}}$. In this case, the measurement $\tilde{\boldsymbol{z}}_{\mathrm{v}}$ is defined as $\tilde{\boldsymbol{v}}_{\mathrm{gb}}^{\mathrm{b}}$, and the measurement equation can be rewritten as

$$\tilde{\boldsymbol{z}}_{\mathrm{v}} = \boldsymbol{C}_{\mathrm{l}}^{\mathrm{b}} \begin{bmatrix} \dot{x} \\ \dot{y} \\ \dot{z} \end{bmatrix} + \boldsymbol{\nu}_{\mathrm{v}} \tag{10.5}$$

where $\boldsymbol{C}_{\mathrm{l}}^{\mathrm{b}}$ is the attitude matrix from $\mathcal{F}_{\mathrm{l}}$ to $\mathcal{F}_{\mathrm{b}}$, and $\boldsymbol{\nu}_{\mathrm{v}}$ is velocity measurement noise. The measurement sensitivity matrix corresponding to Eq. (10.5) can be written as

$$\begin{aligned} \boldsymbol{H} &= \left[\tfrac{\partial \tilde{\boldsymbol{z}}_v}{\partial \boldsymbol{r}} \ \tfrac{\partial \tilde{\boldsymbol{z}}_{\mathrm{v}}}{\partial \boldsymbol{v}} \right] \\ &= \left[\boldsymbol{0}_{3\times 3} \ \boldsymbol{C}_{\mathrm{l}}^{\mathrm{b}} \right] \end{aligned} \tag{10.6}$$

10.3.3 Image-Based Measurement Equation

For optical navigation, Chap. 8 presented measurement equations in the form of LOS vector and pixel, respectively. The measurement equation in the form of LOS vector is

$$\tilde{\boldsymbol{n}}^{\mathrm{s}} = \boldsymbol{C}_{\mathrm{l}}^{\mathrm{s}} \frac{\boldsymbol{R}_{\mathrm{b}}^{\mathrm{l}} - \boldsymbol{r}^{\mathrm{l}}}{\left\| \boldsymbol{R}_{\mathrm{b}}^{\mathrm{l}} - \boldsymbol{r}^{\mathrm{l}} \right\|} + \boldsymbol{\nu} \tag{10.7}$$

where $\boldsymbol{R}_{\mathrm{b}}$ is the position vector of a navigation landmark, $\tilde{\boldsymbol{n}}$ is the LOS vector to the landmark, $\boldsymbol{C}_{\mathrm{l}}^{\mathrm{s}}$ is the matrix of attitude transformation from $\mathcal{F}_{\mathrm{l}}$ to a sensor frame $\mathcal{F}_{\mathrm{s}}$, and $\boldsymbol{\nu}$ is the measurement noise. In the case of a small FOV, the following Gaussian noise approximation can be used [3]

$$\begin{aligned}
&\mathrm{E}\{\boldsymbol{v}\} = \mathbf{0} \\
&\mathrm{E}\{\boldsymbol{v}\boldsymbol{v}^T\} = \sigma^2\Big[\boldsymbol{I}_3 - \big(\boldsymbol{C}_{\mathrm{l}}^{\mathrm{s}}\boldsymbol{n}^{\mathrm{l}}\big)\big(\boldsymbol{C}_{\mathrm{l}}^{\mathrm{s}}\boldsymbol{n}^{\mathrm{l}}\big)^T\Big]
\end{aligned} \tag{10.8}$$

where $\boldsymbol{n}^{\mathrm{l}} = \frac{\boldsymbol{R}_{\mathrm{b}}^{\mathrm{l}} - \boldsymbol{r}^{\mathrm{l}}}{\|\boldsymbol{R}_{\mathrm{b}}^{\mathrm{l}} - \boldsymbol{r}^{\mathrm{l}}\|}$. The measurement sensitivity matrix corresponding to Eq. (10.7) can be written as

$$\begin{aligned}
\boldsymbol{H} &= \begin{bmatrix} \frac{\partial \tilde{\boldsymbol{n}}^{\mathrm{s}}}{\partial \boldsymbol{r}} & \frac{\partial \tilde{\boldsymbol{n}}^{\mathrm{s}}}{\partial \boldsymbol{v}} \end{bmatrix} \\
&= \left[-\frac{1}{R_{\mathrm{br}}}\boldsymbol{C}_{\mathrm{l}}^{\mathrm{s}}\big(\boldsymbol{I}_3 - \boldsymbol{n}^{\mathrm{l}}(\boldsymbol{n}^{\mathrm{l}})^T\big) \quad \mathbf{0}_{3\times 3} \right]
\end{aligned} \tag{10.9}$$

where $R_{\mathrm{br}} = \|\boldsymbol{R}_{\mathrm{b}}^{\mathrm{l}} - \boldsymbol{r}^{\mathrm{l}}\|$.

Denote $\boldsymbol{R}_{\mathrm{b}}^{\mathrm{l}} = [x_d, y_d, z_d]^T$. The measurement equation with the form of pixels can be written as

$$\begin{aligned}
p &= -k_x f \frac{a_{11}(x_d - x) + a_{12}(y_d - y) + a_{13}(z_d - z)}{a_{31}(x_d - x) + a_{32}(y_d - y) + a_{33}(z_d - z)} + p_0 + v_p \\
l &= -k_y f \frac{a_{21}(x_d - x) + a_{22}(y_d - y) + a_{22}(z_d - z)}{a_{31}(x_d - x) + a_{32}(y_d - y) + a_{33}(z_d - z)} + l_0 + v_l
\end{aligned} \tag{10.10}$$

where (p, l) is the projection pixel coordinate of the imaging point on the focal plane; p_0 and l_0 are the image element and image line at the center of navigation sensor; v_p and v_l are measurement noises; $a_{ij}(i = 1, 2, 3;\ j = 1, 2, 3)$ are the corresponding elements in the direction cosine matrix $\boldsymbol{C}_l^s$; k_x and k_y are the conversion from Cartesian coordinate to pixel; and f is focal length.

The measurement sensitivity matrix corresponding to Eq. (10.10) can be written as

$$\boldsymbol{H} = \frac{\partial z}{\partial x} = \begin{bmatrix} \partial p/\partial x \ \partial p/\partial y \ \partial p/\partial z & \mathbf{0}_{2\times 3} \\ \partial l/\partial x \ \partial l/\partial y \ \partial l/\partial z & \end{bmatrix} \tag{10.11}$$

where

$$\begin{aligned}
\partial p/\partial x &= -k_x f \frac{\big(a_{12}a_{31} - a_{11}a_{32}\big)(y_d - y) + \big(a_{13}a_{31} - a_{11}a_{33}\big)(z_d - z)}{\big[a_{31}(x_d - x) + a_{32}(y_d - y) + a_{33}(z_d - z)\big]^2} \\
\partial p/\partial y &= -k_x f \frac{\big(a_{11}a_{32} - a_{12}a_{31}\big)(x_d - x) + \big(a_{13}a_{32} - a_{12}a_{33}\big)(z_d - z)}{\big[a_{31}(x_d - x) + a_{32}(y_d - y) + a_{33}(z_d - z)\big]^2} \\
\partial p/\partial z &= -k_x f \frac{\big(a_{11}a_{33} - a_{13}a_{31}\big)(x_d - x) + \big(a_{12}a_{33} - a_{13}a_{32}\big)(y_d - y)}{\big[a_{31}(x_d - x) + a_{32}(y_d - y) + a_{33}(z_d - z)\big]^2} \\
\partial l/\partial x &= -k_y f \frac{\big(a_{22}a_{31} - a_{21}a_{32}\big)(y_d - y) + \big(a_{23}a_{31} - a_{21}a_{33}\big)(z_d - z)}{\big[a_{31}(x_d - x) + a_{32}(y_d - y) + a_{33}(z_d - z)\big]^2}
\end{aligned}$$

$$\partial l/\partial y = -k_y f \frac{(a_{21}a_{32} - a_{22}a_{31})(x_d - x) + (a_{23}a_{32} - a_{22}a_{33})(z_d - z)}{[a_{31}(x_d - x) + a_{32}(y_d - y) + a_{33}(z_d - z)]^2}$$

$$\partial l/\partial z = -k_y f \frac{(a_{21}a_{33} - a_{23}a_{31})(x_d - x) + (a_{22}a_{33} - a_{23}a_{32})(y_d - y)}{[a_{31}(x_d - x) + a_{32}(y_d - y) + a_{33}(z_d - z)]^2} \quad (10.12)$$

10.4 Observability Analysis

10.4.1 Altimeter-/Velocimeter-Aided Inertial Navigation

Due to the accumulation of initial navigation error and IMU measurement error, the inertial navigation error will gradually increase. To ensure the navigation accuracy, the inertial navigation must be corrected by direct measurement information. This section will analyze the observability of the autonomous navigation system based on IMU + ranging/velocity measurement correction.

(1) Autonomous navigation state equation
The state equation can be written as

$$\dot{\boldsymbol{x}} = \boldsymbol{f}(\boldsymbol{x}) = \begin{bmatrix} \boldsymbol{v} \\ -\frac{\mu_m}{r^3}\boldsymbol{r} - 2\boldsymbol{\omega}_m \times \boldsymbol{v} - \boldsymbol{\omega}_m \times (\boldsymbol{\omega}_m \times \boldsymbol{r}) + \boldsymbol{C}_b^l \boldsymbol{f}^b \end{bmatrix} \quad (10.13)$$

where μ_m is the lunar gravitational constant, $\boldsymbol{\omega}_m$ is the angular velocity of the moon, $\boldsymbol{C}_b^l = (\boldsymbol{C}_l^b)^T$ is the matrix of attitude transformation from $\mathcal{F}_b$ to $\mathcal{F}_l$, and $\boldsymbol{f}^b$ is the non-gravitational acceleration measured by accelerometer.
Define the error state with respect to a reference INS trajectory $\boldsymbol{x}_{ins}$ as $\delta\boldsymbol{x} = \boldsymbol{x} - \boldsymbol{x}_{ins}$, where $\boldsymbol{x}_{ins} = [\boldsymbol{r}_{ins}^T, \boldsymbol{v}_{ins}^T]^T$ and subscript ins indicates a variable derived from the INS. According to Eq. (10.13), the error state equation is given by

$$\frac{d}{dt}(\delta\boldsymbol{x}) = \boldsymbol{A}\delta\boldsymbol{x} \quad (10.14)$$

where

$$\boldsymbol{A} = \left.\frac{\partial \boldsymbol{f}(\boldsymbol{x})}{\partial \boldsymbol{x}}\right|_{\boldsymbol{x}=\boldsymbol{x}_{ins}} = \begin{bmatrix} \boldsymbol{0}_{3\times3} & \boldsymbol{I}_3 \\ \boldsymbol{A}_1 & \boldsymbol{A}_2 \end{bmatrix} \quad (10.15)$$

and $\boldsymbol{A}_1 = \boldsymbol{M} - [\boldsymbol{\omega}_m\times]^2$, $\boldsymbol{A}_2 = -2[\boldsymbol{\omega}_m\times]$, and $\boldsymbol{M}$ is given by

$$
\boldsymbol{M} = -\frac{\mu_{\mathrm{m}}}{r_{\mathrm{ins}}^5}\begin{bmatrix} -2x_{\mathrm{ins}}^2 + y_{\mathrm{ins}}^2 + z_{\mathrm{ins}}^2 & -3x_{\mathrm{ins}}y_{\mathrm{ins}} & -3x_{\mathrm{ins}}z_{\mathrm{ins}} \\ -3x_{\mathrm{ins}}y_{\mathrm{ins}} & x_{\mathrm{ins}}^2 - 2y_{\mathrm{ins}}^2 + z_{\mathrm{ins}}^2 & -3z_{\mathrm{ins}}y_{\mathrm{ins}} \\ -3x_{\mathrm{ins}}z_{\mathrm{ins}} & -3z_{\mathrm{ins}}y_{\mathrm{ins}} & x_{\mathrm{ins}}^2 + y_{\mathrm{ins}}^2 - 2z_{\mathrm{ins}}^2 \end{bmatrix} \tag{10.16}
$$

with x_{ins}, y_{ins} and z_{ins} the three elements of $\boldsymbol{r}_{\mathrm{ins}}$.

(2) Ranging correction

For ranging correction, $\boldsymbol{H}_1$ in Eq. (10.3) can be rewritten as:

$$
\boldsymbol{H}_1 = \begin{bmatrix} \boldsymbol{B} & \boldsymbol{0}_{1\times 3} \end{bmatrix} \tag{10.17}
$$

where $\boldsymbol{B} = [x/r \;\; y/r \;\; z/r]$. Thus, the observability matrix (4.27) can be written as:

$$
\boldsymbol{Q} = \begin{bmatrix} \boldsymbol{B} & \boldsymbol{0}_{1\times 3} \\ \boldsymbol{0}_{1\times 3} & \boldsymbol{B} \\ \boldsymbol{B}\boldsymbol{A}_1 & \boldsymbol{B}\boldsymbol{A}_2 \\ \boldsymbol{B}\boldsymbol{A}_2\boldsymbol{A}_1 & \boldsymbol{B}(\boldsymbol{A}_1 + \boldsymbol{A}_2^2) \\ \boldsymbol{B}(\boldsymbol{A}_1^2 + \boldsymbol{A}_2^2\boldsymbol{A}_1) & \boldsymbol{B}(\boldsymbol{A}_2\boldsymbol{A}_1 + \boldsymbol{A}_1\boldsymbol{A}_2 + \boldsymbol{A}_2^3) \\ \boldsymbol{B}(\boldsymbol{A}_2\boldsymbol{A}_1 + \boldsymbol{A}_1\boldsymbol{A}_2 + \boldsymbol{A}_2^3)\boldsymbol{A}_1 & \boldsymbol{B}(\boldsymbol{A}_1^2 + \boldsymbol{A}_2^2\boldsymbol{A}_1 + \boldsymbol{A}_2\boldsymbol{A}_1\boldsymbol{A}_2 + \boldsymbol{A}_1\boldsymbol{A}_2^2 + \boldsymbol{A}_2^4) \end{bmatrix} \tag{10.18}
$$

The rank of $\boldsymbol{Q}$ is difficult to evaluate. To simplify this problem, neglecting the moon's rotation gives $\boldsymbol{A}_2 = \boldsymbol{0}_{3\times 3}$ and $\boldsymbol{A}_1 = \boldsymbol{M}$. Then, Eq. (10.18) can be simplified as

$$
\boldsymbol{Q} = \begin{bmatrix} \boldsymbol{B} & \boldsymbol{0}_{1\times 3} \\ \boldsymbol{0}_{1\times 3} & \boldsymbol{B} \\ \boldsymbol{B}\boldsymbol{M} & \boldsymbol{0}_{1\times 3} \\ \boldsymbol{0}_{1\times 3} & \boldsymbol{B}\boldsymbol{M} \\ \boldsymbol{B}\boldsymbol{M}^2 & \boldsymbol{0}_{1\times 3} \\ \boldsymbol{0}_{1\times 3} & \boldsymbol{B}\boldsymbol{M}^2 \end{bmatrix} \tag{10.19}
$$

Because $\det(\boldsymbol{M}) > 0$, the rank of $\boldsymbol{Q}$ can be obtained as

$$
\begin{aligned}
\operatorname{rank}(\boldsymbol{Q}) &= \operatorname{rank}\left(\begin{bmatrix} \boldsymbol{B} & \boldsymbol{0}_{1\times 3} \\ \boldsymbol{0}_{1\times 3} & \boldsymbol{B} \\ \boldsymbol{B}\boldsymbol{M} & \boldsymbol{0}_{1\times 3} \\ \boldsymbol{0}_{1\times 3} & \boldsymbol{B}\boldsymbol{M} \\ \boldsymbol{B}\boldsymbol{M}^2 & \boldsymbol{0}_{1\times 3} \\ \boldsymbol{0}_{1\times 3} & \boldsymbol{B}\boldsymbol{M}^2 \end{bmatrix}\right) \\
&= \operatorname{rank}\left(\begin{bmatrix} \boldsymbol{B} & \boldsymbol{0}_{1\times 3} \\ \boldsymbol{0}_{1\times 3} & \boldsymbol{B} \end{bmatrix}\right) = 2
\end{aligned} \tag{10.20}
$$

From the above equation, we see that the range information can only provide the information on two orbital parameters if the influence of the moon rotation is neglecting. According to the measurement equation, range is directly related to the position, so the measuring distance changes with position directly. Therefore, range information is the major contributor to the position correction and the minor contributor to velocity error correction. Since the observation variable used in the navigation algorithm is radial altitude (pointing to the moon center), the navigation correction algorithm that only introduces altitude is insensitive to the position error outside the orbit plane, which, therefore, cannot be corrected. This conclusion will be verified in the later simulation results.

(3) Velocity measurement correction
Since three or more than three beams are available during the descent phase, Eq. (10.5) can be used for observability. According to Chap. 4, the rank of the observability matrix $\boldsymbol{Q}$ for velocity measurement correction is

$$\begin{aligned}\operatorname{rank}(\boldsymbol{Q}) &= \operatorname{rank}\left(\begin{bmatrix} \begin{bmatrix}\mathbf{0}_{3\times3} & \boldsymbol{C}_{\mathrm{l}}^{\mathrm{b}}\end{bmatrix} \\ \begin{bmatrix}\mathbf{0}_{3\times3} & \boldsymbol{C}_{\mathrm{l}}^{\mathrm{b}}\end{bmatrix}\begin{bmatrix}\mathbf{0}_{3\times3} & \boldsymbol{I}_3 \\ \boldsymbol{M}-[\boldsymbol{\omega}_{\mathrm{m}}\times]^2 & -2[\boldsymbol{\omega}_{\mathrm{m}}\times]\end{bmatrix}\end{bmatrix}\right) \\ &= \operatorname{rank}\left(\boldsymbol{C}_{\mathrm{l}}^{\mathrm{b}}\begin{bmatrix}\mathbf{0}_{3\times3} & \boldsymbol{I}_3 \\ \boldsymbol{M}-[\boldsymbol{\omega}_{\mathrm{m}}\times]^2 & -2[\boldsymbol{\omega}_{\mathrm{m}}\times]\end{bmatrix}\right)\end{aligned} \quad (10.21)$$

Since the order of magnitude of the elements in the matrix $\boldsymbol{M}$ is 10^{-6} and that of the elements in the matrix $[\boldsymbol{\omega}_{\mathrm{m}}\times]^2$ is 10^{-12}, the elements in $[\boldsymbol{\omega}_{\mathrm{m}}\times]^2$ are small relative to those in $\boldsymbol{M}$. Accordingly, $\boldsymbol{M}-[\boldsymbol{\omega}_{\mathrm{m}}\times]^2$ is non-singular. Therefore, $\boldsymbol{Q}$ is non-singular, implying that the system is observable.

(1) Qualitative analysis: It can be known from state equation, under the same initial velocity condition, the velocity change will vary with initial position, leading to different velocities. Therefore, the velocity measurement information can correct the position error in inertial navigation system.
(2) Quantitative analysis: According to state equation, the acceleration terms related to position are gravitational acceleration and the acceleration caused by lunar rotation $(-\frac{\mu_{\mathrm{m}}}{r^3}\boldsymbol{r}-\boldsymbol{\omega}_{\mathrm{m}}\times(\boldsymbol{\omega}_{\mathrm{m}}\times\boldsymbol{r}))$. Then, $d\big(-\frac{\mu_{\mathrm{m}}}{r^3}\boldsymbol{r}-\boldsymbol{\omega}_{\mathrm{m}}\times(\boldsymbol{\omega}_{\mathrm{m}}\times\boldsymbol{r})\big)=\big(\boldsymbol{M}-[\boldsymbol{\omega}_{\mathrm{m}}\times]^2\big)d\boldsymbol{r}$, where the order of magnitude of the elements in $\boldsymbol{M}$ is 10^{-6} and that of the elements in $[\boldsymbol{\omega}\times]^2$ is 10^{-12}. Therefore, the velocity change caused by the lander position change within a certain range is very small; that is, the velocity change brought about by the position error within a limited range is very small. As a result, when using the navigation correction algorithm based on velocity measurement, the position correction ability is limited. This conclusion will be verified in the later simulation results.

Based on the above analysis, it can be concluded that the position error cannot be well corrected only by velocity measurement, so range information should be introduced.

Remark If altitude and 3D velocity are corrected by ranging and velocity measurement, the navigation system is absolutely observable because the navigation measurements can be directly converted to the navigation state variables to be estimated.

10.4.2 Optical-Aided Inertial Navigation

Chapter 8 analyzed the geometric observability of optical autonomous navigation based on LOS information. The analysis shows that at least two unparallel LOS vectors are required to achieve the system observability. However, this conclusion does not consider the effect of orbital dynamics. In this chapter, the dynamic observability of optical autonomous navigation based on LOS information will be analyzed by considering the effect of orbital dynamics. Neglecting effects of moon rotation, $\boldsymbol{A}$ in Eq. (10.15) can be simplified as

$$\boldsymbol{A} = \left.\frac{\partial \boldsymbol{f}(\boldsymbol{x})}{\partial \boldsymbol{x}}\right|_{\boldsymbol{x}=\boldsymbol{x}_{\text{ins}}} = \begin{bmatrix} \boldsymbol{0}_{3\times3} & \boldsymbol{I}_3 \\ \boldsymbol{M} & \boldsymbol{0}_{3\times3} \end{bmatrix} \tag{10.22}$$

where $\boldsymbol{M}$ is given by Eq. (10.16). The measurement sensitivity matrix with only one LOS direction is given by Eq. (10.9). Therefore, the observability matrix can be written as

$$\boldsymbol{Q} = \begin{bmatrix} \boldsymbol{H} \\ \boldsymbol{HA} \\ \vdots \\ \boldsymbol{HA}^{n-1} \end{bmatrix} = \begin{bmatrix} \boldsymbol{C}_o & \boldsymbol{0} \\ \boldsymbol{0} & \boldsymbol{C}_o \\ \boldsymbol{C}_o\boldsymbol{M} & \boldsymbol{0} \\ \boldsymbol{0} & \boldsymbol{C}_o\boldsymbol{M} \\ \boldsymbol{C}_o\boldsymbol{M}^2 & \boldsymbol{0} \\ \boldsymbol{0} & \boldsymbol{C}_o\boldsymbol{M}^2 \end{bmatrix} \tag{10.23}$$

Since row exchange does not change the matrix rank,

$$\text{rank}(\boldsymbol{Q}) = \text{rank}\left(\begin{bmatrix} \boldsymbol{C}_o & \boldsymbol{0} \\ \boldsymbol{C}_o\boldsymbol{M} & \boldsymbol{0} \\ \boldsymbol{C}_o\boldsymbol{M}^2 & \boldsymbol{0} \\ \boldsymbol{0} & \boldsymbol{C}_o \\ \boldsymbol{0} & \boldsymbol{C}_o\boldsymbol{M} \\ \boldsymbol{0} & \boldsymbol{C}_o\boldsymbol{M}^2 \end{bmatrix}\right) \tag{10.24}$$

and according to the properties of a partitioned matrix, we have

$$\mathrm{rank}(\boldsymbol{Q}) = 2\mathrm{rank}(\boldsymbol{O}_{o1}) \tag{10.25}$$

where

$$\boldsymbol{O}_{o1} = \begin{bmatrix} \boldsymbol{C}_o \\ \boldsymbol{C}_o\boldsymbol{M} \\ \boldsymbol{C}_o\boldsymbol{M}^2 \end{bmatrix} \tag{10.26}$$

If $\mathrm{rank}(\boldsymbol{O}_{o1}) = 3$, the system is observable. Note that the rank of the matrix $\boldsymbol{C} \in \mathbb{R}^{q\times n}$ $(q \geq n)$ is $n - m$, where m is the maximum number of orthogonal vectors satisfying the equation $\boldsymbol{Cy} = \boldsymbol{0}$ [4]. If a vector $\boldsymbol{y}$ exists such that $\boldsymbol{O}_{o1}\boldsymbol{y} = \boldsymbol{0}$, then $\mathrm{rank}(\boldsymbol{O}_{o1}) = 2$; otherwise, $\mathrm{rank}(\boldsymbol{O}_{o1}) = 3$. From $O_{o1}\boldsymbol{y} = \boldsymbol{0}$, the following equations can be derived

$$\begin{aligned} &\boldsymbol{C}_o\boldsymbol{y} = \boldsymbol{0} \\ &\boldsymbol{C}_o\boldsymbol{G}\boldsymbol{y} = \boldsymbol{0} \\ &\boldsymbol{C}_o\boldsymbol{G}^2\boldsymbol{y} = \boldsymbol{0} \end{aligned} \tag{10.27}$$

The solutions of the above three equations are

$$\begin{aligned} &\boldsymbol{y} = c_1\boldsymbol{n}^{\mathrm{l}} \\ &\boldsymbol{y} = c_2\boldsymbol{M}^{-1}\boldsymbol{n}^{\mathrm{l}} \\ &\boldsymbol{y} = c_3\boldsymbol{M}^{-2}\boldsymbol{n}^{\mathrm{l}} \end{aligned} \tag{10.28}$$

where c_1, c_2, c_3 are nonzero constants. Therefore, the system is unobservable ($\mathrm{rank}(\boldsymbol{O}_{o1}) = 2$) when

$$c_1\boldsymbol{n}^l = c_2\boldsymbol{M}^{-1}\boldsymbol{n}^l = c_3\boldsymbol{M}^{-2}\boldsymbol{n}^l \tag{10.29}$$

In other words, the system is observable when $\boldsymbol{n}^l$, $\boldsymbol{Mn}^l$, and $\boldsymbol{M}^2\boldsymbol{n}^l$ are linearly independent. Because $\det(\boldsymbol{M}) \neq 0$, only the correlation between $\boldsymbol{n}^l$ and $\boldsymbol{Mn}^l$ needs to be considered. Let $\boldsymbol{n}^l = [n_x^l, n_y^l, n_z^l]^T$. Then,

$$\begin{aligned} \boldsymbol{D} &= [\boldsymbol{n}^l, \boldsymbol{Mn}^l] \\ &= \begin{bmatrix} n_x^l & \frac{\mu_m(3x^2-r^2)n_x^l}{r^5} + \frac{3\mu_m xyn_y^l}{r^5} + \frac{3\mu_m xzn_z^l}{r^5} \\ n_y^l & \frac{\mu_m(3y^2-r^2)n_y^l}{r^5} + \frac{3\mu_m xyn_x^l}{r^5} + \frac{3\mu_m yzn_z^l}{r^5} \\ n_z^l & \frac{\mu_m(3z^2-r^2)n_z^l}{r^5} + \frac{3\mu_m xzn_x^l}{r^5} + \frac{3\mu_m yzn_y^l}{r^5} \end{bmatrix} \end{aligned} \tag{10.30}$$

In the above equation, subscripts ins of x, y, z, r are omitted for writing convenience. The elementary column transformation of $\boldsymbol{D}$ is

$$D \rightarrow \begin{bmatrix} n_x^l \, x(xn_x^l + yn_y^l + zn_z^l) \\ n_y^l \, y(xn_x^l + yn_y^l + zn_z^l) \\ n_z^l \, z(xn_x^l + yn_y^l + zn_z^l) \end{bmatrix} \tag{10.31}$$

Obviously, in the case of $\boldsymbol{r}^T\boldsymbol{n}^l = 0$ ($\boldsymbol{r}$ is perpendicular to $\boldsymbol{n}$), $\boldsymbol{n}^l$ and $\boldsymbol{M}\boldsymbol{n}^l$ are linearly dependent and the system is not observable. In addition, if $\boldsymbol{r}^T\boldsymbol{n}^l \neq 0$, a further transformation of $\boldsymbol{D}$ can obtain

$$D \rightarrow \begin{bmatrix} n_x^l \, x \\ n_y^l \, y \\ n_z^l \, z \end{bmatrix} \tag{10.32}$$

It shows that when $\boldsymbol{r}$ and $\boldsymbol{n}^l$ are parallel, the system is unobservable. Therefore, when the lander position vector is neither parallel with nor perpendicular to the LOS direction, the system is observable. In a soft landing scenario, the lander position vector is usually not perpendicular to the LOS direction of a surface landmark. If they are parallel, the lander is descending vertically, and the observed landmark is always below it. Therefore, we can conclude that for the optical-aided inertial autonomous navigation of the soft landing process, the system is unobservable only when the lander descends vertically and the observed landmark is always below it. In other cases, the system is observable.

10.5 Fusion Navigation Method

10.5.1 IMU + Ranging/Velocity Measurement for 3D Position and Velocity Correction

Denote $\delta\hat{\boldsymbol{x}}$ as the estimate of $\delta\boldsymbol{x}$ and $\hat{\boldsymbol{x}}$ as the estimate of $\boldsymbol{x}$. Then, the filter update equation can be written as

$$\begin{cases} \delta\hat{\boldsymbol{x}} = \boldsymbol{K}(\tilde{z} - \hat{z}) \\ \hat{\boldsymbol{x}} = \boldsymbol{x}_{\text{ins}} + \delta\hat{\boldsymbol{x}} \end{cases} \tag{10.33}$$

For ranging correction,

$$\begin{aligned} \tilde{z} &= \tilde{z}_l \\ \hat{z} &= \sqrt{x_{\text{ins}}^2 + y_{\text{ins}}^2 + z_{\text{ins}}^2} - R_m \end{aligned} \tag{10.34}$$

where $\tilde{z}_l$ is the range measurement from Eq. (10.1). For velocity correction, here we assume that the 3D velocity related to the moon can be obtained from velocimeter.

Then,

$$\tilde{z} = \tilde{z}_v$$
$$\hat{z} = \boldsymbol{C}_l^b \begin{bmatrix} \dot{x}_{\text{ins}} \\ \dot{y}_{\text{ins}} \\ \dot{z}_{\text{ins}} \end{bmatrix} \tag{10.35}$$

where $\tilde{z}_v$ is the measured 3D velocity derived from Eq. (10.5). For ranging and velocity correction,

$$\tilde{z} = \left[\tilde{z}_l, \tilde{z}_v^T\right]^T$$
$$\hat{z} = \left[\sqrt{x_{\text{ins}}^2 + y_{\text{ins}}^2 + z_{\text{ins}}^2} - R_m, \left(\boldsymbol{C}_l^b \begin{bmatrix} \dot{x}_{\text{ins}} \\ \dot{y}_{\text{ins}} \\ \dot{z}_{\text{ins}} \end{bmatrix}\right)^T\right]^T \tag{10-3}$$

10.5.2 IMU + Ranging/Velocity Measurement for Altitude and Velocity Correction

The navigation algorithm using IMU + ranging/velocity measurement for altitude and velocity correction is adopted by Chang'e-3 [5, 6]. The algorithm only corrects the lander's altitude and 3D velocity.

1. Altitude correction

For altitude correction, the state equation and measurement equation can be written as

$$\begin{bmatrix} \dot{h} \\ \dot{v}_{\text{ver}} \end{bmatrix} = \begin{bmatrix} 0 & 1 \\ 0 & 0 \end{bmatrix} \begin{bmatrix} h \\ v_{\text{ver}} \end{bmatrix} + \begin{bmatrix} 0 \\ 1 \end{bmatrix} (f_{\text{ver}} + g)$$
$$\tilde{h} = \begin{bmatrix} 1 & 0 \end{bmatrix} \begin{bmatrix} h \\ v_{\text{ver}} \end{bmatrix} + \nu_h \tag{10.37}$$

where subscript ver represents vertical component of a quantity, ν_h is the altitude calculation error of the altimeter, and $\tilde{h}$ is the altitude derived by the altimeter.

Equation (10.37) is a LTI system, and the correction for the Kalman filter is given by

$$\begin{bmatrix} \delta h \\ \delta v_{\text{ver}} \end{bmatrix} = \boldsymbol{K}_h(\tilde{h} - h_{\text{ins}}) \tag{10.38}$$

where $\boldsymbol{K}_h$ is the Kalman gain. Once altitude is updated, the inertial position update equation can be written as

$$\boldsymbol{r}_{\text{ins}} = \boldsymbol{r}_{\text{ins}} + \frac{\boldsymbol{r}_{\text{ins}}}{r_{\text{ins}}}\delta h \tag{10.39}$$

For a m-beam altimeter, Eq. (10.39) can be rewritten as

$$\boldsymbol{r}_{\text{ins}} = \boldsymbol{r}_{\text{ins}} + \frac{\boldsymbol{r}_{\text{ins}}}{r_{\text{ins}}}\sum_{i}^{m} \beta_i \delta h_{(i)} \tag{10.40}$$

where $\delta h_{(i)}$ is the altitude correction of the ith beam and β_i is information distribution factor (refer to the knowledge on federated filtering in Chap. 3), determined by the trace of the error covariance and satisfying $\sum_i^m \beta_i = 1$.

2. Velocity correction

Consider velocity correction along the ith beam. The filter state is chosen as the velocity along the beam. Then, the state and measurement equations can be written as

$$\begin{aligned} \dot{v}_{p,(i)} &\approx 0 \cdot v_{p,(i)} + w_{vp} \\ \tilde{v}_{p,(i)} &= v_{p,(i)} + \nu_{v(i)} \end{aligned} \tag{10.41}$$

where w_{vp} is the system noise and $\nu_{v(i)}$ is the velocity measurement error along the beam direction. The velocity correction for the Kalman filter can be written as

$$\delta v_{p,(i)} = K_{v,(i)}(\tilde{v}_{p,(i)} - v_{p,(i),\text{ins}}) \tag{10.42}$$

where $v_{p,,(i)\text{ins}}$ is the projection (in ith beam direction) of the velocity calculated by INS and $K_{v,(i)}$ is the Kalman gain.

In actual landing process, the number of available velocity measurement beams is not fixed. According to Eqs. (10.4) and (10.42), the following equation can be derived

$$\begin{bmatrix} \delta v_{p,(1)} \\ \vdots \\ \delta v_{p,(n)} \end{bmatrix} = \boldsymbol{H}\delta \boldsymbol{v}_{\text{gb}}^{\text{b}} + \boldsymbol{\nu}_{v,1\sim n} \tag{10.43}$$

where n is the number of available beams, $\boldsymbol{\nu}_{v,1\sim n}$ is the measurement noise vector, and $\boldsymbol{H} = \left[\boldsymbol{p}_{(1)}^{\text{b}}, \boldsymbol{p}_{(2)}^{\text{b}}, \ldots, \boldsymbol{p}_{(n)}^{\text{b}}\right]^T$.

Depending on the number of available beams, the following correction strategy is used [5–7].

(1) If $n = 0$, no correction is made.
(2) If $n = 1$, only the velocity along the beam direction is corrected:

$$\delta \boldsymbol{v}_{\mathrm{gb}}^{\mathrm{b}} = \delta v_p \boldsymbol{p}^{\mathrm{b}} \tag{10.44}$$

where $\boldsymbol{p}^{\mathrm{b}}$ is the direction vector of the available beam expressed in the body frame.

(3) If $n = 2$, only the velocity projected in the plane formed by these two beams is corrected:

Firstly, a new body frame $\mathcal{F}_{\mathrm{b1}}$ is constructed such that the two available beams on its $x - y$ plane, x_{b1} is coincided with the pointing direction of the first beam y_{b1} is orthogonal with x_{b1} and z_{b_1} is selected to form a right-handed a coordinate system. plane. Define

$$\boldsymbol{H}_{\mathrm{b1}} = \boldsymbol{H}\boldsymbol{C}_{\mathrm{b1}}^{\mathrm{b}} \tag{10.45}$$

Obviously, $\boldsymbol{H}_{\mathrm{b1}}$ can be written as $\boldsymbol{H}_{\mathrm{b1}} = [\boldsymbol{H}_{\mathrm{b1,1}}, \boldsymbol{0}_{2\times 1}]$ where $\boldsymbol{H}_{\mathrm{b1,1}} \in \mathbb{R}^{2\times 2}$. According to Eq. (10.43), the velocity correction in $\mathcal{F}_{\mathrm{b1}}$, denoted as $\delta \boldsymbol{v}_{gb}^{b}$, is given by

$$\delta \boldsymbol{v}_{\mathrm{gb}}^{\mathrm{b1}} = \begin{bmatrix} \left(\boldsymbol{H}_{\mathrm{b1,1}}^T \boldsymbol{H}_{\mathrm{b1,1}}\right)^{-1} \boldsymbol{H}_{\mathrm{b1,1}}^T \begin{bmatrix} \delta v_{p,(1)} \\ \delta v_{p,(2)} \end{bmatrix} \\ 0 \end{bmatrix} \tag{10.46}$$

and $\delta \boldsymbol{v}_{\mathrm{gb}}^{\mathrm{b}} = \boldsymbol{C}_{\mathrm{b1}}^{\mathrm{b}} \delta \boldsymbol{v}_{\mathrm{gb}}^{\mathrm{b1}}$.

(4) If $n \geq 3$, the 3D velocity correction is obtained by a least square algorithm:

$$\delta \boldsymbol{v}_{\mathrm{gb}}^{\mathrm{b}} = (\boldsymbol{H}^T \boldsymbol{H})^{-1} \boldsymbol{H}^T \begin{bmatrix} \delta v_{p,(1)} \\ \vdots \\ \delta v_{p,(n)} \end{bmatrix} \tag{10.47}$$

Once $\delta \boldsymbol{v}_{\mathrm{gb}}$ is obtained, the inertial velocity can be corrected as

$$\boldsymbol{v}_{\mathrm{ins}} = \boldsymbol{v}_{\mathrm{ins}} + \boldsymbol{C}_{\mathrm{b}}^{i} \delta \boldsymbol{v}_{\mathrm{gb}}^{\mathrm{b}} \tag{10.48}$$

It should be noted that Chang'e-4 also uses IMU + ranging/velocity measurement to correct its altitude and velocity. But the landing site of Chang'e-4 is on the far side of the moon, rather than in the Sinus Iridum on the near side of the moon, where the Chang'e-3 landed. This site is a relatively rugged terrain dominated by highlands and spread with impact craters and ringlike mountains, with few large lunar seas. Such complex terrain conditions bring new challenges to the navigation during landing. To solve the problem of sensor data exception caused by rugged lunar terrain, the Chang'e-4 has established a fault-tolerant computing framework for intelligent autonomous navigation on the basis of the navigation algorithms of Chang'e-3, thereby realizing the rapid screening and isolation of abnormal measurements and the robust fusion of multi-beam measurement information and significantly improving the reliability of landing navigation. Interested readers may refer to [8].

10.5.3 IMU + Imaging for 3D Position and Velocity Correction

Note that optical image information can better determine the lander position relative to the landing site so as to not only improve the navigation accuracy but also autonomously identify the obstacles, and this section studies the fusion navigation method using IMU + imaging for 3D position and velocity correction. Before discussing the navigation filtering, image recognition and terrain matching are introduced, which consists of three steps: construction of landing topography map, image matching, and feature point tracking.

The topographic map of the areas where the lander passes through and lands is constructed by using the images and orbital information captured on the fly-around orbit. Then, by matching the images stored on the lander with those taken by optical imaging sensor, some corresponding feature points can be acquired. From them, certain feature points can be selected. With the aid of topographic map, their position can be determined. In order to quickly obtain the feature point information, the feature points should be tracked after being identified. Once the feature points overflow, the images and terrain are matched again.

After completing the image recognition and terrain matching, navigation filter can be used for correction. The correction equation is given by Eq. (10.33), where $\tilde{z} = [p, l]^T$ is the measured landmark pixel; $\hat{z} = [\hat{p}, \hat{l}]^T$ is the predicted landmark pixel given by the following equation

$$\begin{aligned}\hat{p} &= -k_x f \frac{a_{11}(x_d - \hat{x}) + a_{12}(y_d - \hat{y}) + a_{13}(z_d - \hat{z})}{a_{31}(x_d - \hat{x}) + a_{32}(y_d - \hat{y}) + a_{33}(z_d - \hat{z})} \\ \hat{l} &= -k_y f \frac{a_{21}(x_d - \hat{x}) + a_{22}(y_d - \hat{y}) + a_{23}(z_d - \hat{z})}{a_{31}(x_d - \hat{x}) + a_{32}(y_d - \hat{y}) + a_{33}(z_d - \hat{z})}\end{aligned} \tag{10.50}$$

The implementation process of the correction algorithm based on image feature points is as follows: First, calculate the pixels corresponding to feature points: (1) Calculate the direction vectors of feature points relative to the lander by using the position information of feature points and the predicted lander position; (2) rotate the relative direction vectors to sensor coordinate system by using the attitude information of observation epoch and the installation matrix of optical imaging sensor; (3)

project the sensor parameters to the focal plane of camera; (4) convert the focal plane projection into image elements and lines, and calculate the pixels corresponding to feature points. Then, input the feature point pixels obtained by image matching and the calculated feature point pixels into the navigation filter to correct and update the lander's orbit.

10.6 Simulation Application Examples

10.6.1 Lunar Soft Landing Navigation

In this section, consider a lunar soft landing navigation scenario. The initial navigation errors are as follows: (1) position error: 1 km (altitude) and (2) velocity errors: 0.1 m/s (in the descending direction), 1 m/s (in the heading direction) and 0.1 m/s (in the lateral direction). The sampling period of IMU is 0.1 s, and the output periods of altimeter and velocimeter are 1 s. For the IMU, the random noise of accelerometer measurement error is set as $5 \times 10^{-5} g_0$ (1σ), and the random noise of gyro measurement error is set as $0.02°/\sqrt{h}$ (1σ). In addition, a misalignment of $40''$ is assumed. The altimeter and velocimeter measurement accuracies are shown in Table 10.1.

If only ranging correction is adopted, the ranging information is introduced from an altitude of 10 km to correct inertial navigation results. Simulation results are shown in Figs. 10.5 and 10.6. It is observed that the altimeter-aided inertial navigation can correct the errors of altitude, and radial and heading velocities.

If only velocity measurement correction is adopted, the measurements of velocimeter are introduced from an altitude of 10 km to correct the 3D position and velocity. Simulation results are shown in Figs. 10.7 and 10.8. It can be seen that the velocimeter-aided inertial navigation algorithm can correct the velocity error very well and the position error to some extent, but the final position estimation error is still quite large.

If both ranging correction and velocity measurement correction are adopted, the measurements of altimeter and velocimeter are introduced from an altitude of 10 km to correct inertial navigation. Simulation results are shown in Figs. 10.9 and 10.10. The terminal position estimation accuracy is about 100 m, and the velocity estimation accuracy is less than 0.1 m/s, implying that the velocity can be corrected whereas the observability of the 3D position correction is weak.

Table 10.1 Altimeter and velocimeter measurement accuracy (H—LOS distance and V—velocity)

Sensor	Measurement accuracy (1σ)
Velocimeter	0.2 + 0.001 V m/s (V < 200 m/s) 0.2 + 0.005 V m/s (200 m/s < V < 1000 m/s)
Altimeter	0.5 m (H < 2000 m) 0.5 m + 0.0005 H (2000 m < H < 20000 m)

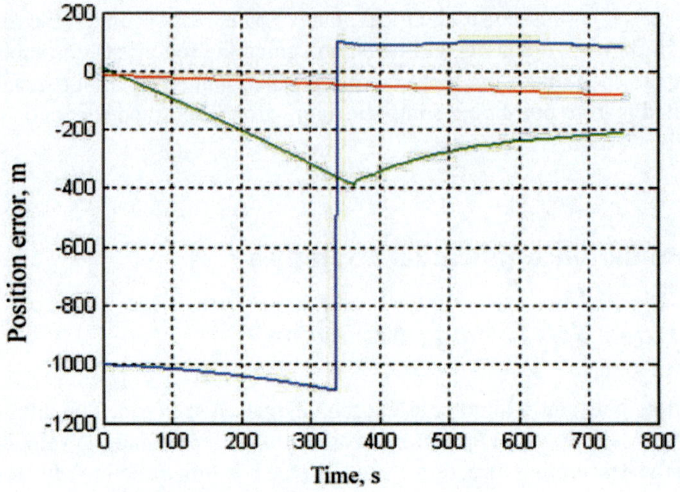

Fig. 10.5 3D position error (ranging correction)

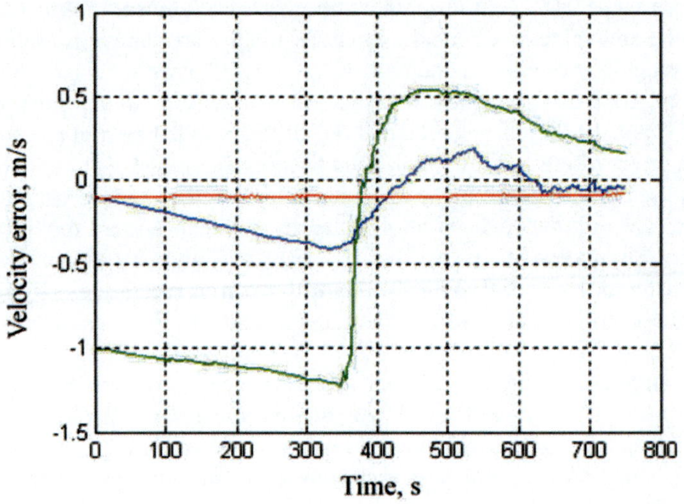

Fig. 10.6 3D velocity error (ranging correction)

In the autonomous navigation of Chang'e-3's lunar soft landing mission, IMU + ranging/velocity measurement are used to correct the altitude and 3D velocity. The flight process and navigation results of Chang'e-3 are shown below [5, 7]. In the lunar powered descent process, the Chang'e-3 lander experienced 685 s powered flight starting from the altitude of 15 km above the lunar surface at the deceleration generated by 7500 N braking engine. After completing the flight phases

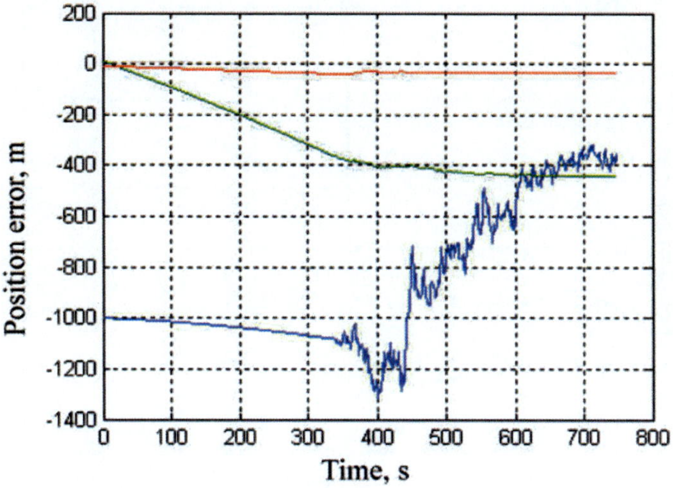

Fig. 10.7 3D position error (velocity measurement correction)

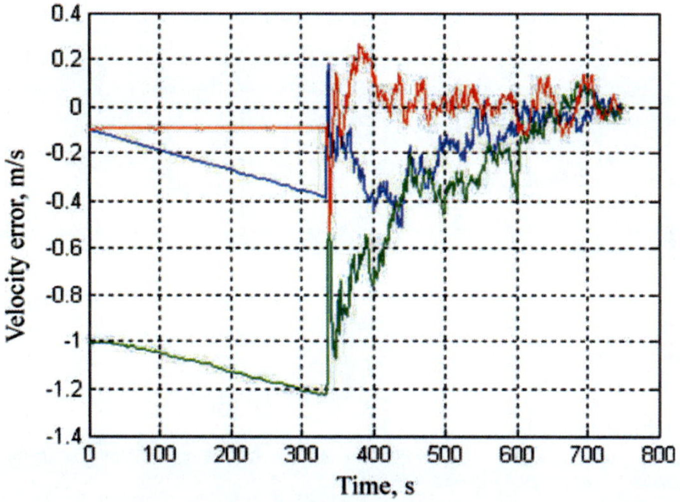

Fig. 10.8 3D velocity error (velocity measurement correction)

such as main braking, rapid adjustment, approaching, hovering, obstacle avoidance, and slow descent, it finally succeeded in soft landing in the Sinus Iridum of lunar surface at 21:11:17, December 14, 2013 (Beijing time). The whole flight was completely autonomous, without any ground intervention. In this process, autonomous navigation system played an important role.

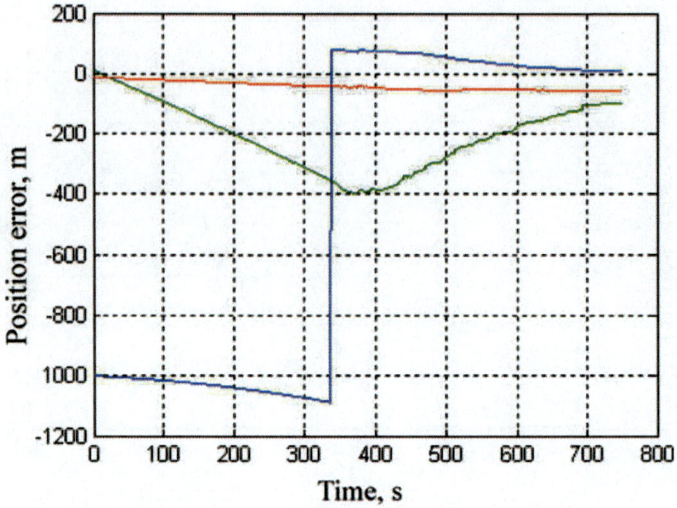

Fig. 10.9 3D position error (ranging + velocity measurement correction)

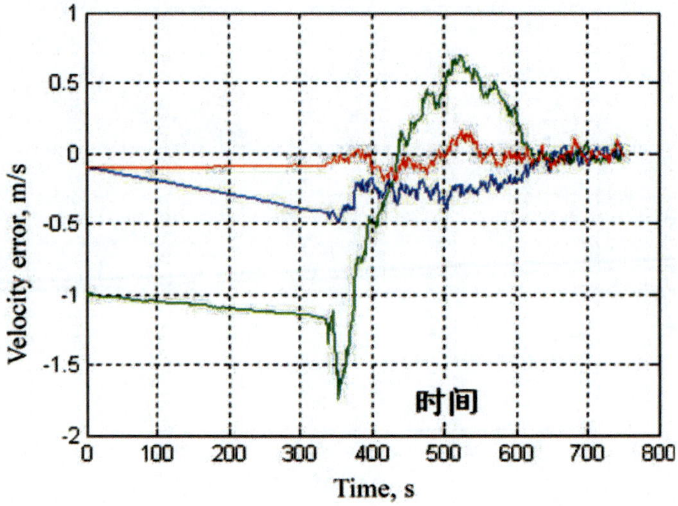

Fig. 10.10 3D velocity error (ranging + velocity measurement correction)

The Chang'e-3 is equipped with an inertial measurement unit, a laser altimeter, a microwave altimeter, and a microwave velocimeter to achieve autonomous navigation during the powered descent. The powered descent phase of this soft landing mission adopted the altimeter-/velocimeter-aided inertial navigation. At the beginning of main braking phase, only inertial autonomous navigation algorithm was used. To ensure safety in the altitude direction, overcome the influence of initial navigation errors and terrain uncertainty, and obtain an accurate altitude relative to lunar surface in the main braking phase, ranging correction was introduced to the navigation algorithm at the end of this phase. After entering the approaching phase, the lander began to keep a relatively stable attitude relative to the moon, which constituted a favorable velocity measurement condition so that the velocity measurement accuracy gradually improved. At the same time, more and more accurate velocity estimates were required to achieve guidance, so velocity measurement correction was introduced to the navigation algorithm. Note that the lunar dust caused by engine plume might affect the measuring performance of velocimeter and altimeters, and only inertial autonomous navigation algorithm should be adopted once the slow descent began.

The final navigation result is shown in Figs. 10.11, 10.12, and 10.13. The

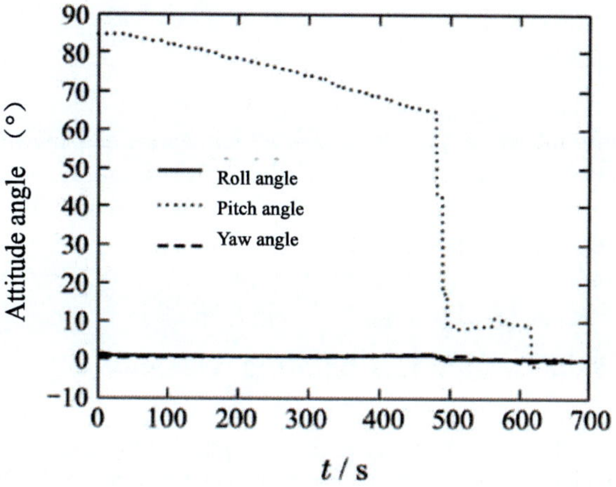

Fig. 10.11 Navigation attitude angles of the Chang'e-3

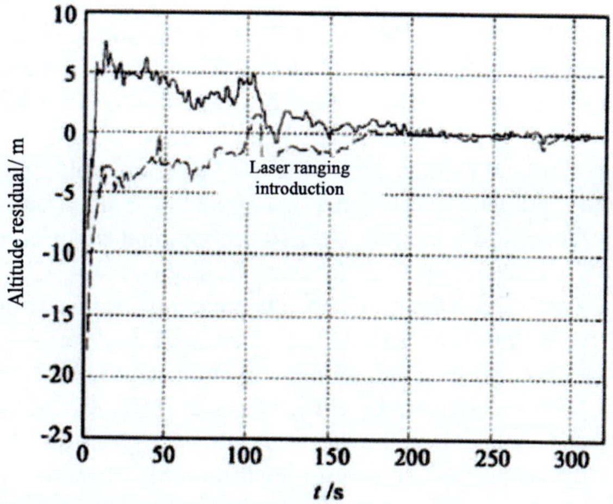

Fig. 10.12 Navigation altitude residual

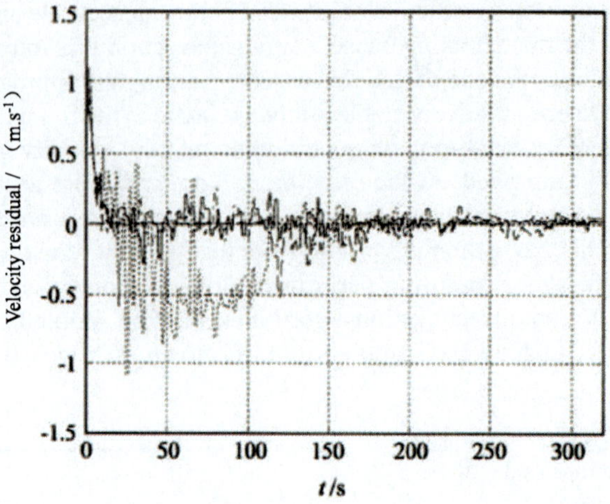

Fig. 10.13 Navigation velocity residual along beam direction

introducing of the altimeter-/velocimeter-aided inertial navigation provides high-precision estimates such as the lander's altitude and velocity relative to lunar surface, thus meeting the requirements for high-precision obstacle avoidance guidance and lunar landing safety.

10.6.2 Mars Soft Landing Navigation

Mars is a celestial body with atmosphere. In this section, a Mars soft landing mission is taken as an example to simulate the autonomous navigation in this mission. Figure 10.14 shows thc nominal Mars landing trajectory. A ballistic flight is assumed during the atmospheric entry phase, where the initial altitude is 126.1 km, the initial velocity (relative to Mars) is 6.75 km/s, the initial longitude and latitude are 0, and the flight path angle is $-14.4°$. When descending to an altitude of 10 km, the parachute descents begin. At the altitude of about 3 km, the lander's velocity is reduced to 50 m/s. The process of descending from the altitude of 3 km to the Martian surface is the terminal landing phase. At this stage, an orbital engine is used for deceleration, and a polynomial guidance is adopted. The polynomial guidance assumes that the distance to the landing site is a polynomial function of time and that the velocity when descending to the Martian surface is 0.

For the INS, the initial absolute position error is 3 km(3σ), initial absolute velocity error is 2 m/s (3σ), initial absolute attitude angle error is 0.05° (3σ), the random noise of accelerometer measurement error is set as $5 \times 10^{-4} g_0$ (3σ), and the random noise of gyro measurement error is set as $0.015°/\sqrt{h}$ rad/h (3σ). The inertial navigation error results during the atmospheric entry phase are shown in Fig. 10.15. As shown in the figure, the maximum position error of inertial navigation is 600 m, and the

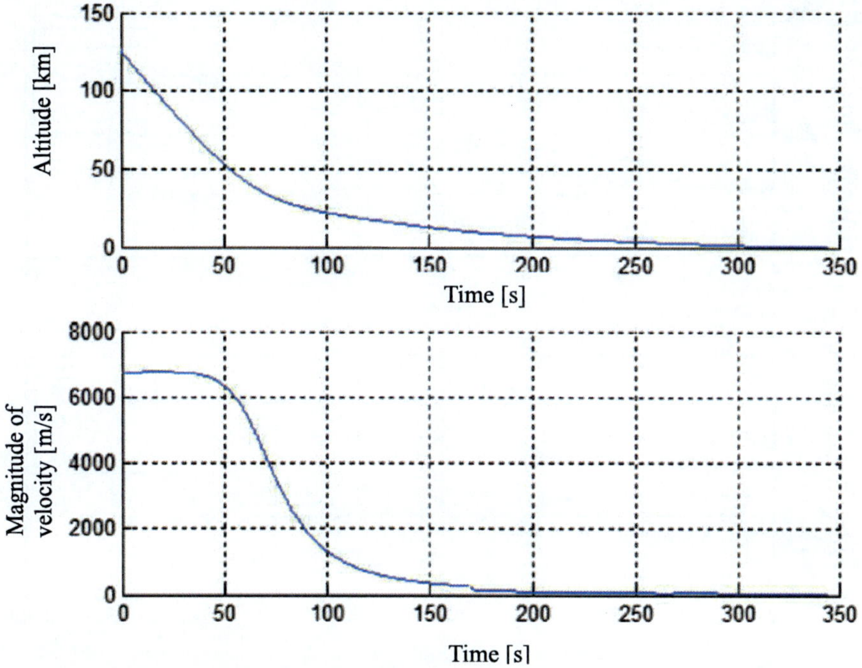

Fig. 10.14 Nominal trajectory of entering the entire landing process

maximum velocity error is 4 m/s. Therefore, at the subsequent stages, optical images and range/velocity measurements are required to correct INS errors.

At parachute phase, once the heat shield is separated, an optical camera is used to image the surface landmarks and then correct inertial navigation errors. The navigation simulation results after the filtering convergence are shown in Fig. 10.16. It can be seen that the optical navigation can effectively correct the inertial navigation position, but with the limited effect of velocity correction.

In the terminal landing phase, velocity measurements are introduced in order to improve the velocity estimation accuracy. The corresponding navigation simulation results are shown in Fig. 10.17. It can be seen from the figure that, after adding the velocity measurement, the velocity can be well corrected, and the velocity estimation accuracy improved. The landing position accuracy is 0.1 m, and the velocity error is 0.2 m/s.

10.7 Brief Summary

This chapter analyzes the requirements for soft landing on the celestial bodies with and without atmosphere and studies the composition and workflow of soft landing

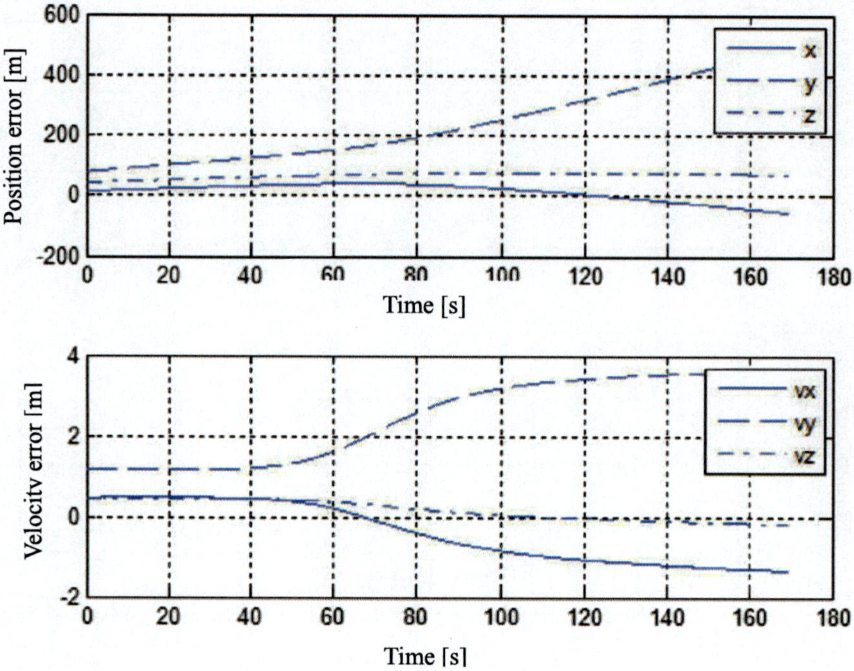

Fig. 10.15 INS navigation errors during the atmospheric entry phase

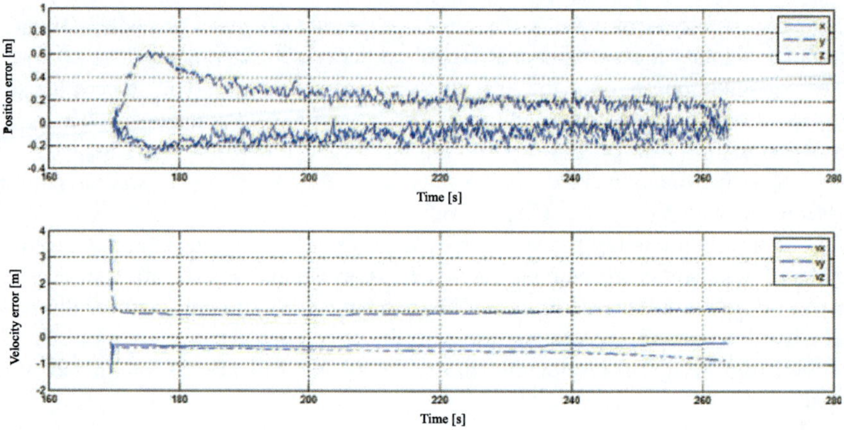

Fig. 10.16 Navigation errors during the parachute descent phase

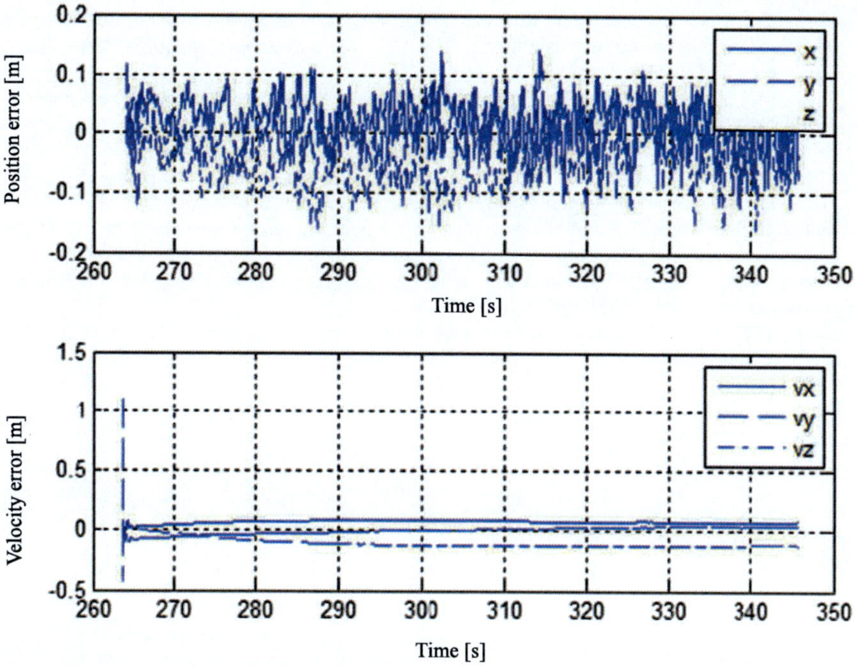

Fig. 10.17 Position and velocity errors in the terminal landing phase

autonomous navigation system. For soft landing on a celestial body without atmosphere, this chapter investigates the autonomous navigation method based on the fusion of IMU + range and velocity measurements/optics. It presents the autonomous navigation methods using IMU + ranging/velocity measurement for 3D position and velocity correction, using IMU + image for 3D position and velocity correction, and using IMU + ranging/velocity measurement for altitude and velocity correction, while conducting the corresponding observability analysis. Among them, the first two methods are analyzed by mathematical simulation. The third method is the actual autonomous navigation scheme of Chang'e-3, to which the corresponding flight results are also presented. Mathematical simulation analysis and actual flight results have demonstrated the rationality and effectiveness of the proposed soft landing navigation method.

References

1. Wong, E.C., J. Masciarelli, and G. Singh. 2002. Autonomous guidance and control design for hazard avoidance and safe landing on Mars. In *AIAA atmospheric flight mechanics conference and exhibit*.

2. Neveu, D., J. De Lafontaine, and K. Lebel. 2005. Validation of autonomous hazard-avoidance Mars landing via closed-loop simulations. In *AIAA Modeling and Simulation Technologies Conference and Exhibit.*
3. Shuster, M.D., and S.D. Oh. 1981. Three-axis attitude determination from vector observations. *Journal of Guidance, Control and Dynamics* 4 (1): 70–77.
4. Hildebrand, F.B. 2012. *Methods of applied mathematics*. Courier Corporation.
5. Zhang, H.H., et al. 2014. Guidance navigation and control for Chang'E-3 powered descent. *SCIENTIA SINICA Technologica* 44 (4): 377–384.
6. Zhang, H.H., et al. 2014. Autonomous navigation for powered descent phase of Chang'E-3 lunar lander. *Control Theory & Applications* 31 (12): 1686–1694.
7. Huang, X.Y., et al. 2014. Autonomous navigation and guidance for Chang'e-3 soft landing. *Journal of Deep Space Exploration* 1 (1): 52–59.
8. Zhang, H.H., et al., 2019. Guidance navigation and control for Chang'E-4 lander. SCIENTIA SINICA Technologica.

Chapter 11
Simulation Testing Techniques for Autonomous Navigation Based on Multi-source Information Fusion

This chapter introduces semi-physical ground test techniques of the autonomous navigation based on multi-source information fusion, including scheme design, system composition and test cases. Two semi-physical test facilities will be presented: one is optical and pulsar integrated autonomous navigation test facility for the transfer, approaching and orbiting phases of deep space exploration missions, and the other is altimeter and velocimeter aided inertial navigation system (INS) test facility for the hovering, obstacle avoidance and slow descent phases of lunar soft landing missions.

11.1 Ground Test of Optical and Pulsar Integrated Autonomous Navigation

11.1.1 Scheme Design

Consider ground simulation test of optical and pulsar integrated autonomous navigation for the transfer, approaching and orbiting phases of deep space exploration missions. The hardware includes an autonomous navigation camera, a celestial body simulator, a pulsar signal simulator, an asteroid/star simulator and a 3-axis turntable, and the software consists of the autonomous navigation algorithm based on information fusion and a ground simulation control algorithm.

During the transfer phase, the asteroid/star simulator generates point light sources for distant asteroids and their background stars, imaged by navigation camera. Then through image processing and calculation, the navigation computer acquires the asteroid LOS direction as the observation of optical autonomous navigation. During the target-approaching phase, the celestial body simulator generates global images of target body, which can be imaged by navigation camera for image collection. Then through image processing and calculation, the navigation computer acquires

D. Wang et al., *Spacecraft Autonomous Navigation Technologies Based on Multi-source Information Fusion*, Space Science and Technologies, https://doi.org/10.1007/978-981-15-4879-6_11

the direction vector of target body center and the apparent radius as the measurements of optical autonomous navigation. During the orbiting phase, the large FOV celestial body simulator generates local images of the target body (including its edge), so as to provide measurement targets to the adjustable-observation navigation camera for image collection. Then through image processing and calculation, the navigation computer acquires the direction vector of the center of the target body and the apparent radius as the measurements of optical autonomous navigation. During the transfer, approaching and orbiting phases, the pulsar signal simulator generates pulsar photons according to the orbit information. Then the receiver of pulsar signal processes a large number of photons through the profile folding and phase comparison to obtain the measurements necessary for pulsar navigation. Based on the information fusion technique, optical navigation information and pulsar navigation information is fused and processed to obtain the autonomous navigation results.

11.1.2 System Composition

11.1.2.1 Transfer Phase

The autonomous navigation testing system for transfer phase is composed of an autonomous navigation camera, a dynamic asteroid/star simulator, a 3-axis turntable, a X-ray pulsar signal simulator, an image acquisition computer, a navigation computer, a simulation control and evaluation computer and other components,

The test system composition of the transfer phase is shown in Fig. 11.1, which are described as follows:

(1) Simulation control and evaluation computer
(2) Autonomous navigation camera
This camera takes images generated by the asteroid/star simulator and then send the images to image acquisition computer. This camera is an optical lens structure with two reflectors and aperture corrector. It uses two internal hoods and one external hood to eliminate the influence of stray light satisfactorily. The prototype of the autonomous navigation camera is shown in Fig. 11.2, and the main technical indexes are as follows:

- Focal length: 953 mm
- FOV: $0.8° \times 0.8°$
- Sensitive magnitude: 12 Mv
- Random measurement error: $0.5''$ (1σ).

(3) X-ray pulsar signal simulator
As shown in Fig. 11.3, the main task of the X-ray pulsar signal simulator is to generate the corresponding pulse signals according to the command of the signal simulation computer, and then send measurements such as the observation

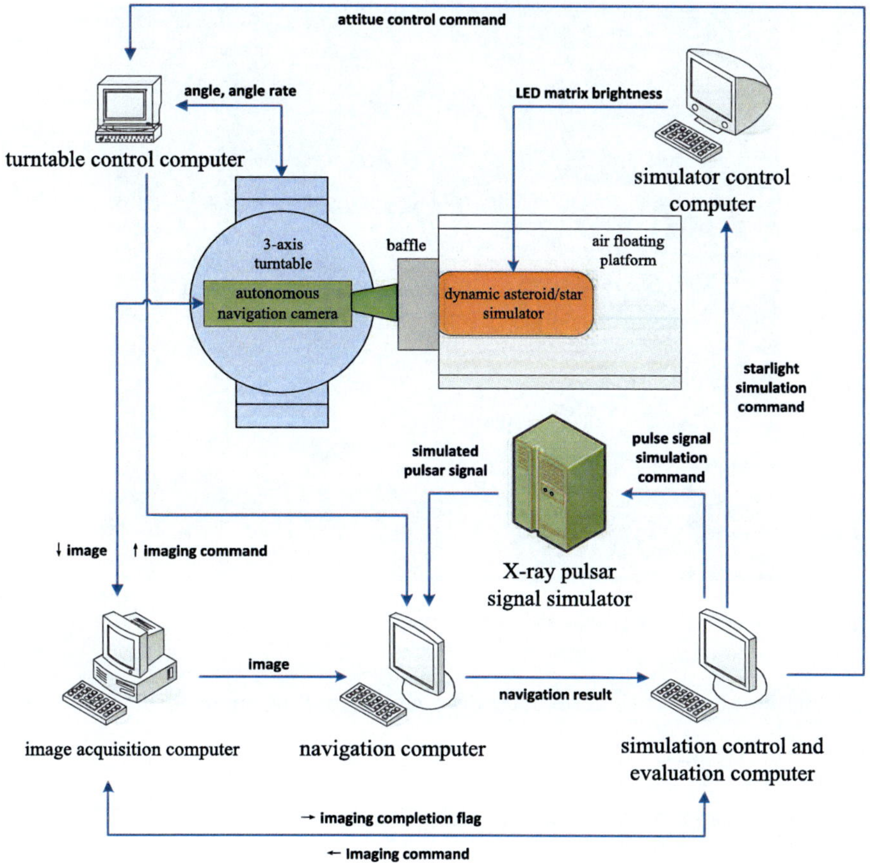

Fig. 11.1 Composition of autonomous navigation testing system for transfer phase

time, the time when pulse signal arrives at the spacecraft and the solar system barycenter, respectively, to the navigation computer.

(4) Dynamic asteroid/star simulator

As shown in Fig. 11.4, the asteroid/star simulator is designed to generate a simulated star map at infinity according to the FOV of navigation camera in the celestial sphere coordinate system. It mainly consists of variable target standard source and optical projection system. To reduce the occupied space caused by long focal length, a reflective optical projection system has been designed. By using attenuator, the point targets of different apparent magnitudes can be simulated. To ensure the projection accuracy of star map, both a small-object simulator and a variable target standard source are installed on a high-precision air floating platform. To ensure the accuracy and installation/adjustment convenience, the structure of test equipment is installed and fixed in sections, which are connected through soft connection.

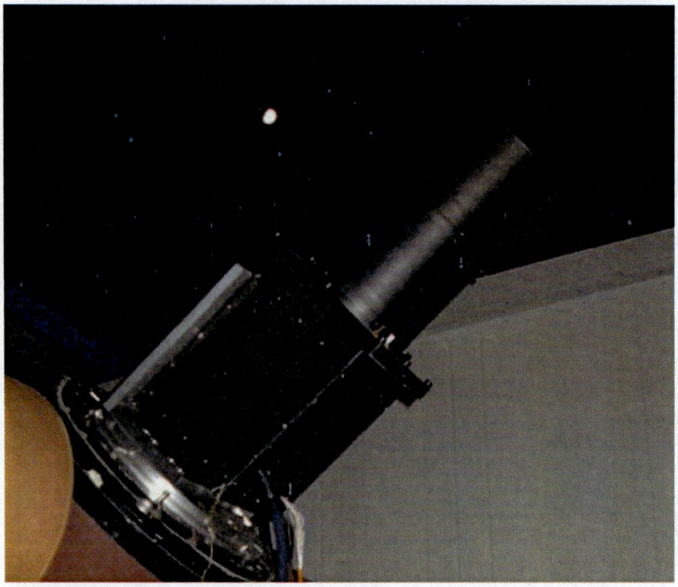

Fig. 11.2 Autonomous navigation camera prototype

The main technical indexes of the dynamic asteroid/star simulator are as follows:

- Focal length: 5 m
- FOV: 1.2° × 1.2°
- Entrance-pupil diameter: 200 mm
- Wave aberration: 1.2° × 1.2°, and $\lambda/10$ (1σ) in FOV
- Maximum relative distortion: 0.000422%
- speckle uniformity: $\leq$0.89 μm
- Star-point projection accuracy: 0.2″.

(5) High-precision 3-axis turntable
As shown in Fig. 11.5, the high-precision 3-axis turntable simulates the impact of spacecraft attitude disturbance on camera imaging during the long-time exposure. Its relative attitude transformation is within 1.0°.
The main technical indexes of the 3-axis turntable are as follows:

- Angle range: 360° full-rotation
- Control accuracy of angular position: better than 1″
- Repeated accuracy: better than 0.5″
- Range of rotation speed: ±0.00001°/s to ±30°/s
- Rotation speed accuracy: 0.01% for $\omega > 1°$/s, and 0.0001°/s for $\omega \leq 1°$/s.

Fig. 11.3 X-ray pulsar signal simulator

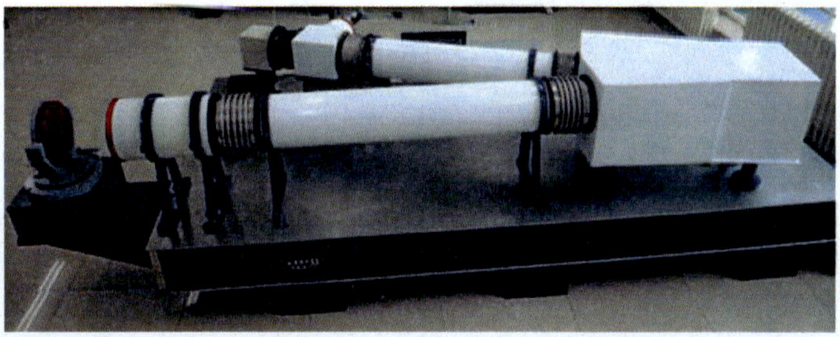

Fig. 11.4 Dynamic asteroid/star simulator

11.1.2.2 Approach Phase

The autonomous navigation testing system for approaching phase is composed of an adjustable-observation navigation camera, a dynamic celestial body simulator, a X-ray pulsar signal simulator, an image acquisition computer, a navigation computer, a simulation control and evaluation computer and other components.

Fig. 11.5 High-precision 3-axis turntable

The test system composition of the approach phase is shown in Fig. 11.6, which are described as follows:

(1) Adjustable-observation navigation camera

A prototype of the adjustable-observation navigation camera is shown in Fig. 11.7. It takes images generated by the celestial body simulator according to the image acquisition command and then send the images to the image acquisition computer.

The main technical indexes of the adjustable-observation navigation camera prototype are as follows:

- Range of FOV angle: $\varphi 7.8°$ to $\varphi 30.1°$
- Object-approaching measurement accuracy (for an 8° FOV): $0.0039°(3\sigma)$
- IFOV resolution (for a 30° FOV): $0.0017°(3\sigma)$

(2) Dynamic celestial body simulator.

As shown in Fig. 11.8, the dynamic celestial body simulator simulates the target body's image when approaching the target. By receiving the command from image simulation software, the simulator can display the simulated images of solar system objects in any shape and size and under any light condition on the internal LCD in real time. This simulator consists of a half-field display system, a splicing optical system and a projection optical system.

The main technical indexes of the dynamic celestial body simulator are as follows:

- Resolution: 1920×2160
- FOV: $10° \times 10°$
- System accuracy:$7.5''$.

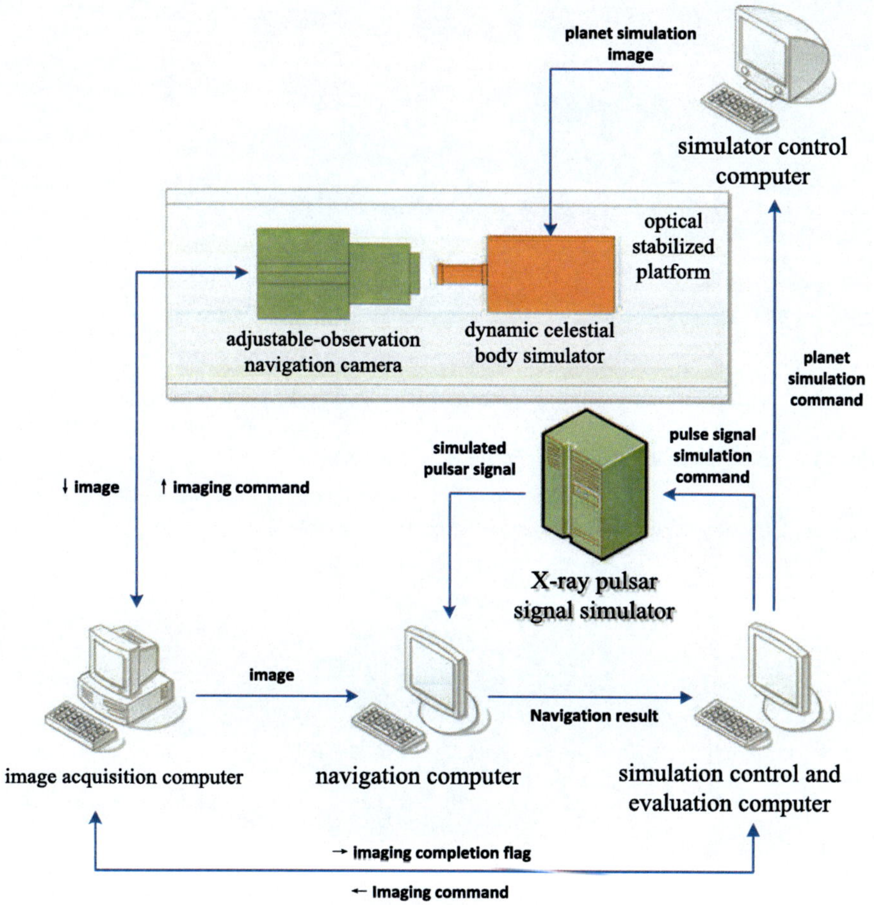

Fig. 11.6 Composition of autonomous navigation testing system for approaching phase

11.1.2.3 Orbiting Phase

The test system composition of the orbiting phase is similar with the approaching phase, but the dynamic celestial body simulator is replaced with a large FOV celestial body simulator, as shown in Fig. 11.9

The large FOV celestial body simulator simulates images of the central celestial object when observing it at close range, as shown in Fig. 11.10. The area illuminated by the simulator simulates the edge profile of the central celestial object for the adjustable-observation navigation camera to collect.

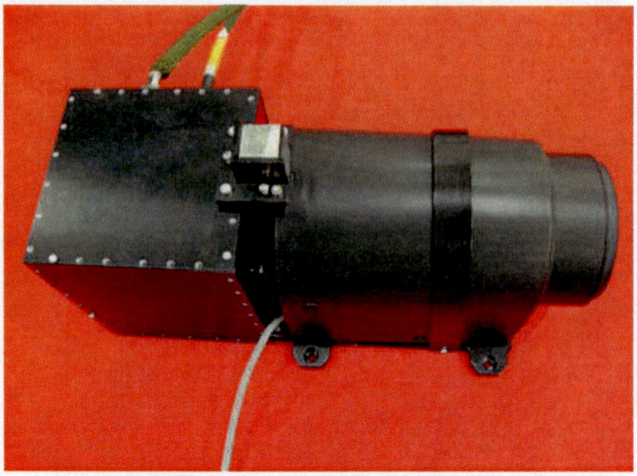

Fig. 11.7 Prototype of adjustable-observation navigation camera

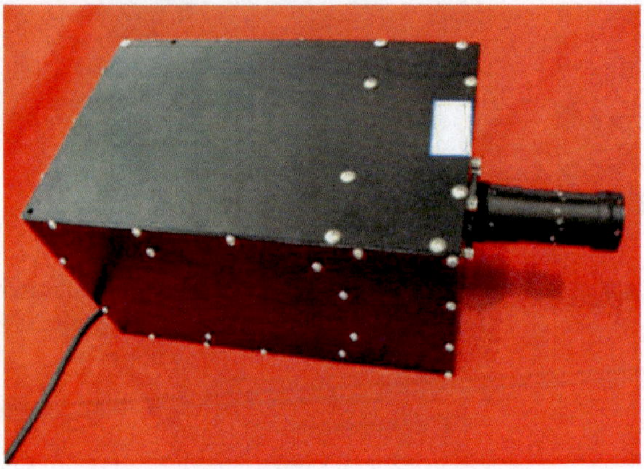

Fig. 11.8 dynamic celestial body simulator

11.1.3 Test Cases

11.1.3.1 Transfer Phase

In this section, the Earth-Mars transfer phase is taken as an example to carry out the simulation test for optical and pulsar integrated autonomous navigation technique. The initial position and velocity errors for each axis are 1000 km and 5 m/s, respectively. The exposure time of navigation camera is set as 2 s. Figure 11.11 shows a picture of the test system.

The asteroid simulation image with background stars taken by autonomous navigation camera under the condition of attitude disturbance is shown in Fig. 11.12.

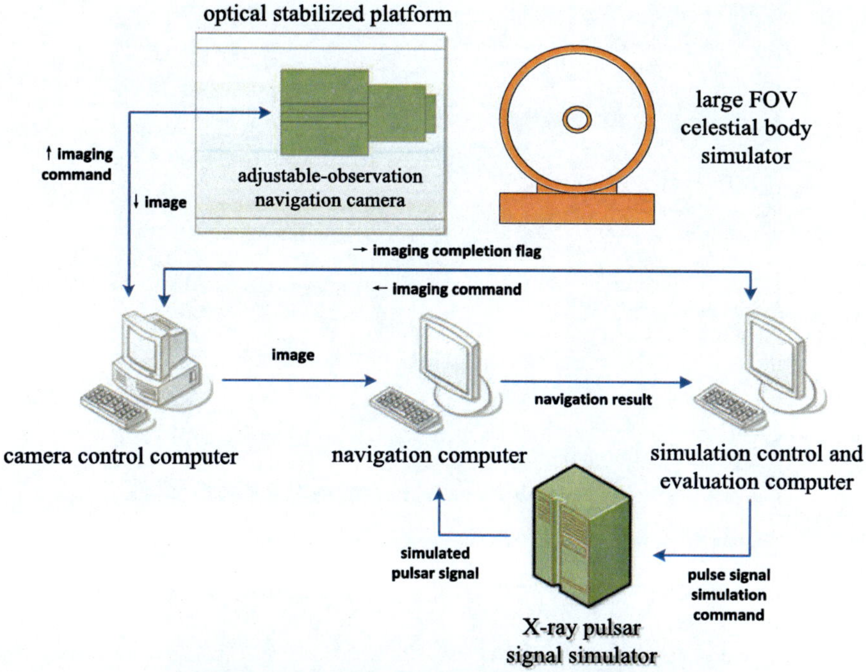

Fig. 11.9 Composition of autonomous navigation testing system for orbiting phase

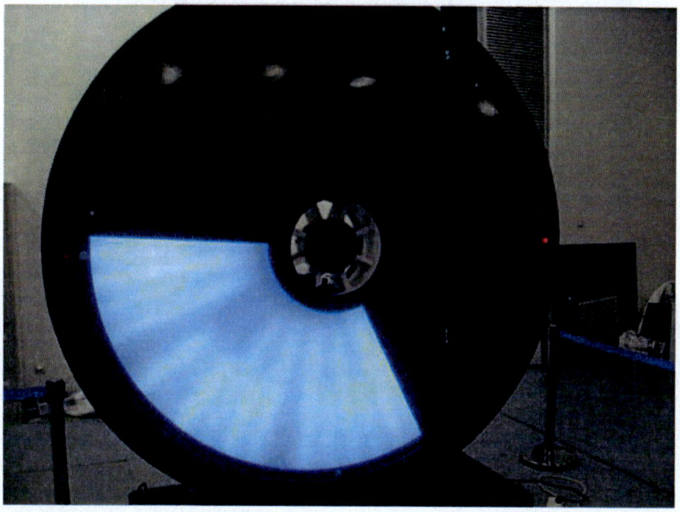

Fig. 11.10 Large FOV celestial body simulator

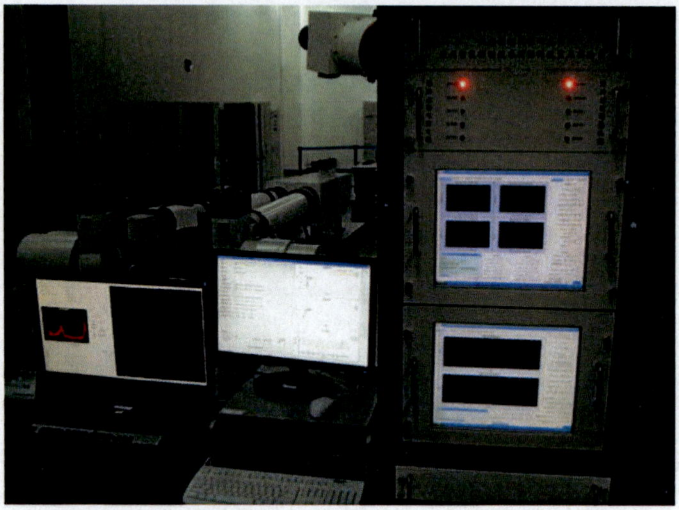

Fig. 11.11 Picture of the autonomous navigation testing system for transfer phase

Fig. 11.12 Asteroid simulation image with background stars taken by autonomous navigation camera

Table 11.1 Accuracy results of optical autonomous navigation and optical and pulsar integrated autonomous navigation in the transfer phase

Navigation accuracy (1σ)	Results of optical autonomous navigation	Results of optical and pulsar integrated autonomous navigation
Position (km)	90.2356	0.3292
Velocity (m/s)	0.0532	0.0437

The navigation accuracy results of optical autonomous navigation in the Earth-Mars transfer phase and those of optical and pulsar integrated autonomous navigation that are finally obtained are shown in Table 11.1. It can be seen that, the information fusion technique has significantly improved the estimation accuracy of spacecraft position and velocity in the transfer phase.

11.1.3.2 Approaching Phase

Take the Mars-approaching phase as an example to carry out the simulation test for optical and pulsar integrated autonomous navigation technique. The initial position and velocity errors for each axis are 500 km and 10 m/s, respectively. The initial FOV of adjustable-observation navigation camera is set as an 8° narrow FOV. Figure 11.13 shows a picture of the test system.

The image of Mars taken by the adjustable-observation navigation camera aiming at the dynamic celestial body simulator is shown in Fig. 11.14.

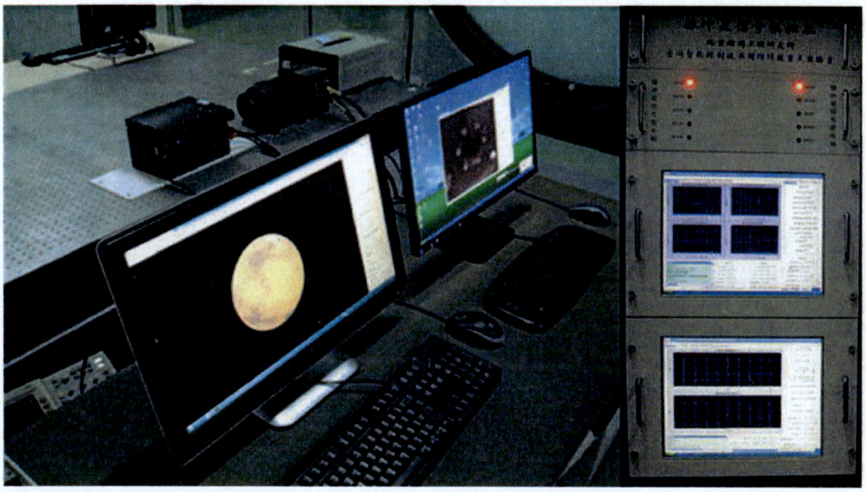

Fig. 11.13 Picture of the autonomous navigation testing system for approaching phase

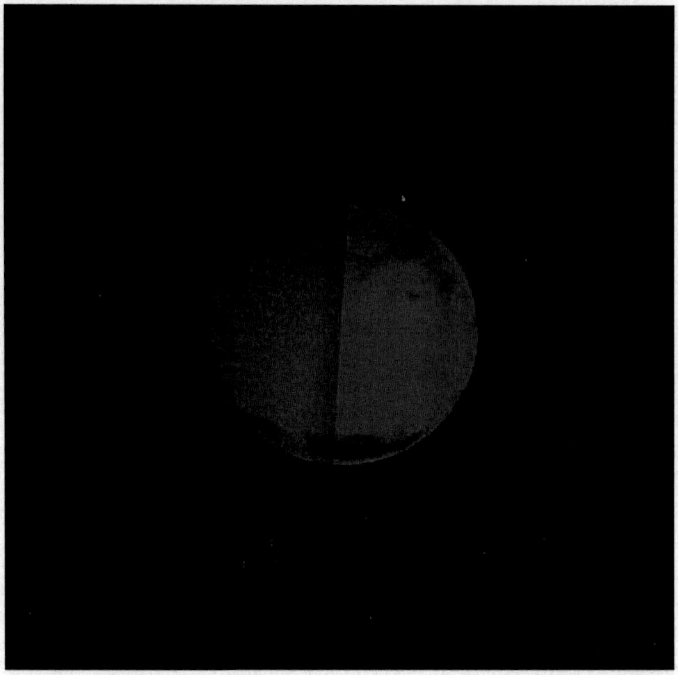

Fig. 11.14 An image taken by adjustable-observation navigation camera in the approaching phase

Table 11.2 Accuracy results of optical autonomous navigation and optical and pulsar integrated autonomous navigation in the approaching phase

Navigation accuracy (1σ)	Results of optical autonomous navigation	Results of optical and pulsar integrated autonomous navigation
Position (km)	1.4802	0.3080
Velocity (m/s)	0.3815	0.1461

The navigation accuracy results of optical autonomous navigation in the Mars-approaching phase and those of optical and pulsar integrated autonomous navigation that are finally obtained are shown in Table 11.2. It can be seen that, the information fusion technique has significantly improved the estimation accuracy of spacecraft position and velocity in the approaching phase.

11.1.3.3 Orbiting Phase

Take the Mars-orbiting phase as an example to carry out the simulation test for optical and pulsar integrated autonomous navigation technique. The initial position and velocity errors for each axis are 100 km and 5 m/s, respectively. The adjustable-observation navigation camera is set as a 30° large FOV aiming at the luminous part of large FOV celestial body simulator and confining the Mars edge to the FOV. Figure 11.15 shows a picture of the test system.

In the orbiting phase, the simulation image of local Mars edge taken by adjustable-observation navigation camera aiming at the large FOV celestial body simulator is shown in Fig. 11.16.

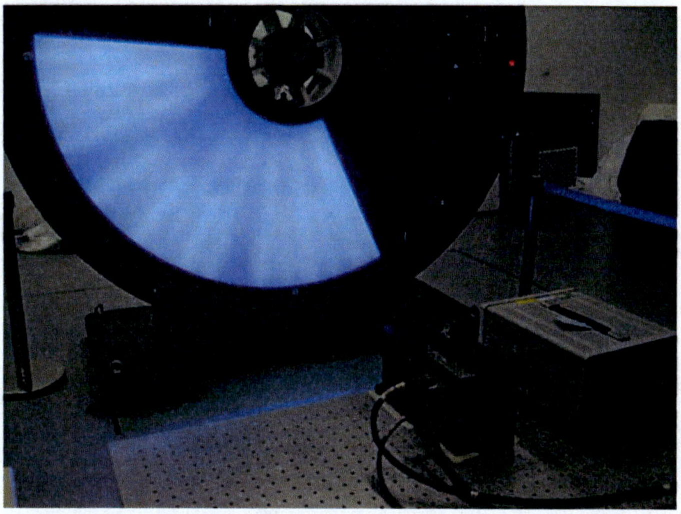

Fig. 11.15 Picture of the autonomous navigation testing system for orbiting phase

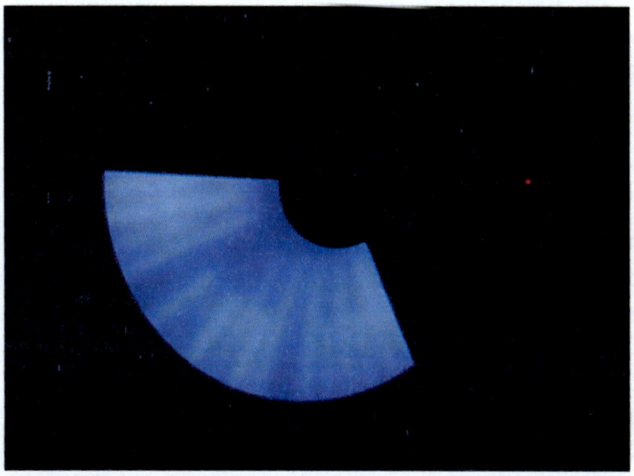

Fig. 11.16 Simulation image of local mars edge taken by adjustable-observation navigation camera in the orbiting phase

Table 11.3 Accuracy results of optical autonomous navigation and optical and pulsar integrated autonomous navigation in the orbiting phase

Navigation accuracy (1σ)	Results of optical autonomous navigation	Results of optical and pulsar integrated autonomous navigation
Position (km)	1.3058	0.2886
Velocity (m/s)	0.2718	0.0234

The navigation accuracy results of optical autonomous navigation in the orbiting phase and those of optical and pulsar integrated autonomous navigation that are finally obtained are shown in Table 11.3. It can be seen that, the information fusion technique has significantly improved the estimation accuracy of spacecraft position and velocity in the orbiting phase.

11.2 Ground Test of Altimeter and Velocimeter Aided Inertial Navigation

11.2.1 Scheme Design

Consider the hovering, obstacle avoidance and slow descent phases of lunar soft landing missions. As shown in Fig. 11.17, the closed-loop ground test scheme of altimeter and velocimeter aided inertial navigation are as follows. A six-degree-of-freedom (6-DOF) motion simulation system simulates the landing orbit and attitude motion of the lander. An inertial measurement unit (IMU) and laser altimeter are mounted on the rotation platform. A radar altimeter/velocimeter and a microwave target simulator is placed on the ground. A lunar surface sand table on the ground provides relative measurement target to the laser altimeter. The dynamics simulation computer is responsible for translation and rotation command of the 6-DOF motion simulation system. The ground test master control computer controls and checks the test system, sets navigation conditions, and analyzes and evaluates navigation results.

Due to space limitation, it is impossible for the radar altimeter/velocimeter to measure the slant range and range rate related to the lunar surface sand table, so a microwave target simulator is configured to generate measurements. The sensor triggers the microwave target simulator through an electrical signal. According to the relative distance and velocity calculated by the kinematics model in real-time, the microwave target simulator simulates the waves reflected by lunar surface under different altitudes and velocities. Then through the appropriate processing (such as delay), the waves are then transmitted to the radar altimeter/velocimeter. In contrast, the laser altimeter can directly measure the slant range between lander (namely rotation platform) and lunar surface sand table.

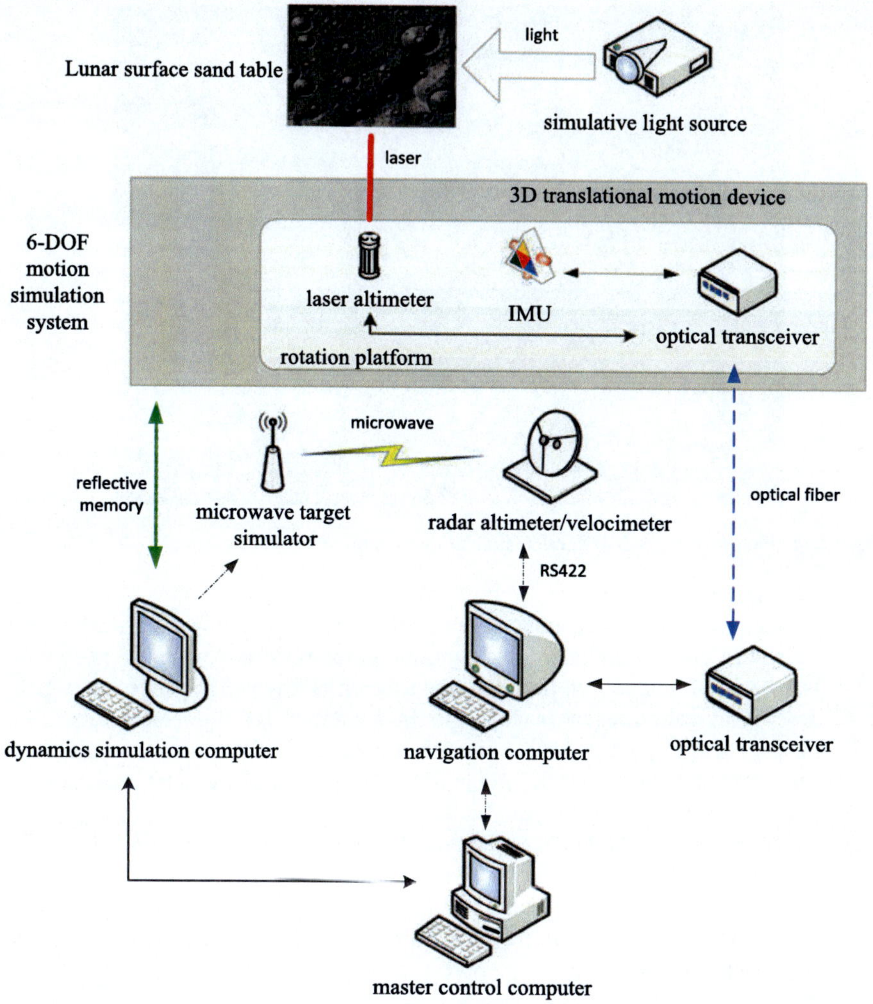

Fig. 11.17 Test scheme and working principle of altimeter and velocimeter aided inertial navigation for landing phase

11.2.2 System Composition

The composition of the altimeter and velocimeter aided inertial navigation testing system is as follows:

(1) 6-DOF motion simulation system
As shown in Fig. 11.18, the 6-DOF motion simulation system is composed of a 3D translational motion device and a 3D rotation platform. The rotation platform is a 3-axis turntable, which is mounted on the translation motion

Fig. 11.18 Picture of the 6-DOF motion simulation system

device. Its longitudinal translational subsystem is composed of four guide rails. The cross beam supported by three guide rails moves transversely in a translational manner; while the vertical beam is driven up and down by ball screw. Navigation sensors such as the IMU and the laser altimeter are installed on the rotation platform that simulates the lander's rotation motion. Driving the platform to move on the 3D guide rail, lander's 3D translational motion related to the lunar surface can be simulated.

(2) Lunar surface sand table
As shown in Fig. 11.19, the lunar surface sand table consists of a sand table and its supporting mechanism. The main function of the lunar surface sand table is simulating the lunar terrain features and the reflection characteristics of visible and laser light.

(3) Simulative light source
The simulative light source simulates illumination conditions on the lunar surface. It consists of four stand-alone units placed on a special lifting platform, with two units in a group. Through a 10 m × 15 m glass screen, light is reflected to the sand table to ensure its complete illumination. Each unit is composed of xenon lamps and an optical system.

(4) IMU
The IMU, consisting of three orthogonally mounted optical gyros and three orthogonally mounted quartz accelerometers, measures specific force and angular velocity.

(5) Laser altimeter
The laser altimeter transmits the modulated laser to the sand table to measure the slant range.

Fig. 11.19 Picture of the lunar surface sand table

(6) Radar altimeter/velocimeter
This sensor has three measuring antennas whose pointing directions are not coplanar. By receiving analog echo signals from the microwave target simulator, it can measure the slant range and slant rate relative to virtual lunar surface.

(7) Microwave target simulator
The microwave target simulator simulates the echo signal of the microwave measurement beam hitting the sand table.

(8) Master control computer
The master control computer is equipped with a ground test control software, which is mainly used for the ground management. Its functions include: test initialization, start-stop control, state monitoring during testing, storage and processing of test data, and data display. This computer communicates with navigation simulation computer and dynamics simulation computer through TCP/IP network.

(9) Navigation computer
The navigation computer collects measurements of IMU, laser altimeter and radar altimeter/velocimeters and processes measurements through the information-fusion navigation filtering algorithm to estimate the attitude, position and velocity of the lander.

(10) Dynamics simulation computer
The dynamics simulation computer mainly has two functions: calculating the lander kinematics and the measurement model of radar altimeter/velocimeter, and controlling the motion of the 6-DOF motion simulation system in real time

according to a given command. The dynamics simulation computer communicates and exchanges data with the 6-DOF motion simulation system through the reflection memory network interface, and with ground test control computer through the TCP/IP protocol.

(11) Optical transceiver
Optical transceiver is mainly used for the real-time data transmission (e.g., IMU and laser altimeter measurements) between the landing navigation sensors on rotation platform and the navigation computer under the platform. Two optical transceivers are configured. The one connected with the navigation computer and placed under the platform is called host optical transceiver; while the other connected with the landing navigation camera and placed on the load plate of rotation platform is called slave optical transceiver. The transmission between the two optical transceivers is achieved through optical fiber so as to meet the timing sequence and real time communication requirements.

11.2.3 Test Case

In this case, the hovering, obstacle avoidance and slow descent phases of a lunar soft landing is considered. At beginning, the 6-DOF motion simulation device is located 45 m away from the sand table. The spacecraft initial attitude and position navigation errors are set to zero, and the initial velocity navigation error is set as 0.1 m/s. The layout of the test system facility is shown in Fig. 11.20.

Steps of altimeter and velocimeter aided inertial navigation test are as follows:

(1) Check the devices on rotation platform and the ground devices to ensure that they are in good mechanical and electrical conditions.
(2) Power up the equipment according to their power-up requirements, and keep them in standby mode until they meet the working conditions.
(3) The 6-DOF motion simulation system moves to the designated position according to the test requirements, and switches to the remote control mode.
(4) The ground test control computer initializes the dynamics simulator and navigation computer.
(5) The dynamics simulation computer calculates the initial startup program in a cycle of 8 ms according to the initialization parameters. The dynamics data is then transmitted through the reflective memory to the control computer of the 6-DOF motion simulation system, driving the 3D translational motion device and rotation platform to move as required.
(6) At the same time, the 6-DOF motion simulation system transmits its actual 6-DOF motion data through the reflective memory to the dynamics simulation computer.
(7) The dynamics simulation computer determines whether the dynamics data meets initial dynamic environment conditions or not. If so, it will send a signal to the ground test control computer to formally start the test. Otherwise, it repeats steps (5)–(7) until the dynamics condition is satisfied.

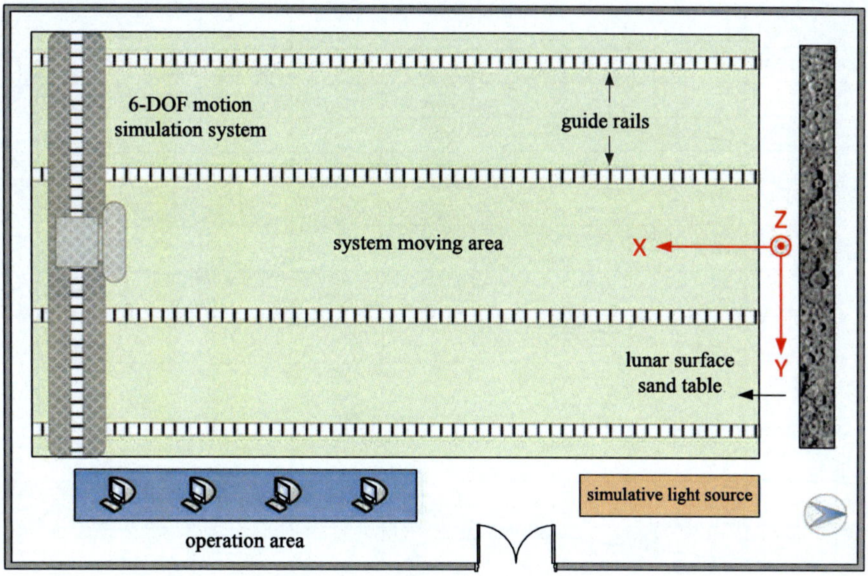

Fig. 11.20 Ground test site layout for the altimeter and velocimeter aided INS

(8) Once the formal test begins, the dynamics simulation computer calculates the current position, velocity and attitude in a cycle of 8 ms. The dynamics data is then transmitted through the reflective memory to the control computer of the 6-DOF motion simulation system, where the data drives the 3D translational motion device and rotation platform to move as required. At the same time, the measurement model of radar altimeter/velocimeter in the dynamics simulation computer sends the corresponding data to the microwave target simulator through calculation. In addition, 16 dynamics data packages obtained in every 8 ms are transmitted via high-speed Ethernet to the ground test control computer for monitor, data analysis, and display.

(9) The laser altimeter installed on the rotation platform begins to measure the slant range to the sand table. Meanwhile, the IMU measures the angle/velocity increments of the platform in motion every 32 ms. The two devices transmit the measurement data to the navigation computer through optical transceivers and optical fiber every 128 ms.

(10) Meanwhile, the radar altimeter/velocimeter placed under the platform transmits the slant range and velocity data to the navigation computer through a RS422 interface every 128 ms.

(11) The navigation computer receives measurements from the IMU, the laser altimeter and the radar altimeter/velocimeter, and calculates navigation solutions every 128 ms. Navigation results are then transmitted to the ground test control computer via Ethernet.

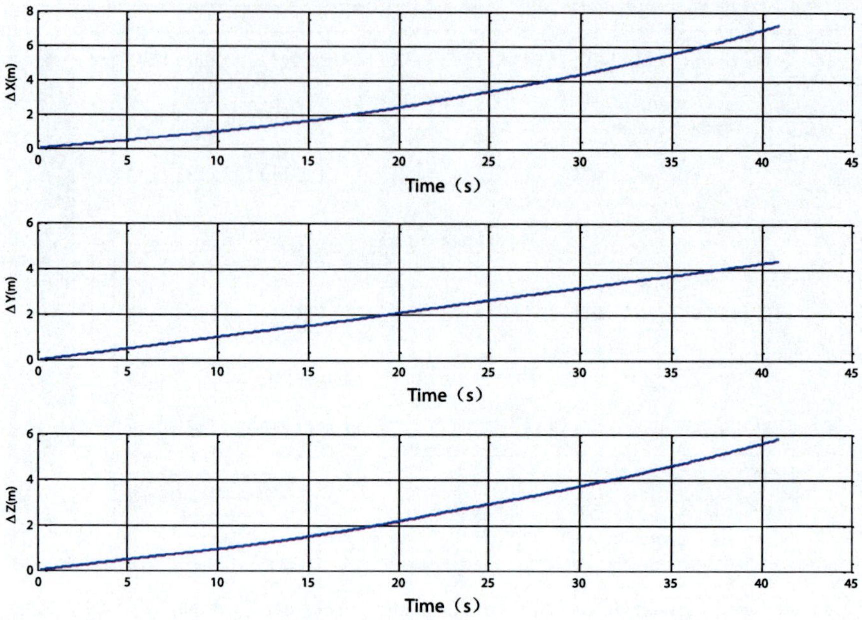

Fig. 11.21 INS position estimation error

(12) The dynamics simulator calculates the X-direction distance of the 6-DOF motion simulation system every 8 ms and determines whether the termination or deceleration conditions are satisfied or not. If the deceleration condition is satisfied, the 6-DOF system can switch to the deceleration program. Otherwise, repeats steps (8)–(11).

(13) In the deceleration program, the dynamics simulator controls the braking of the 6-DOF system until there is no translational and rotational motions.

The strapdown inertial navigation results are shown in Figs. 11.21 and 11.22. The altimeter and velocimeter aided inertial navigation results are shown in Figs. 11.23, 11.24, and Table 11.4.

It can be seen that the inertial navigation errors increase over time and altimeter/velocimeter aided inertial navigation has significantly improved the estimation accuracy of spacecraft position and velocity in the hovering, obstacle avoidance and slow descent phases of lunar soft landing.

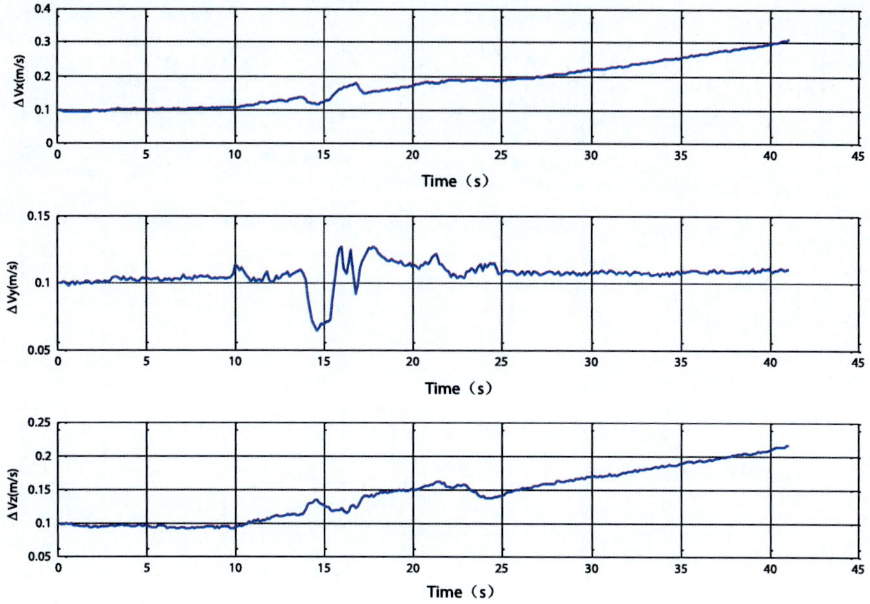

Fig. 11.22 INS velocity estimation error

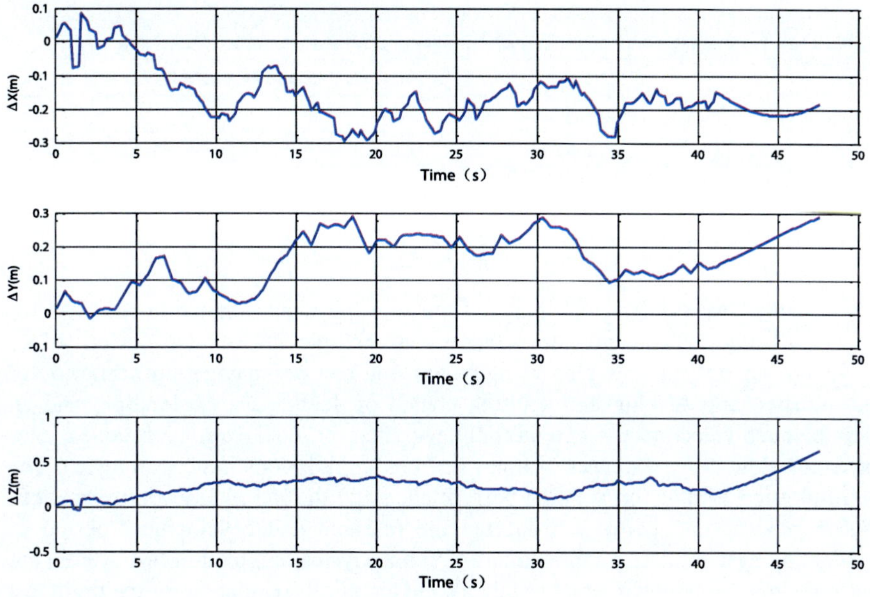

Fig. 11.23 Altimeter and velocimeter aided INS position estimation error

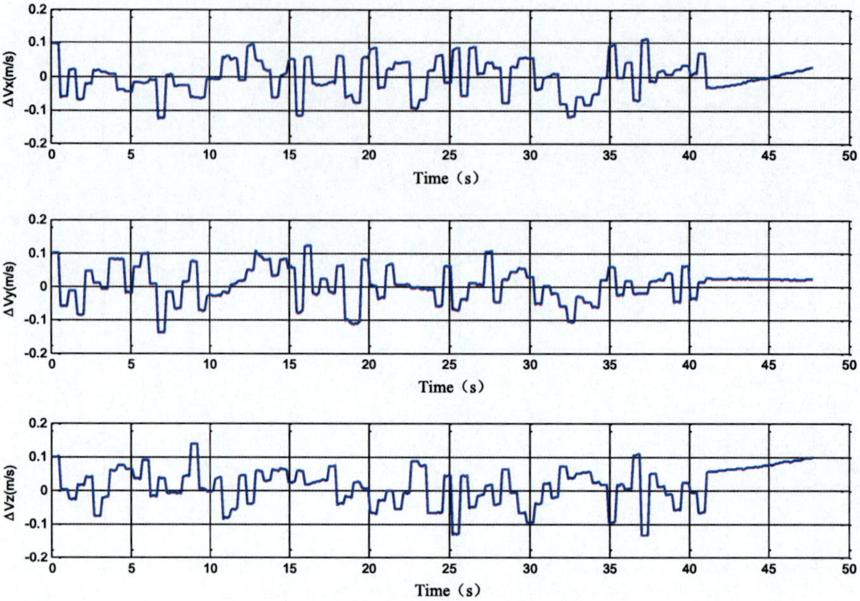

Fig. 11.24 Altimeter and velocimeter aided INS velocity estimation error

Table 11.4 Comparison between INS and altimeter and velocimeter aided INS result

	INS	Altimeter/velocimeter aided INS
Position (m)	10.252	0.705
Velocity (m/s)	0.383	0.105

11.3 Brief Summary

This chapter has studied and analyzed the semi-physical simulation schemes for spacecraft autonomous navigation based on multiple information fusion. More specially, an optical and pulsar integrated autonomous navigation scenario for the transfer, approaching and orbiting phases of deep space exploration, and an altimeter and velocimeter aided inertial navigation for the hovering, obstacle avoidance and slow descent phases of lunar soft landing missions have been tested. The semi-physical simulation verified performance and the rationality of information-fusion navigation scheme and shorten the research and development period of navigation system. The establishment of semi-physical simulation test system for information-fusion navigation has laid a solid technical foundation for the flight test of autonomous navigation technique and its practical engineering application.

Chapter 12
Prospect for Multi-source Information Fusion Navigation

Autonomous navigation is the core technology of spacecraft autonomous operation, but a single navigation mode cannot meet the requirements for long-time operation, high accuracy, and high reliability. The autonomous navigation technique based on multi-source information fusion can realize the advantage complementation among subsystems and achieve the performance better than that of subsystems. Therefore, it is an effective way to improve the performance of autonomous navigation. On the one hand, the fusion navigation system integrates the navigation information from each subsystem to realize the mutual correction of systematic errors and to achieve better navigation performance. On the other hand, this system obtains redundant measurement information, which can improve the system adaptability to the environment and the reliability of the whole system.

With the emergence of new navigation concepts and the development of modern signal processing, computing technology, artificial intelligence, and other technologies, information fusion has gradually become an indispensable technology in autonomous navigation system. To achieve high performance, many practical problems need to be addressed, including fusion navigation scheme, filter and fusion algorithms, sensor choice and configuration, data synchronization, and system computation capability. In terms of theoretical method, it is of great significance for the development of fusion navigation technique to improve the basic theory of information fusion and the data processing method and to design the filter algorithm in engineering application practically. This chapter takes autonomous navigation technique based on multi-source information fusion as the research object and briefly introduces its development trend and research hotspots from three aspects: information fusion navigation scheme, information fusion technology, and fused navigation sensors.

D. Wang et al., *Spacecraft Autonomous Navigation Technologies Based on Multi-source Information Fusion*, Space Science and Technologies, https://doi.org/10.1007/978-981-15-4879-6_12

12.1 Development of Multi-source Information Fusion Navigation Scheme

The ultimate development trend of spacecraft autonomous navigation technology is high autonomy. An important feature of high autonomy is to achieve fault detect, isolate, and recover autonomously, and the system is required to have a certain redundancy capacity, which can be achieved through fusion navigation technique. In addition, future deep-space exploration missions would be more complex, which also puts forward new requirements for multi-source information fusion based navigation. Next, development of several typical multi-source information fusion schemes will be introduced.

12.1.1 Fusion Navigation Based on Optical Measurement

In deep-space exploration missions, optical autonomous navigation technique, which can realize the navigation of spacecraft relative to target celestial body, is still the mainstream navigation method in the future. But it is susceptible to the influence of the target's light conditions and other factors, so its absolute navigation accuracy is not high. In addition, optical autonomous navigation cannot correct onboard clock errors for a long-time flight mission. If clock error diverges, the autonomous survival ability of spacecraft will be affected. Therefore, it is necessary to develop a fusion navigation technique based on optical measurement. As an emerging autonomous navigation technique, pulsar navigation has great development prospects once fused with optical autonomous navigation. On the one hand, optical navigation can provide accurate priori information for pulsar navigation to facilitate the establishment of measurement pulse profile. On the other hand, pulsar navigation can achieve absolute navigation and correct onboard clock errors to eliminate the divergence of clock time difference in optical navigation, so the performance of the whole navigation system can be improved.

12.1.2 Fusion Navigation Based on Inertial Measurement

Inertial autonomous navigation is still the main navigation mode in powered flight phases of deep-space exploration missions in the future. Since the inertial navigation errors are accumulated with time, it is necessary to study the fusion navigation technique based on inertial measurement unit (IMU).

The fusion navigation technique based on IMU has been successfully applied in the powered descent segment of soft landing, but not in the atmospheric entry phase for extraterrestrial bodies. This is because in the process of atmospheric entry, high-speed friction between spacecraft and atmosphere will cause high temperature on the contact surface of the spacecraft. In order to protect spacecraft from being damaged by high temperature, the spacecraft is generally installed in the heat shield. Due to the isolation effect of heat shield, very few sensors are available at the entry stage. In addition to IMU, only the radio communication equipment with special bands and flush air data system can work in this special environment. Among them, the radio communication equipment can communicate with other wireless beacons for velocity measurement and ranging. The flush air data system is mainly used to calculate dynamic pressure, Mach number, angle of attack, and other key atmospheric parameters by measuring the pressure distribution on the spacecraft surface. In order to achieve pin-point soft landing, the navigation accuracy at the entry stage cannot be ignored. It is necessary to combine IMU with available sensors (radio sensors or flush air data system) to study inertial-aided navigation system [1]. For this particular navigation system, the influence of various error sources on the performance of navigation system should be comprehensively analyzed. In addition, some robust information fusion algorithm shall be studied to improve the reliability of navigation system.

In addition, future extraterrestrial-body (large or small) landing missions require the ability of highly autonomous relative navigation in the terminal landing segment, as well as the ability to quickly avoid obstacles and select safe landing sites in unknown environments. In terms of sensor configuration, the sensors shall be capable of both autonomous navigation and obstacle detection. Available sensors include Doppler velocity rangefinder, laser rangefinder, and LIDAR [2, 3]. In terms of navigation scheme, inertial and optical autonomous navigation will be an important development direction for pin-point landing missions. In addition, artificial wireless beacons can be released to further improve the performance of autonomous navigation system.

12.1.3 Autonomous Navigation Based on Deep-Space Navigation Constellation

At present, most of the earth satellites rely on GNSS constellation for autonomous navigation. This technique can be further extended to deep-space exploration missions, for example, by establishing the corresponding deep-space navigation constellation. In 2015, NASA proposed multi-autonomous positioning system (MAPS) [4, 5], planning to combine the existing "Mars Reconnaissance Orbiter" (MRO) and Mars Odyssey spacecraft with ground-based TT&C system to initially establish a Mars navigation network. This network, which will be combined with future commercial Mars communication orbiters, is expected to realize the communication interactions and navigation information sharing/updating among multiple spacecrafts through the increasing communication bandwidth and broadcast information.

Researchers also proposed concept of satellite network [6]. Spacecrafts in the satellite network can be classified into two types: reference and user spacecrafts. Reference spacecrafts realize autonomous navigation by observing pulsars, thus providing the navigation reference for the whole network. A user spacecraft is equipped with the inter-satellite ranging device compatible with the satellite network, relying on the communication with reference spacecrafts and other user spacecrafts to determine its own position. Different from GNSS navigation constellation, the configuration and member spacecrafts of satellite network are flexible and variable. Any spacecraft configured with inter-satellite link measurement equipment compatible with the network can join the network, while any spacecraft in the network can provide navigation services for other spacecrafts.

Deep-space navigation constellation is a typical distributed fusion architecture. With the increasing of spacecrafts, it can be predicted that the autonomous navigation technique based on deep-space navigation constellation will also be one of the important development directions in the future.

12.2 Development of Information Fusion Technology

12.2.1 Development of Fusion Architecture

Centralized fusion has many problems, such as heavy computation burden and limited fault tolerance capability. Whereas distributed fusion has significant advantages over centralized fusion in terms of fault tolerance and computing efficiency.

Distributed fusion architecture with a global fusion center is mainly used in the autonomous navigation scenarios for a single spacecraft, such as in the descent and landing segments. This architecture has several local information fusion units and a global information fusion center—the latter fuse the former's information.

Distributed fusion architecture without global fusion centers is mainly used in the autonomous navigation scenarios where multiple spacecrafts are flying cooperatively, such as the constellation satellite navigation based on inter-satellite measurement and the autonomous navigation for formation flight. In this architecture, each sensor is connected with other sensors through nodes, in order to output its fusion measurements. In the absence of a central processing unit, the system output no longer depends on central fusion unit, but by any node. Therefore, the damage of a single sensor or node will be no big deal for the whole system. However, the communication and fusion algorithm of this fusion architecture are complex, as its nodes can be connected with all the other nodes or with only some of them.

12.2.2 Development of Filter and Fusion Algorithms

The main difficulties of future deep-space exploration are high environmental complexity and unpredictability and the requirement of high mission autonomy. Therefore, from the perspective of filter and fusion algorithm, the information fusion algorithm with good self-adaptation to uncertain factors such as environment and aerodynamics and high on-orbit operation efficiency is one of the development directions. Besides, in order to achieve a high degree of autonomy, the filtering and fusion algorithm with strong robustness, good fault tolerance, and fast convergence is also an inevitable development direction.

1. Robust filtering and H_∞ filtering

Robust filtering is designed to ensure the availability and minimization of the upper bound of variance of estimation error under the condition of model uncertainty. H_∞ filtering is designed to ensure the boundedness of the ratio of the norm of estimation error to that of model error (uncertainty). Robust filtering (or H_∞ filtering) is based on uncertain model. If the model uncertainty in the actual system is consistent with the expression of uncertainty in the model, then robust filtering can obtain the estimation accuracy better than traditional Kalman filtering. The reason is that the uncertain model contains the priori knowledge on the uncertainty of actual system model and describes the actual system more accurately, while traditional Kalman filtering fails to consider the influence of model uncertainty in the design process. How to establish an appropriate uncertainty model for describing the uncertainty in actual systems reasonably and effectively is a problem that should be solved to apply robust filtering in engineering practice.

It should be noted that, if the expression of uncertainty in the model is not consistent with actual system, robust filter algorithm in place of Kalman filtering may reduce the estimation accuracy. For example, if the unknown disturbance exists as model uncertainty only in few cases, the uncertain model will be inconsistent with actual system in most cases. Thus, the robust filter algorithm based on uncertain model is often not the best choice for designers. To this end, the combination of robust filtering with multi-model filtering is helpful to solve the above problem.

2. Multi-model adaptive filtering

Multiple model adaptive estimation (MMAE) is another filtering method capable of handling the model uncertainty. To design the MMAE algorithm, a model set composed of a series of models is often used to describe the model uncertainty. The design of model set has an important influence on the performance of MMAE algorithm. In the MMAE algorithm, a filter is designed for each model in the model set, and each filter obtains its own state estimate through parallel solving. The total state estimate of MMAE algorithm is the weighted sum of the state estimate of each parallel filter. The weight of each parallel filter is obtained by calculating the measurement residual (innovation) and reflects the degree of similarity between each filter model and actual system model. By optimizing the model set and designing the adaptive law, the model consistent with actual system can play a major role in the state estimation. In this case, multi-model filtering can achieve better estimation accuracy than Kalman filtering.

The development of MMAE has gone through three generations. The first generation is static multi-model filtering algorithm, where each parallel filter carries out recursive estimation centering on a fixed model set. The second generation is interacting multiple model (IMM) filtering algorithm, where each parallel filter exchanges information with others. The third generation is variable-structure multiple model (VSMM) filtering algorithm. Its feature is that the model set used to design the parallel filtering is time-varying and can be adapted to environmental change. With the development of MMAE algorithm, its ability to handle the problem of model uncertainty has been continuously improved.

MMAE algorithm can be used to identify the time when the unknown disturbance in the system starts and ends and its effects, and then handle this through appropriate strategies. For example, when an unknown disturbance occurs, the model contributing to a larger noise variance matrix will be dominant. On the contrary, when there is no unknown disturbance, the model contributing to a smaller noise variance matrix will be dominant. The MMAE algorithm can adaptively switch to different models and make a compromise between them. This makes it better than Kalman filtering and robust filtering when dealing with the uncertainty of engine jet thrust or the change of sensor measurement characteristics due to environmental influence. The optimization design of model set and the analysis of its influence on filtering performance are the main concerns in multi-model filter research.

3. Fusion algorithm

Owing to the absolute advantage of distributed fusion architecture, the distributed fusion algorithm will be the main development direction of future fusion algorithm. But the common orbital dynamics model and measurement coupling bring unknown correlations to local estimation. In addition, the sampling frequency asynchrony between different sensors, communication delay, and other factors makes it difficult to design the distributed fusion algorithm. Covariance intersection (CI) fusion algorithm can effectively solve the unknown correlations in local estimation and provide a suboptimal fusion estimate. Typical solutions to the asynchronous delay of sensors include: (1) designing a time synchronization device; (2) augmenting states at different time into a new state for estimation. These methods provide ideas for the design of fusion algorithm. For typical fusion navigation systems, the design of distributed fusion algorithm considering the unknown correlations between subsystems, the asynchronous sensors delay will be an important research direction in the future.

In addition, with the continuous improvement of onboard computing capability, the intelligent fusion algorithm based on neural network will be developed and applied to the autonomous navigation of spacecrafts. This is because the intelligent fusion algorithm has strong fault tolerance to improve the autonomous survival of the system and has a strong ability to handle the problems of nonlinearity and uncertainty.

From the perspective of information source, the current fusion algorithm is mainly to fuse the observation information and state equation of multiple sensors. With the development of estimation theory, future fusion algorithm will not stop here, for example, by fusing multiple models (the above-mentioned MMAE) or system constraints (constraint filtering).

12.3 Development of Fused Autonomous Navigation Sensors

Sensors in fused autonomous navigation systems will become miniaturized, reusable, integrated, and intelligent. The target objects in deep-space exploration are very far away from the earth, imposing even more stringent weight limits on the probes [7]. The USA and the ESA are actively developing various miniature solar sensors, stellar compasses, video sensors, and gyroscopes. The microminiaturization technique can not only reduce the weight of onboard functional components, but also bring about the overall miniaturization and low power consumption of related subsystems, thus greatly reducing the weight, size, and power consumption of the entire satellite. It can be predicted that with the deepening of micro-nano research, the MEMS and micro-nano techniques will surely play an increasingly important role in the future development of space technology. Moreover, by integrating multiple sensors with different functions in autonomous navigation, systematic errors among sensors can be effectively reduced.

In terms of reusability, navigation sensors and optical load on a spacecraft are often shared by different flight stages, where they may play different roles. For example, when flying by the "Wild-2" comet, the "Stardust" probe used its navigation sensor to map the comet's nucleus in three dimensions. The landing navigation sensor of "Hayabusa" probe also undertook the task of imaging and observing the asteroid surface [8]. In addition, the deep-space probes launched by the USA, such as "Voyager 1" and "Deep Space 1," often use payload cameras as navigation sensors to complete the task of optical autonomous navigation.

In deep-space missions, optical imaging sensors are required to image and observe the celestial bodies at different distances. For example, the "Altair" probe used in the US lunar return program is equipped with an integrated navigation sensor capable of observing stars, solar planets, and lunar landmarks [9]. To meet the requirements of imaging navigation in multiple mission segments (e.g.,, approaching, orbiting, and flyby), the varifocal and FOV-adjustable technique can also be used. This technique is also a feasible means to image and measure different targets, such as stars, weak asteroids, and near-earth objects (NEOs). Another typical example is to improve the ultraviolet sensor through the opto-mechatronic integration design. Only one solid-state sensor is used to simultaneously image the star and target object, so as to provide triaxial attitude data and autonomous navigation information at the same time that can effectively reduce the influence of relative installation error on the calculation of autonomous orbit.

To improve the system's autonomous survival and reduce the communication bandwidth, future sensors will develop toward the intelligentization direction. Each sensor uses its own processor to process original measurement data and directly exchanges information with other sensor nodes to complete the final decision-making and fusion.

12.4 Conclusion

There is no end to scientific research. High accuracy and reliability of navigation and positioning is always the goal. With the continuous implementation of deep-space exploration missions, new opportunities and challenges have emerged in the development of various deep-space exploration technologies, and higher requirements have also been put forward for autonomous navigation technology. With great advantages in accuracy, reliability, and other aspects, autonomous navigation technology based on multi-source information fusion is bound to flourish and make greater breakthroughs in the future, playing an important role in deep-space exploration missions.

References

1. Wang, D., M. Li, and X. Huang. 2016. Review of the mars atmospheric entry autonomous navigation technology. *Aerospace Control and Application* 42 (5): 1–7.
2. Wang, D.Y., and X. Huang. 2009. Survey of autonomous navigation and control for deep space exploration. *Aerospace Control and Application* 35 (3): 6–12.
3. Wang, D.Y., C. Xu, and X. Huang. 2016. Overview of autonomous navigation based on sequential images for planetary landing. *Journal of Harbin Institute of Technology* 48 (4): 1–12.
4. Anzalone, E., J. Chuang, and C. Olsen. 2014. Unified simulation and analysis framework for deep space navigation design. In *Guidance, Navigation, and Control Conference*, Breckenridge, CO.
5. Anzalone, E., et al. 2015. Multi-spacecraft autonomous positioning system: LEO demo development. In *29th Annual AIAA/USU Conference on Small Satellites*, Logan, UT.
6. Zheng, W., L. Zhang, and Y.D. Wang. 2017. Design of deep space autonomous navigation system based on spacecraft networking. *Journal of Deep Space Exploration* 4 (1): 31–37.
7. Wu, W.R., D. Wang, and X. Ning. 2011. *Autonomous Navigation Principle and Technology for Deep Space Spacecraft*. Beijing China Astronautic Publishing House.
8. Yoshimitsu, T., et al. 2009. Hayabusa-final autonomous descent and landing based on target marker tracking. *Acta Astronautica* 65 (5): 657–665.
9. Riedel, J.E., et al. 2010. Optical navigation plan and strategy for the Lunar Lander Altair. In *AIAA Guidance, Navigation, and Control Conference*, Toronto, Ontario, Canada.

Appendix A
Constants and Unit Conversion

A.1 Constants

Meter (m), kilogram (kg), and second (s) are the units of length, mass, and time, respectively, in the international system of units (SI).

1. Velocity of light: 299,792,458 m/s
2. Equatorial radius of the earth: 6,378,140 m
3. Polar radius of the earth: 6,356,775 m
4. Average radius of the earth: 6,371,004 m
5. Equator circumference of the earth: 40,075.13 km
6. Gravitational acceleration on the earth surface: 9.8061 m/s^2
7. Average orbital velocity of the earth: 29.79 km/s
8. Earth gravitational constant: 3.986005×10^{14} m^3/s^2
9. The flattening of the earth ellipsoid: 0.00335281
10. Earth mass: 5.9742×10^{24} kg
11. Mean sun–earth distance (1 AU): $1.49597870 \times 10^{11}$ m
12. Actual distance from the sun to the earth: 1.4710×10^{11}–1.5210×10^{11} m
13. Heliocentric gravitational constant: $1.32712438 \times 10^{20}$ m^3/s^2
14. Solar mass: 1.9891×10^{30} kg
15. Solar radius: 696,265 km
16. Apparent magnitude of the sun: −26.74
17. Gravitational acceleration on the solar surface: 274 m/s^2
18. Lunar mass: 7.3506×10^{22} kg
19. Lunar diameter: 3476.4 km
20. The flattening of the moon ellipsoid: 0.0012
21. Mean earth–moon distance: 384,401 km
22. Average lunar orbital velocity: 1 km/s
23. Full-moon brightness: −12.7
24. Gravitational acceleration on the lunar surface: 1.62 m/s^2
25. Mars gravitational constant: 4.28284×10^{13} m^3/s^2
26. Mars equatorial radius: 3,396,190 m

D. Wang et al., *Spacecraft Autonomous Navigation Technologies Based on Multi-source Information Fusion*, Space Science and Technologies, https://doi.org/10.1007/978-981-15-4879-6

27. Mars polar radius: 3,376,200 m
28. 1 light-year (ly): 9.4605536×10^{15} m
29. 1 parsec (pc): 3.085678×10^{16} m (1pc = 1AU/tan(1″)).

A.2 Unit Conversion

A.2.1 Time Conversion

1. 1 Julian century = 100 Julian days (d) [1]
2. 1 Julian year (yr) = 365.2500 Julian days (d) = 8766 hours (hr) = 31,557,600 seconds (s)
3. 1 second (s) = 10^3 milliseconds (ms) = 10^6 microseconds (μs) = 10^9 nanoseconds (ns) = 10^{12} picoseconds (ps) = 10^{15} femtoseconds (fs) = 10^{18} attoseconds.

A.2.2 Angle Conversion

1. 1 radian (rad) = 57.29577951 degrees (°)
2. 1 degree (°) = 60 minutes of arc (′) = 3600 seconds of arc (″)
3. 1 second of arc (″) = 10^3 milliseconds of arc (mas) = 10^6 microseconds of arc (μas).

Appendix B
Matrix-Related Knowledge

B.1 Matrix Trace

The sum of the diagonal elements of the square matrix $\boldsymbol{A} \in \mathbb{R}^{n \times n}$ is called the trace of $\boldsymbol{A}$, denoted as $\mathrm{Tr}[\boldsymbol{A}]$, and given by the following formula

$$\mathrm{Tr}[\boldsymbol{A}] = \sum_{i=1}^{n} a_{ii} \tag{B.1}$$

where a_{ii} is the ith diagonal element of $\boldsymbol{A}$. The formulas related to trace are as follows [4]:

$$\begin{aligned}
\mathrm{Tr}[\boldsymbol{A}^{\mathrm{T}}] &= \mathrm{Tr}[\boldsymbol{A}] \\
\mathrm{Tr}[\boldsymbol{A}] &= \sum_{i=1}^{n} \lambda_i(\boldsymbol{A}) \\
\mathrm{Tr}[\alpha \boldsymbol{A}] &= \alpha \mathrm{Tr}[\boldsymbol{A}] \\
\mathrm{Tr}[\boldsymbol{A} + \boldsymbol{B}) &= \mathrm{Tr}[\boldsymbol{B} + \boldsymbol{A}] \\
\mathrm{Tr}[\boldsymbol{A}\boldsymbol{B}) &= \mathrm{Tr}[\boldsymbol{B}\boldsymbol{A}] \\
\mathrm{Tr}[\boldsymbol{A}\boldsymbol{B}\boldsymbol{C}\boldsymbol{D}) &= \mathrm{Tr}[\boldsymbol{B}\boldsymbol{C}\boldsymbol{D}\boldsymbol{A}] = \mathrm{Tr}[\boldsymbol{C}\boldsymbol{D}\boldsymbol{A}\boldsymbol{B}] = \mathrm{Tr}[\boldsymbol{D}\boldsymbol{A}\boldsymbol{B}\boldsymbol{C}]
\end{aligned} \tag{B.2}$$

where $\boldsymbol{A}, \boldsymbol{B}, \boldsymbol{C}, \boldsymbol{D} \in \mathbb{R}^{n \times n}$; $\lambda_i(\boldsymbol{A})$ is the ith eigenvalue of $\boldsymbol{A}$. If $\boldsymbol{x}, \boldsymbol{y} \in \mathbb{R}^n$, then

$$\begin{aligned}
\mathrm{Tr}[\boldsymbol{x}\boldsymbol{y}^{\mathrm{T}}] &= \boldsymbol{x}^{\mathrm{T}}\boldsymbol{y} \\
\mathrm{Tr}[\boldsymbol{A}\boldsymbol{y}\boldsymbol{x}^{\mathrm{T}}] &= \mathrm{Tr}[\boldsymbol{x}^{\mathrm{T}}\boldsymbol{A}\boldsymbol{y}]
\end{aligned} \tag{B.3}$$

D. Wang et al., *Spacecraft Autonomous Navigation Technologies Based on Multi-source Information Fusion*, Space Science and Technologies, https://doi.org/10.1007/978-981-15-4879-6

B.2 Kronecker Operator

Let $\boldsymbol{A} \in \mathbb{R}^{m \times n}$, $\boldsymbol{B} \in \mathbb{R}^{p \times q}$. Then, the Kronecker product of $\boldsymbol{A}$ and $\boldsymbol{B}$ is denoted as $\boldsymbol{A} \otimes \boldsymbol{B}$, and given by the following equation [5]

$$\boldsymbol{A} \otimes \boldsymbol{B} \triangleq \begin{bmatrix} a_{11}\boldsymbol{B} & \cdots & a_{1n}\boldsymbol{B} \\ \vdots & & \vdots \\ a_{m1}\boldsymbol{B} & \cdots & a_{mn}\boldsymbol{B} \end{bmatrix} \in \mathbb{R}^{mp \times nq} \tag{B.4}$$

where $\otimes$ is called Kronecker operator, a_{ij} is the (row i, column j) element of $\boldsymbol{A}$.

B.3 vec Operator

Let $\boldsymbol{A} \in \mathbb{R}^{m \times n}$. $\boldsymbol{a}_j \in \mathbb{R}^m$ is the column j of $\boldsymbol{A}$. Then, $\text{vec}(\boldsymbol{A})$ is defined as

$$\text{vec}(\boldsymbol{A}) \triangleq \begin{bmatrix} \boldsymbol{a}_1 \\ \vdots \\ \boldsymbol{a}_n \end{bmatrix} \in \mathbb{R}^{mn \times 1} \tag{B.5}$$

The formulas related to vec are as follows [5]:

$$\begin{aligned} \text{vec}(\boldsymbol{a}\boldsymbol{b}^{\mathrm{T}}) &= \boldsymbol{b} \otimes \boldsymbol{a} \\ \text{vec}(\boldsymbol{A}\boldsymbol{B}\boldsymbol{C}) &= (\boldsymbol{C}^{\mathrm{T}} \otimes \boldsymbol{A})\text{vec}(\boldsymbol{B}) \\ \text{Tr}[\boldsymbol{A}\boldsymbol{B}\boldsymbol{C}\boldsymbol{D}] &= \left(\text{vec}(\boldsymbol{D}^{\mathrm{T}})\right)^{\mathrm{T}}(\boldsymbol{C}^{T} \otimes \boldsymbol{A})\text{vec}(\boldsymbol{B}) = (\text{vec}(\boldsymbol{D}))^{\mathrm{T}}(\boldsymbol{A} \otimes \boldsymbol{C}^{\mathrm{T}})\text{vec}(\boldsymbol{B}^{\mathrm{T}}) \end{aligned} \tag{B.6}$$

B.4]

Matrix Calculus [4, 6]

1. Partial derivative of function: Let $f(\boldsymbol{x}) : \mathbb{R}^n \to \mathbb{R}$, where $\boldsymbol{x} = [x_1, \ldots, x_n]^{\mathrm{T}} \in \mathbb{R}^n$, and $\mathbb{R}^n \to \mathbb{R}$ represents the mapping of $\mathbb{R}^n$ onto $\mathbb{R}$. Define the partial derivative of f with respect to x as a row vector, that is,

$$\frac{\partial f}{\partial \boldsymbol{x}} \triangleq \left[\frac{\partial f}{\partial x_1} \cdots \frac{\partial f}{\partial x_n}\right] \tag{B.7}$$

Note: some partial derivatives in this book are also expressed by the gradient operator ∇, for example, $\frac{\partial f}{\partial \boldsymbol{x}}$ is expressed by $\nabla_{\boldsymbol{x}} f$.

2. Partial derivative of vector function: Let $\boldsymbol{x} \in \mathbb{R}^n, \boldsymbol{z} \in \mathbb{R}^k$. Then, the partial derivative of vector function is defined as

$$\frac{\partial \boldsymbol{x}}{\partial \boldsymbol{z}} = \begin{bmatrix} \frac{\partial x_1}{\partial \boldsymbol{z}} \\ \vdots \\ \frac{\partial x_n}{\partial \boldsymbol{z}} \end{bmatrix} = \begin{bmatrix} \frac{\partial x_1}{\partial z_1} & \frac{\partial x_1}{\partial z_2} & \cdots & \frac{\partial x_1}{\partial z_k} \\ \frac{\partial x_2}{\partial z_1} & \frac{\partial x_2}{\partial z_2} & \cdots & \frac{\partial x_2}{\partial z_k} \\ \vdots & \vdots & \ddots & \vdots \\ \frac{\partial x_n}{\partial z_1} & \frac{\partial x_n}{\partial z_2} & \cdots & \frac{\partial x_n}{\partial z_k} \end{bmatrix} \in \mathbb{R}^{n \times k} \tag{B.8}$$

Note: In some literatures, $\frac{\partial \boldsymbol{x}}{\partial \boldsymbol{z}}$ is defined as the transpose matrix on the right side of Eq. (B.8).

3. Partial derivative of matrix with respect to scalar: Let $\boldsymbol{A} \in \mathbb{R}^{m \times n}, \alpha \in \mathbb{R}$. Then, the partial derivative of matrix with respect to a scalar is defined as

$$\frac{\partial \boldsymbol{A}}{\partial \alpha} = \begin{bmatrix} \frac{\partial a_{11}}{\partial \alpha} & \frac{\partial a_{12}}{\partial \alpha} & \cdots & \frac{\partial a_{1n}}{\partial \alpha} \\ \frac{\partial a_{21}}{\partial \alpha} & \frac{\partial a_{22}}{\partial \alpha} & \cdots & \frac{\partial a_{2n}}{\partial \alpha} \\ \vdots & \vdots & & \vdots \\ \frac{\partial a_{m1}}{\partial \alpha} & \frac{\partial a_{11}}{\partial \alpha} & \cdots & \frac{\partial a_{mn}}{\partial \alpha} \end{bmatrix} \tag{B.9}$$

4. Chain rule: If $\boldsymbol{x}, \boldsymbol{y}, \boldsymbol{z}$ are all vectors fields, then

$$\frac{\partial \boldsymbol{x}}{\partial \boldsymbol{z}} = \frac{\partial \boldsymbol{x}}{\partial \boldsymbol{y}} \frac{\partial \boldsymbol{y}}{\partial \boldsymbol{z}} \tag{B.10}$$

Note: If $\frac{\partial \boldsymbol{x}}{\partial \boldsymbol{z}}$ is defined as the transpose matrix on the right side of Eq. (B.8), then the chain rule can be written as $\frac{\partial \boldsymbol{x}}{\partial \boldsymbol{z}} = \frac{\partial \boldsymbol{y}}{\partial \boldsymbol{z}} \frac{\partial \boldsymbol{x}}{\partial \boldsymbol{y}}$.

5. If $f(\boldsymbol{x}) : \mathbb{R}^n \to \mathbb{R}$ and $z = z(\boldsymbol{x})$, then

$$\frac{f(\boldsymbol{x})z}{\boldsymbol{x}} = z\frac{\partial f(\boldsymbol{x})}{\partial \boldsymbol{x}} + f(\boldsymbol{x})\frac{\partial z}{\partial \boldsymbol{x}} \tag{B.11}$$

If $\boldsymbol{y} = \boldsymbol{y}(z) \in \mathbb{R}^m, \boldsymbol{x} = \boldsymbol{x}(z) \in \mathbb{R}^n, \boldsymbol{A} \in \mathbb{R}^{m \times n}$, then

$$\frac{\partial \boldsymbol{y}^{\mathrm{T}} \boldsymbol{A} \boldsymbol{x}}{\partial z} = \boldsymbol{x}^{\mathrm{T}} \boldsymbol{A}^{\mathrm{T}} \frac{\partial \boldsymbol{y}}{\partial z} + \boldsymbol{y}^{\mathrm{T}} \boldsymbol{A} \frac{\partial \boldsymbol{x}}{\partial z} \tag{B.12}$$

6. Partial derivatives of several common vector operations:

$$\frac{\partial \boldsymbol{a}^{\mathrm{T}} \boldsymbol{x}}{\partial \boldsymbol{x}} = \frac{\partial \boldsymbol{x}^{\mathrm{T}} \boldsymbol{a}}{\partial \boldsymbol{x}} = \boldsymbol{a}^{\mathrm{T}}$$
$$\frac{\partial \|\boldsymbol{x}\|^2}{\partial \boldsymbol{x}} = \frac{\boldsymbol{x}^{\mathrm{T}} \boldsymbol{x}}{\partial \boldsymbol{x}} = 2\boldsymbol{x}^{\mathrm{T}}$$
$$\frac{\partial \|\boldsymbol{x} - \boldsymbol{a}\|}{\partial \boldsymbol{x}} = \frac{(\boldsymbol{x} - \boldsymbol{a})^{\mathrm{T}}}{|\boldsymbol{x} - \boldsymbol{a}|}$$

$$\begin{aligned} \frac{\partial 1/\|\boldsymbol{x}-\boldsymbol{a}\|}{\partial \boldsymbol{x}} &= -\frac{(\boldsymbol{x}-\boldsymbol{a})^{\mathrm{T}}}{|\boldsymbol{x}-\boldsymbol{a}|^{3}} \\ \frac{\partial \frac{\boldsymbol{x}-\boldsymbol{a}}{\|\boldsymbol{x}-\boldsymbol{a}\|}}{\partial \boldsymbol{x}} &= \frac{1}{\|\boldsymbol{x}-\boldsymbol{a}\|}\left(\boldsymbol{I}-\frac{\boldsymbol{x}-\boldsymbol{a}}{|\boldsymbol{x}-\boldsymbol{a}|}\frac{(\boldsymbol{x}-\boldsymbol{a})^{\mathrm{T}}}{|\boldsymbol{x}-\boldsymbol{a}|}\right) \\ \frac{\partial \boldsymbol{A}\boldsymbol{x}}{\partial \boldsymbol{x}} &= \boldsymbol{A} \end{aligned} \tag{B.13}$$

where $\boldsymbol{a}, \boldsymbol{x} \in \mathbb{R}^n$; $\boldsymbol{A} \in \mathbb{R}^{m\times n}$; $\|\cdot\|$ represent the 2-norm of a vector.

7. Partial derivative of trace with respect to a matrix:

$$\begin{aligned} \frac{\partial}{\partial \boldsymbol{A}}\mathrm{Tr}[\boldsymbol{A}^{\alpha}] &= \alpha(\boldsymbol{A}^{\alpha-1})^{\mathrm{T}} \\ \frac{\partial}{\partial \boldsymbol{A}}\mathrm{Tr}[\boldsymbol{C}\boldsymbol{A}^{-1}\boldsymbol{B}] &= -\boldsymbol{A}^{-\mathrm{T}}\boldsymbol{C}\boldsymbol{B}\boldsymbol{A}^{-\mathrm{T}} \\ \frac{\partial}{\partial \boldsymbol{A}}\mathrm{Tr}[\boldsymbol{C}\boldsymbol{A}\boldsymbol{B}\boldsymbol{A}^{\mathrm{T}}\boldsymbol{D}] &= \boldsymbol{C}^{\mathrm{T}}\boldsymbol{D}^{\mathrm{T}}\boldsymbol{A}\boldsymbol{B}^{\mathrm{T}} + \boldsymbol{D}\boldsymbol{C}\boldsymbol{A}\boldsymbol{B} \\ \frac{\partial}{\partial \boldsymbol{A}}\mathrm{Tr}[\boldsymbol{C}^{\mathrm{T}}\boldsymbol{A}\boldsymbol{B}^{\mathrm{T}}] &= \frac{\partial}{\partial \boldsymbol{A}}\mathrm{Tr}[\boldsymbol{B}\boldsymbol{A}^{\mathrm{T}}\boldsymbol{C}] = \boldsymbol{C}\boldsymbol{B} \end{aligned} \tag{B.14}$$

8. Partial derivative of a quadratic or cubic form with respect to a vector or matrix:

$$\begin{aligned} \frac{\partial}{\partial \boldsymbol{x}}(\boldsymbol{A}\boldsymbol{x}+\boldsymbol{b})^{\mathrm{T}}\boldsymbol{C}(\boldsymbol{D}\boldsymbol{x}+\boldsymbol{e}) &= \boldsymbol{A}^{\mathrm{T}}\boldsymbol{C}(\boldsymbol{D}\boldsymbol{x}+\boldsymbol{e}) + \boldsymbol{D}^{\mathrm{T}}\boldsymbol{C}^{\mathrm{T}}(\boldsymbol{A}\boldsymbol{x}+\boldsymbol{b}) \\ \frac{\partial}{\partial \boldsymbol{A}}(\boldsymbol{A}\boldsymbol{a}+\boldsymbol{b})^{\mathrm{T}}\boldsymbol{C}(\boldsymbol{A}\boldsymbol{a}+\boldsymbol{b}) &= (\boldsymbol{C}+\boldsymbol{C}^{\mathrm{T}})(\boldsymbol{A}\boldsymbol{a}+\boldsymbol{b})\boldsymbol{a}^{\mathrm{T}} \\ \frac{\partial}{\partial \boldsymbol{x}}(\boldsymbol{x}^{\mathrm{T}}\boldsymbol{A}\boldsymbol{x}\boldsymbol{x}^{\mathrm{T}}) &= (\boldsymbol{A}+\boldsymbol{A}^{\mathrm{T}})\boldsymbol{x}\boldsymbol{x}^{\mathrm{T}} + (\boldsymbol{x}^{\mathrm{T}}\boldsymbol{A}\boldsymbol{x})\boldsymbol{I} \end{aligned} \tag{B.15}$$

9. Partial derivative operation related to determinant:

$$\begin{aligned} \frac{\partial}{\partial \boldsymbol{A}}\det(\boldsymbol{A}) &= \frac{\partial}{\partial \boldsymbol{A}}\det(\boldsymbol{A}^{\mathrm{T}}) = [\mathrm{adj}(\boldsymbol{A})]^{\mathrm{T}} \\ \frac{\partial}{\partial \boldsymbol{A}}\det(\boldsymbol{C}\boldsymbol{A}\boldsymbol{B}) &= \det(\boldsymbol{C}\boldsymbol{A}\boldsymbol{B})\boldsymbol{A}^{-\mathrm{T}} \\ \frac{\partial}{\partial \boldsymbol{A}}\ln[\det(\boldsymbol{C}\boldsymbol{A}\boldsymbol{B})] &= \boldsymbol{A}^{-\mathrm{T}} \\ \frac{\partial}{\partial \boldsymbol{A}}\det(\boldsymbol{A}^{\alpha}) &= \alpha\det(\boldsymbol{A}^{\alpha})\boldsymbol{A}^{-\mathrm{T}} \\ \frac{\partial}{\partial \boldsymbol{A}}\ln\left[\det(\boldsymbol{A}^{\alpha})\right] &= \alpha\boldsymbol{A}^{-\mathrm{T}} \\ \frac{\partial}{\partial \boldsymbol{A}}\det(\boldsymbol{A}^{\mathrm{T}}\boldsymbol{C}\boldsymbol{A}) &= \det(\boldsymbol{A}^{\mathrm{T}}\boldsymbol{C}\boldsymbol{A})(\boldsymbol{C}+\boldsymbol{C}^{\mathrm{T}})\boldsymbol{A}(\boldsymbol{A}^{\mathrm{T}}\boldsymbol{C}\boldsymbol{A})^{-1} \\ \frac{\partial}{\partial \boldsymbol{A}}\ln\left[\det(\boldsymbol{A}^{\mathrm{T}}\boldsymbol{C}\boldsymbol{A})\right] &= (\boldsymbol{C}+\boldsymbol{C}^{\mathrm{T}})\boldsymbol{A}(\boldsymbol{A}^{\mathrm{T}}\boldsymbol{C}\boldsymbol{A})^{-1} \end{aligned} \tag{B.16}$$

where adj(A) is the adjoint matrix of A.

10. Partial derivative operation related to matrix inversion:

$$\frac{\partial \boldsymbol{A}^{-1}}{\partial \alpha} = -\boldsymbol{A}^{-1}\frac{\partial \boldsymbol{A}}{\partial \alpha}\boldsymbol{A}^{-1}$$
$$\frac{\partial \boldsymbol{a}^{\mathrm{T}}\boldsymbol{A}^{-1}\boldsymbol{b}}{\partial \boldsymbol{A}} = -\boldsymbol{A}^{-\mathrm{T}}\boldsymbol{a}\boldsymbol{b}^{\mathrm{T}}\boldsymbol{A}^{-\mathrm{T}} \tag{B.17}$$

B.5 Cross-Product Algorithm

If $\boldsymbol{x} = [x_1, x_2, x_3]^{\mathrm{T}} \in \mathbb{R}^3$, the cross-product operator of $\boldsymbol{x}$ can be defined as

$$[\boldsymbol{x}\times] \triangleq \begin{bmatrix} 0 & -x_3 & x_2 \\ x_3 & 0 & -x_1 \\ -x_2 & x_1 & 0 \end{bmatrix} \tag{B.18}$$

Some properties of cross-product operator are as follows:

$$[\boldsymbol{x}\times]\boldsymbol{y} = \boldsymbol{x} \times \boldsymbol{y} = -\boldsymbol{y} \times \boldsymbol{x}$$
$$\boldsymbol{x}^{\mathrm{T}}(\boldsymbol{y} \times \boldsymbol{z}) = (\boldsymbol{x} \times \boldsymbol{y})^{\mathrm{T}}\boldsymbol{z}$$
$$[\boldsymbol{x}\times]^2 = -(\boldsymbol{x}^T\boldsymbol{x})\boldsymbol{I} + \boldsymbol{x}\boldsymbol{x}^{\mathrm{T}} \tag{B.19}$$

where $\boldsymbol{y}, \boldsymbol{z} \in \mathbb{R}^3$.

B.6 Matrix Equalities

Sherman–Morrison lemma:

$$(\boldsymbol{I} + \boldsymbol{A}\boldsymbol{B})^{-1} = \boldsymbol{I} - \boldsymbol{A}(\boldsymbol{I} + \boldsymbol{B}\boldsymbol{A})^{-1}\boldsymbol{B} \tag{B.20}$$

Matrix inversion lemma:

$$(\boldsymbol{A} + \boldsymbol{B}\boldsymbol{C}\boldsymbol{D})^{-1} = \boldsymbol{A}^{-1} - \boldsymbol{A}^{-1}\boldsymbol{B}(\boldsymbol{D}\boldsymbol{A}^{-1}\boldsymbol{B} + \boldsymbol{C}^{-1})\boldsymbol{D}\boldsymbol{A}^{-1} \tag{B.21}$$

Lemma B.1

$$(\boldsymbol{P}^{-1}+\boldsymbol{H}^{\mathrm{T}}\boldsymbol{R}^{-1}\boldsymbol{H})^{-1}\boldsymbol{H}^{\mathrm{T}}\boldsymbol{R}^{-1} = \boldsymbol{P}\boldsymbol{H}^{\mathrm{T}}(\boldsymbol{H}\boldsymbol{P}\boldsymbol{H}^{\mathrm{T}} + \boldsymbol{R})^{-1} \tag{B.22}$$

Proof At first, from Eq. (B.21), it can be derived that

$$\boldsymbol{P}\boldsymbol{H}^{T}(\boldsymbol{H}\boldsymbol{P}\boldsymbol{H}^{\mathrm{T}} + \boldsymbol{R})^{-1} = \boldsymbol{P}\boldsymbol{H}^{\mathrm{T}}[\boldsymbol{R}^{-1} - \boldsymbol{R}^{-1}\boldsymbol{H}(\boldsymbol{P}^{-1} + \boldsymbol{H}^{\mathrm{T}}\boldsymbol{R}^{-1}\boldsymbol{H})^{-1}\boldsymbol{H}^{\mathrm{T}}\boldsymbol{R}^{-1}]$$

$$
\begin{aligned}
&= \boldsymbol{P}\big[\boldsymbol{H}^{\mathrm{T}}\boldsymbol{R}^{-1} - \boldsymbol{H}^{\mathrm{T}}\boldsymbol{R}^{-1}\boldsymbol{H}(\boldsymbol{P}^{-1} + \boldsymbol{H}^{\mathrm{T}}\boldsymbol{R}^{-1}\boldsymbol{H})^{-1}\boldsymbol{H}^{\mathrm{T}}\boldsymbol{R}^{-1}\big] \\
&= \boldsymbol{P}\big[\boldsymbol{I} - \boldsymbol{H}^{\mathrm{T}}\boldsymbol{R}^{-1}\boldsymbol{H}(\boldsymbol{P}^{-1} + \boldsymbol{H}^{\mathrm{T}}\boldsymbol{R}^{-1}\boldsymbol{H})^{-1}\big]\boldsymbol{H}^{\mathrm{T}}\boldsymbol{R}^{-1} \\
&= \boldsymbol{P}\big[\boldsymbol{I} - \boldsymbol{H}^{\mathrm{T}}\boldsymbol{R}^{-1}\boldsymbol{H}(\boldsymbol{I} + \boldsymbol{H}^{\mathrm{T}}\boldsymbol{R}^{-1}\boldsymbol{H})^{-1}\boldsymbol{P}\big]\boldsymbol{H}^{\mathrm{T}}\boldsymbol{R}^{-1}
\end{aligned} \tag{B.23}
$$

then from Eq. (B.20), it can be derived that

$$
\begin{aligned}
\boldsymbol{P}\boldsymbol{H}^{\mathrm{T}}(\boldsymbol{H}\boldsymbol{P}\boldsymbol{H}^{\mathrm{T}} + \boldsymbol{R})^{-1} &= \boldsymbol{P}\big[\boldsymbol{I} - \boldsymbol{H}^{\mathrm{T}}\boldsymbol{R}^{-1}\boldsymbol{H}(\boldsymbol{I} + \boldsymbol{H}^{\mathrm{T}}\boldsymbol{R}^{-1}\boldsymbol{H})^{-1}\boldsymbol{P}\big]\boldsymbol{H}^{\mathrm{T}}\boldsymbol{R}^{-1} \\
&= \boldsymbol{P}(\boldsymbol{I}+\boldsymbol{H}^{\mathrm{T}}\boldsymbol{R}^{-1}\boldsymbol{H}\boldsymbol{P})^{-1}\boldsymbol{H}^{\mathrm{T}}\boldsymbol{R}^{-1} \\
&= (\boldsymbol{P}^{-1}+\boldsymbol{H}^{\mathrm{T}}\boldsymbol{R}^{-1}\boldsymbol{H})^{-1}\boldsymbol{H}^{\mathrm{T}}\boldsymbol{R}^{-1}
\end{aligned} \tag{B.24}
$$

which completes the proof.

Remark Equation (B.22) is often used to derive different forms of Kalman filtering.

B.7 Matrix Inequalities

Definition B.1 In the real number domain, $\boldsymbol{M} \in \mathbb{R}^{n\times n}$ is a positive definite matrix, if and only if for $\boldsymbol{x} \in \mathbb{R}^n$ ($\boldsymbol{x} \neq 0$), the following inequality is hold

$$
\boldsymbol{x}^{\mathrm{T}}\boldsymbol{M}\boldsymbol{x} > 0 \tag{B.25}
$$

Definition B.2 In the real number domain, $\boldsymbol{M} \in \mathbb{R}^{n\times n}$ is a positive semi-definite matrix, if and only if for $\boldsymbol{x} \in \mathbb{R}^n$ $\boldsymbol{x} \neq 0$, the following inequality is hold

$$
\boldsymbol{x}^{\mathrm{T}}\boldsymbol{M}\boldsymbol{x} \geq 0 \tag{B.26}
$$

Theorem B.1 *Suppose that the symmetric matrix*$\mathcal{A}$*can be written into a partitioned form*

$$
\mathcal{A} \triangleq \begin{bmatrix} \boldsymbol{A} & \boldsymbol{C} \\ \boldsymbol{C}^{\mathrm{T}} & \boldsymbol{B} \end{bmatrix} \tag{B.27}
$$

where$\boldsymbol{A}$, $\boldsymbol{B}$*are all square matrices. Then* [7]

(1) $\mathcal{A} > \mathbf{0}$, *if and only if*

$$
\begin{aligned}
&\boldsymbol{A} > 0 \\
&\boldsymbol{B} > \boldsymbol{C}^{\mathrm{T}}\boldsymbol{A}^{-1}\boldsymbol{C}
\end{aligned} \tag{B.28}
$$

(2) $\mathcal{A} > \mathbf{0}$, *if and only if*

$$\begin{aligned} &\boldsymbol{B} > \mathbf{0} \\ &\boldsymbol{A} > \boldsymbol{C}\boldsymbol{B}^{-1}\boldsymbol{C}^{\mathrm{T}} \end{aligned} \tag{B.29}$$

(3) $\boldsymbol{A} > 0$, *then* $\mathcal{A} \geq \mathbf{0}$, *if and only if*

$$\boldsymbol{B} \geq \boldsymbol{C}^{\mathrm{T}}\boldsymbol{A}^{-1}\boldsymbol{C} \tag{B.30}$$

(4) $\boldsymbol{B} > 0$, *then* $\mathcal{A} \geq \mathbf{0}$, *if and only if*

$$\boldsymbol{A} \geq \boldsymbol{C}\boldsymbol{B}^{-1}\boldsymbol{C}^{\mathrm{T}} \tag{B.31}$$

Corollary *As far as* Eq. (B.27) *is concerned, if* $\mathcal{A} > \mathbf{0}$, *then the following inequality will be hold for* $\omega \in [0, 1]$

$$\begin{bmatrix} \boldsymbol{A} & \boldsymbol{C} \\ \boldsymbol{C}^{\mathrm{T}} & \boldsymbol{B} \end{bmatrix} \leq \begin{bmatrix} \frac{1}{\omega}\boldsymbol{A} & \mathbf{0} \\ \mathbf{0} & \frac{1}{1-\omega}\boldsymbol{B} \end{bmatrix} \tag{B.32}$$

Proof $\mathcal{A} > \mathbf{0}$, so

$$\begin{aligned} &\boldsymbol{A} > \mathbf{0} \\ &\boldsymbol{B} > \boldsymbol{C}^{\mathrm{T}}\boldsymbol{A}^{-1}\boldsymbol{C} \end{aligned} \tag{B.33}$$

While

$$\begin{bmatrix} \frac{1}{\omega}\boldsymbol{A} & \mathbf{0} \\ \mathbf{0} & \frac{1}{1-\omega}\boldsymbol{B} \end{bmatrix} - \begin{bmatrix} \boldsymbol{A} & \boldsymbol{C} \\ \boldsymbol{C}^{\mathrm{T}} & \boldsymbol{B} \end{bmatrix} = \begin{bmatrix} \frac{1-\omega}{\omega}\boldsymbol{A} & -\boldsymbol{C} \\ -\boldsymbol{C}^{\mathrm{T}} & \frac{\omega}{1-\omega}\boldsymbol{B} \end{bmatrix} \tag{B.34}$$

If $\omega \in [0, 1]$, then

$$\begin{aligned} &\frac{1-\omega}{\omega}\boldsymbol{A} \geq 0 \\ &\frac{\omega}{1-\omega}\boldsymbol{B} - \boldsymbol{C}^{\mathrm{T}}\left(\frac{1-\omega}{\omega}\boldsymbol{A}\right)^{-1}\boldsymbol{C} = \frac{\omega}{1-\omega}(\boldsymbol{B} - \boldsymbol{C}^{\mathrm{T}}\boldsymbol{A}^{-1}\boldsymbol{C}) \geq 0 \end{aligned} \tag{B.35}$$

If $\omega \in (0, 1)$, then it can be derived from Eq. (B.28) that

$$\begin{bmatrix} \frac{1-\omega}{\omega}\boldsymbol{A} & -\boldsymbol{C} \\ -\boldsymbol{C}^{\mathrm{T}} & \frac{\omega}{1-\omega}\boldsymbol{B} \end{bmatrix} > \mathbf{0} \tag{B.36}$$

that is,

$$\begin{bmatrix} \boldsymbol{A} & \boldsymbol{C} \\ \boldsymbol{C}^{\mathrm{T}} & \boldsymbol{B} \end{bmatrix} < \begin{bmatrix} \frac{1}{\omega}\boldsymbol{A} & \boldsymbol{0} \\ \boldsymbol{0} & \frac{1}{1-\omega}\boldsymbol{B} \end{bmatrix} \tag{B.37}$$

If $\omega = 0, 1$, then

$$\begin{bmatrix} \boldsymbol{A} & \boldsymbol{C} \\ \boldsymbol{C}^{\mathrm{T}} & \boldsymbol{B} \end{bmatrix} \leq \begin{bmatrix} \frac{1}{\omega}\boldsymbol{A} & \boldsymbol{0} \\ \boldsymbol{0} & \frac{1}{1-\omega}\boldsymbol{B} \end{bmatrix} \tag{B.38}$$

So this proposition is proved.
Note:

(1) Equation (B.32) can be used to derive the covariance intersection (CI) algorithm.
(2) Equation (B.32) is a 2×2 block matrix, which can be further generalized to the case of $m \times m$ $(m > 2)$ block matrix.

Corollary *If $\boldsymbol{Q} \geq 0$, then*

$$\begin{bmatrix} \boldsymbol{Q} & \cdots & \boldsymbol{Q} \\ \vdots & \ddots & \\ \boldsymbol{Q} & & \boldsymbol{Q} \end{bmatrix} \leq \begin{bmatrix} \gamma_1 \boldsymbol{Q} & \cdots & \boldsymbol{0} \\ \vdots & \ddots & \\ \boldsymbol{0} & & \gamma_N \boldsymbol{Q} \end{bmatrix}, \tag{B.39}$$

where

$$\begin{aligned} &\frac{1}{\gamma_1} + \cdots + \frac{1}{\gamma_N} = 1 \\ &0 \leq 1/\gamma_i \leq 1 \\ &1 \leq \gamma_i \leq \infty \end{aligned} \tag{B.40}$$

Note:
Equation (B.39) can be used to derive the federated filtering algorithm.

Appendix C
Probability-Related Knowledge

C.1 Basic Concepts

C.1.1 Probability Axioms

The sample space for random events is denoted as Ω, a subset of which is called event. If the real function $P(A)$ satisfies the following 3 axioms [8, 9] for each event A in Ω, then P is called probability function and $P(A)$ is called the probability of the event A.

Axiom 1 (non-negativity): $P(A) \geq 0$.
Axiom 2 (normalization): The probability of the entire sample space is 1, i.e., $P(\Omega) = 1$.
Axiom 3 (countable additivity): For countable events $\{A_i\}_{i\in\mathcal{N}}$ that are pair-wise mutually exclusive, therc is $\sum_{i=1}^{\infty} P(A_i) = P\left(\bigcup_{i=1}^{\infty} A_i\right)$, where $\mathcal{N}$ is the set of natural numbers, $\cup$ represents the union of the set, the mutex of events A_i and A_j means $A_i \cap A_j = \emptyset$, and $\emptyset$ represents the empty set.

C.1.2 Joint Probability and Conditional Probability

Definition C.1 The probability that several events occur together is called **joint probability**. For example, the joint probability of the events A and B can be expressed as $P(A \cap B)$ or $P(AB)$, $P(A, B)$.

Definition C.2 Suppose that A and B are two events in the sample space, where $P(B) > 0$. So, on the condition that the event B has occurred, the occurrence probability of the event A is called the **conditional probability** of the event A under the condition of B occurrence, denoted as $P(A|B)$, and given by the following equation

D. Wang et al., *Spacecraft Autonomous Navigation Technologies Based on Multi-source Information Fusion*, Space Science and Technologies, https://doi.org/10.1007/978-981-15-4879-6

$$P(A|B) = \frac{P(AB)}{P(B)} \tag{C.1}$$

Multiplication formula of conditional probability:

$$P(AB) = P(B|A)P(A) = P(A|B)P(B) \tag{C.2}$$

Equation (C.2) can be further generalized to multiple events: For $n \geq 2$, if $P(A_1 A_2 \ldots A_{n-1}) > 0$, then:

$$\begin{aligned} P(A_1 A_2 \ldots A_n) &= P(A_n | A_1 A_2 \ldots A_{n-1}) P(A_1 A_2 \ldots A_{n-1}) \\ &= P(A_1) P(A_2|A_1) P(A_3|A_1 A_2) \ldots P(A_n|A_1 A_2 \ldots A_{n-1}) \end{aligned} \tag{C.3}$$

C.1.3 Bayes Rule and Total Probability Formula

Definition C.3 If the events $\{B_1, B_2, \ldots, B_n\}$ satisfy

$$\begin{aligned} &B_i \cap B_j = \emptyset, \quad \forall i \neq j \\ &B_1 \cup B_2 \cdots \cup B_n = \Omega \\ &P(B_i) > 0 \end{aligned} \tag{C.4}$$

Then $\{B_1, B_2, \ldots, B_n\}$ is called a **partition** of the sample space Ω.
If $\{B_1, B_2, \ldots, B_n\}$ is a partition of the sample space Ω, then

$$\begin{aligned} P(A) &= \sum_{i=1}^{n} P(AB_i) \\ &= \sum_{i=1}^{n} P(A|B_i)P(B_i) \end{aligned} \tag{C.5}$$

Equation (C.5) is called **total probability formula.**
If $\{B_1, B_2, \ldots, B_n\}$ is a partition of the sample space Ω, then

$$\begin{aligned} P(B|A) &= \frac{P(A|B)P(B)}{P(A)} \\ &= \frac{P(A|B)P(B)}{\sum_{i=1}^{n} P(A|B_i)P(B_i)} \end{aligned} \tag{C.6}$$

Equation (C.6) is called Bayes rule.

C.1.4 Independence and Conditional Independence

Definition C.4 If the two events A and B satisfy the following equation, then A and B are independent

$$P(AB) = P(A)P(B) \tag{C.7}$$

Definition C.5 Under the given condition C, if A and B satisfy

$$P(AB|C) = P(A|C)P(B|C) \tag{C.8}$$

A and B are claimed to be **conditionally independent** under the given condition C.

C.2 A Single Random Variable

C.2.1 Distribution Function and Density Function

Definition C.6 The cumulative distribution function (CDF) of a single random variable x is denoted as $F_x(x)$ and defined as

$$F_x(x) \triangleq P(x \le x) \tag{C.9}$$

Usually, $F_x(x)$ is simplified as $F(x)$.

Definition C.7 The derivative of the CDF of a single random variable x is called probability density function (PDF) $p_x(x)$ and simplified as $p(x)$, i.e.,

$$p(x) \triangleq \frac{\mathrm{d}F(x)}{\mathrm{d}x} \tag{C.10}$$

In particular, if x is a discrete random variable with the sample space of $\{x_1, x_2, \ldots\}$, then

$$p(x) = \sum_i p_i \delta(x - x_i), \tag{C.11}$$

where $\delta(x - x_i)$ is δ function, equal to 1 if $x = x_i$ and 0 in other cases; $p_i = P(x = x_i)$ is called the probability mass function (PMF) of a discrete random variable.

The properties of PDF are:

$$p(x) \ge 0 \quad \forall -\infty < x < \infty$$

$$\int_{-\infty}^{\infty} p(x)\mathrm{d}x = 1 \tag{C.12}$$

For a discrete-value random variable, the second equation in (C.12) means $\sum_i p_i = 1$.

C.2.2 Conditional Distribution

Definition C.8 Suppose that the event B has occurred. The conditional distribution function of the random variable x is defined as

$$F_x(x|B) \triangleq P(x \leq x|B) = \frac{P(x \leq x, B)}{P(B)} \tag{C.13}$$

Usually, $F_x(x|B)$ is simplified as $F(x|B)$. Similarly, the conditional probability density function is defined as the derivative of $F(x|B)$, namely

$$p(x|B) = \frac{\mathrm{d}F(x|B)}{\mathrm{d}x} \tag{C.14}$$

C.2.3 Mean and Variance

Definition C.9 The **mean (also known as expectation)** of the random variable x is denoted as $\bar{x}$ and defined as follows

$$\bar{x} \triangleq E\{x\} = \int_{-\infty}^{\infty} xp(x)\mathrm{d}x \tag{C.15}$$

For a discrete random variable, the substitution of Eq. (C.11) into Eq. (C.15) and the utilization of

$$\int_{-\infty}^{\infty} x\delta(x - x_i)\mathrm{d}x = x_i \tag{C.16}$$

can obtain

$$\bar{x} \triangleq \sum_i x_i p_i \tag{C.17}$$

Definition C.10 The **variance** of the random variable x is denoted as σ_x^2 and defined as follows:

$$\sigma_x^2 \triangleq \mathrm{var}(x) = E\{(x - \bar{x})^2\} = \int_{-\infty}^{\infty} (x - \bar{x})^2 p(x)\mathrm{d}x, \tag{C.18}$$

where σ_x is called standard deviation. If x is a discrete random variable, then $\sigma_x^2 = \sum_i (x_i - \bar{x})^2 p_i$.

The relation between variance and mean is

$$\mathrm{var}(x) + E\{x\}^2 = E\{x^2\} \tag{C.19}$$

C.3 Binary Random Variable

C.3.1 Joint Distribution Function and Density Function

Definition C.11 The **joint distribution function** of the random variables x, y is defined as

$$F_{x,y}(x, y) = P(x \leq x, y \leq y) \tag{C.20}$$

Usually, $F_{x,y}(x, y)$ is simplified as $F(x, y)$.

Definition C.12 The **joint probability density function** of the random variables x, y is defined as

$$p(x, y) = \frac{\partial F(x, y)}{\partial x \partial y} \tag{C.21}$$

In the multivariate random variables, the statistical property of each random variable is called marginal property. Thus, for the binary random variables x, y, $F(x)$ and $p(x)$ are the marginal cumulative distribution function (marginal CDF) and **marginal probability density function** (marginal PDF) of x, respectively, while satisfying

$$p(x) = \int_{-\infty}^{\infty} p(x, y)\mathrm{d}y \tag{C.22}$$

Similarly,

$$p(y) = \int_{-\infty}^{\infty} p(x, y)\mathrm{d}x \tag{C.23}$$

C.3.2 Conditional Distribution

The marginal PDFs of the random variables x, y are $p_x(x)$ and $p_y(y)$, and then the conditional PDF of y in the case of $x = \chi$ will be

$$p_{y|x}(\gamma|\chi) = p_{y|x}(\gamma|x = \chi) = \frac{p(x, y)}{p(x)} \tag{C.24}$$

Similarly, if $y = \gamma$, the conditional PDF of x will be

$$p_{x|y}(\chi|\gamma) = p_{x|y}(\chi|y = \gamma) = \frac{p(x, y)}{p(y)} \tag{C.25}$$

Usually, $p_{\gamma|\chi}(\gamma|\chi)$ and $p_{x|y}(\chi|\gamma)$ are simplified as $p(y|x)$ and $p(x|y)$. The conditional mean is given by the following equation

$$\begin{aligned} E\{x|y\} &= \int_{-\infty}^{\infty} xp(x|y)\mathrm{d}x \\ E\{y|x\} &= \int_{-\infty}^{\infty} yp(y|x)\mathrm{d}y \end{aligned} \tag{C.26}$$

The PDF form of the Bayes rule (C.6) is given by

$$\begin{aligned} p(x|y) &= \frac{p(x|y)p(x)}{p(y)} \\ &= \frac{p(x|y)p(x)}{\sum_{-\infty}^{\infty} p(x, y)\mathrm{d}x} \end{aligned} \tag{C.27}$$

C.3.3 Relationship Between Covariance and Two Random Variables

The cross-covariance of the two random variables x, y is

$$C_{xy} \triangleq E\{(x-\bar{x})(y-\bar{y})\} \tag{C.28}$$

The correlation coefficient of the two random variables x, y is

$$\rho_{xy} = \frac{C_{xy}}{\sigma_x \sigma_y} \tag{C.29}$$

If $C_{xy} = 0$, the two random variables x, y are not related; if $E\{xy\} = 0$, they are orthogonal.

Note:

(1) C_{xy} is also denoted as $\mathrm{cov}(x, y)$.
(2) $C_{xy} = 0$ means $E\{xy\} = E\{x\}E\{y\}$.
(3) If x, y are the same random variable, then $C_{xx} = \mathrm{var}(x)$.

C.4 Random Vector

C.4.1 Joint Distribution Function and Density Function

Definition C.13 Multiple one-dimensional random variables $\{x_1, \ldots, x_n\}$ can be written into one random vector $\boldsymbol{x} = [x_1, \ldots, x_n]^{\mathrm{T}}$, whose joint CDF is defined as

$$F_{\boldsymbol{x}}(\mathbf{x}) = P(x_1 \le x_1, \ldots, x_n \le x_n) \tag{C.30}$$

Usually, $F_{\boldsymbol{x}}(\mathbf{x})$ is written as $F(\boldsymbol{x})$.

Definition C.14 The **joint PDF** of the random vector $\boldsymbol{x} = [x_1, \ldots, x_n]^{\mathrm{T}}$ is defined as

$$p_{\boldsymbol{x}}(\mathrm{x}) = \frac{\partial^n F(x_1, \ldots, x_n)}{\partial x_1 \ldots \partial x_n} \tag{C.31}$$

Usually, $p_{\boldsymbol{x}}(\mathbf{x})$ is simplified as $p(\boldsymbol{x})$.

The marginal PDF of a random variable in the random vector can be derived from the joint PDF integral, for example

$$p(x_1) = \int\int\int p(\boldsymbol{x})\mathrm{d}x_2, \ldots, \mathrm{d}x_n \tag{C.32}$$

Definition C.15 If a set of random variables $x_1, \ldots, x_n$ satisfy

$$p(x_1, x_2, \ldots, x_n) = p(x_1)p(x_2)\ldots p(x_n) \tag{C.33}$$

they are mutually (statistically) independent.

C.4.2 Formulas Related to Conditional Probability

The random vectors $\boldsymbol{x}$, $\boldsymbol{y}$, $\boldsymbol{z}$ have the following relations

$$p(\boldsymbol{x}, \mathrm{z}) = p(\boldsymbol{x}|\mathrm{z})p(\mathrm{z}) = p(\mathrm{z}|\boldsymbol{x})p(\boldsymbol{x}) \tag{C.34}$$

$$p(\boldsymbol{x}) = \int p(\boldsymbol{x}, \mathrm{z})\mathrm{dz} = \int p(\boldsymbol{x}|\mathrm{z})p(\mathrm{z})\mathrm{dz} \tag{C.35}$$

$$\int p(\boldsymbol{x}|\mathrm{z})\mathrm{d}\boldsymbol{x} = 1 \tag{C.36}$$

The integrals in Eqs. (C.35) and (C.36) are multi-vector integrals.

$$p(\boldsymbol{x}, \boldsymbol{y}|\mathrm{z}) = p(\boldsymbol{x}|\boldsymbol{y}, \mathrm{z})p(\boldsymbol{y}|\mathrm{z}) \tag{C.37}$$

From Eq. (C.34), the PDF form of Bayes formula can be derived as shown below

$$p(\boldsymbol{x}|\mathrm{z}) = \frac{p(\mathrm{z}|\boldsymbol{x})p(\boldsymbol{x})}{p(\mathrm{z})} \tag{C.38}$$

Its conditional mean satisfies

$$E\{E\{\boldsymbol{x}|\mathrm{z}\}\} = E\{\boldsymbol{x}\} \tag{C.39}$$

C.4.3 Statistical Properties of a Single Random Vector

Mean value: The mean value of multi-dimensional continuous random variables $\boldsymbol{x}$ is denoted as $\bar{\boldsymbol{x}}$ and defined as follows

$$\bar{\boldsymbol{x}} \triangleq E\{\boldsymbol{x}\} = \int_{-\infty}^{\infty} \boldsymbol{x}\, p(\boldsymbol{x})\mathrm{d}\boldsymbol{x} \tag{C.40}$$

The integral symbol in Eq. (C.40) represents multiple integral.

Autocorrelation matrix and autocovariance matrix: The autocorrelation matrix and autocovariance matrix of x are denoted as $\boldsymbol{R}_x$ and $\boldsymbol{P}_x$, and defined as

$$\begin{aligned} \boldsymbol{R}_x &\triangleq E\{\boldsymbol{x}\boldsymbol{x}^{\mathrm{T}}\} \\ \boldsymbol{P}_x &\triangleq \mathrm{var}(\boldsymbol{x}) = E\{(\boldsymbol{x}-\bar{\boldsymbol{x}})(\boldsymbol{x}-\bar{\boldsymbol{x}})^{\mathrm{T}}\} \end{aligned} \tag{C.41}$$

Sometimes, autocovariance matrix is also called "variance matrix" for short. The relation between $\boldsymbol{R}_x$ and $\boldsymbol{P}_x$ is as follows:

$$\boldsymbol{R}_x = \boldsymbol{P}_x + \overline{\boldsymbol{x}\boldsymbol{x}}^{\mathrm{T}} \tag{C.42}$$

Note:

(1) For a random vector $\boldsymbol{x}$, $\mathrm{var}(\boldsymbol{x}) = \mathrm{cov}(\boldsymbol{x}, \boldsymbol{x})$.
(2) Without causing confusion, the subscript of $\boldsymbol{P}_x$ can be omitted so that $\boldsymbol{P}_x$ can be denoted as $\boldsymbol{P}$.

C.4.4 Statistical Properties of Two Random Vectors

Autocorrelation matrix and autocovariance matrix: For two sets of multi-dimensional random variables $\boldsymbol{x}$ and y, their autocorrelation matrix and autocovariance matrix are denoted as $\boldsymbol{R}_{xy}$ and $\boldsymbol{P}_{xy}$, and defined as follows

$$\begin{aligned} \boldsymbol{R}_{xy} &\triangleq E\{\boldsymbol{x}\boldsymbol{y}^{\mathrm{T}}\} \\ \boldsymbol{P}_{xy} &\triangleq E\{(\boldsymbol{x}-\bar{\boldsymbol{x}})(\boldsymbol{y}-\bar{\boldsymbol{y}})^{\mathrm{T}}\} \end{aligned} \tag{C.43}$$

The relation between the two matrices is:

$$\boldsymbol{R}_{xy} = \boldsymbol{P}_{xy} + \overline{\boldsymbol{x}\boldsymbol{y}}^{\mathrm{T}} \tag{C.44}$$

Statistical irrelevance: If the cross-covariance matrix $\boldsymbol{P}_{xy}$ of the two sets of multi-dimensional random variables x and y is zero, x and y are claimed to be statistically uncorrelated.

Orthogonality: If the cross-correlation matrix $\boldsymbol{R}_{xy}$ of the two sets of multi-dimensional random variables x and y is zero, x and y are claimed to be orthogonal.

Note: In some literatures, $\boldsymbol{P}_x$ is also expressed by $\boldsymbol{C}_x$, and $\boldsymbol{P}_{xy}$ is expressed by $\boldsymbol{C}_{xy}$.

C.5 Gaussian Random Variable

C.5.1 Definition

The probability density function (PDF) of a Gaussian random vector $\boldsymbol{x}$ is

$$\begin{aligned} p(\boldsymbol{x}) &= \mathcal{N}(\boldsymbol{x}; \bar{\boldsymbol{x}}, \boldsymbol{P}_x) \\ &\triangleq \frac{1}{\det(2\pi \boldsymbol{P}_x)^{1/2}} \exp\left[-\frac{1}{2}(\boldsymbol{x}-\bar{\boldsymbol{x}})^{\mathrm{T}} \boldsymbol{P}_x^{-1} (\boldsymbol{x}-\bar{\boldsymbol{x}})\right] \end{aligned} \tag{C.45}$$

where $\mathcal{N}(\cdot)$ denotes a Gaussian PDF, $\det(\cdot)$ is a determinant operator, $\exp[\cdot]$ is an exponential function, $\bar{\boldsymbol{x}}$ and $\boldsymbol{P}_x$ are the mean and variance of $\boldsymbol{x}$, respectively.

Important properties of Gaussian random vectors:

1. The PDF of a Gaussian random variable is completely dependent on its mean and covariance matrices.
2. The mean value point of a Gaussian random vector is the peak point of its PDF.
3. If the components of a Gaussian random variable are statistically unrelated to each other, they are also statistically independent.
4. The linear combination of Gaussian random variables is also a Gaussian random variable. If y and x satisfy

$$\boldsymbol{y} = \boldsymbol{A}\boldsymbol{x} + \boldsymbol{b} \tag{C.46}$$

and $p(\boldsymbol{x}) = \mathcal{N}(\boldsymbol{x}; \bar{\boldsymbol{x}}, \boldsymbol{P}_x)$, then $p(\boldsymbol{y}) = \mathcal{N}(\boldsymbol{y}; \bar{\boldsymbol{y}}, \boldsymbol{P}_y)$, where

$$\begin{aligned} \bar{\boldsymbol{y}} &= \boldsymbol{A}\bar{\boldsymbol{x}} + \boldsymbol{b} \\ \boldsymbol{P}_y &= \boldsymbol{A}\boldsymbol{P}_x\boldsymbol{A}^{\mathrm{T}} \end{aligned} \tag{C.47}$$

C.5.2 Joint Gaussian Distribution

Definition Two random vectors $\boldsymbol{x}$ and $\boldsymbol{z}$ are joint Gaussian distribution vectors if $\boldsymbol{y} \triangleq [\boldsymbol{x}^{\mathrm{T}}, \boldsymbol{z}^{\mathrm{T}}]^{\mathrm{T}}$ is a Gaussian random variable [10].

Theorem C.1 *If* $\boldsymbol{x}$ and $\boldsymbol{z}$ satisfy the joint Gaussian distribution, $p(\boldsymbol{x}|\boldsymbol{z})$ is also a Gaussian PDF and can be given by the following equation

$$p(\boldsymbol{x}|\boldsymbol{z}) = \mathcal{N}(\boldsymbol{x}; \bar{\boldsymbol{x}}_{x|z}, \boldsymbol{P}_{xx|z}) \tag{C.48}$$

where

$$\begin{aligned}\bar{\boldsymbol{x}}_{x|z} &\triangleq E\{\boldsymbol{x}|\boldsymbol{z}\} = \bar{\boldsymbol{x}} + \boldsymbol{P}_{xz}\boldsymbol{P}_{zz}^{-1}(\boldsymbol{z} - \bar{\boldsymbol{z}}) \\ \boldsymbol{P}_{xx|z} &\triangleq E\{(\boldsymbol{x} - E\{\boldsymbol{x}|\boldsymbol{z}\})(\boldsymbol{x} - E\{\boldsymbol{x}|\boldsymbol{z}\})^{\mathrm{T}}|\boldsymbol{z}\} \\ &= \boldsymbol{P}_{xx} - \boldsymbol{P}_{xz}\boldsymbol{P}_{zz}^{-1}\boldsymbol{P}_{zx} \end{aligned} \tag{C.49}$$

The proof follows from [10, 11]: Define $\boldsymbol{y} \triangleq [\boldsymbol{x}^{\mathrm{T}}, \boldsymbol{z}^{\mathrm{T}}]^{\mathrm{T}}$. If $\boldsymbol{x}$ and $\boldsymbol{z}$ satisfy the joint Gaussian distribution, then

$$p(\boldsymbol{y}) = \frac{1}{\det(2\pi\boldsymbol{P}_{yy})^{1/2}} \exp\left[-\frac{1}{2}(\boldsymbol{y} - \bar{\boldsymbol{y}})^{\mathrm{T}}\boldsymbol{P}_{yy}^{-1}(\boldsymbol{y} - \bar{\boldsymbol{y}})\right] \tag{C.50}$$

and $\boldsymbol{z}$ also satisfies the Gaussian distribution. Its PDF can be written as

$$p(\boldsymbol{z}) = \frac{1}{\det(2\pi\boldsymbol{P}_{zz})^{1/2}} \exp\left[-\frac{1}{2}(\boldsymbol{z} - \bar{\boldsymbol{z}})^{\mathrm{T}}\boldsymbol{P}_{zz}^{-1}(\boldsymbol{z} - \bar{\boldsymbol{z}})\right] \tag{C.51}$$

Because

$$\boldsymbol{P}_{yy} = \begin{bmatrix} \boldsymbol{P}_{xx} & \boldsymbol{P}_{xz} \\ \boldsymbol{P}_{zx} & \boldsymbol{P}_{zz} \end{bmatrix} \tag{C.52}$$

$\boldsymbol{P}_{yy}$ can be converted to obtain

$$\begin{bmatrix} \boldsymbol{I} & -\boldsymbol{P}_{xz}\boldsymbol{P}_{zz}^{-1} \\ \boldsymbol{0} & \boldsymbol{I} \end{bmatrix} \boldsymbol{P}_{yy} \begin{bmatrix} \boldsymbol{I} & \boldsymbol{0} \\ -\boldsymbol{P}_{zz}^{-1}\boldsymbol{P}_{zx} & \boldsymbol{I} \end{bmatrix} = \begin{bmatrix} \boldsymbol{P}_{xx} - \boldsymbol{P}_{xz}\boldsymbol{P}_{zz}^{-1}\boldsymbol{P}_{zx} & \boldsymbol{0} \\ \boldsymbol{0} & \boldsymbol{P}_{zz} \end{bmatrix} \tag{C.53}$$

so

$$\begin{aligned} \boldsymbol{P}_{yy}^{-1} &= \begin{bmatrix} \boldsymbol{I} & \boldsymbol{0} \\ -\boldsymbol{P}_{zz}^{-1}\boldsymbol{P}_{zx} & \boldsymbol{I} \end{bmatrix} \begin{bmatrix} \boldsymbol{P}_{xx} - \boldsymbol{P}_{xz}\boldsymbol{P}_{zz}^{-1}\boldsymbol{P}_{zx} & \boldsymbol{0} \\ \boldsymbol{0} & \boldsymbol{P}_{zz} \end{bmatrix}^{-1} \begin{bmatrix} \boldsymbol{I} & -\boldsymbol{P}_{xz}\boldsymbol{P}_{zz}^{-1} \\ \boldsymbol{0} & \boldsymbol{I} \end{bmatrix} \\ \det(\boldsymbol{P}_{yy}) &= \det(\boldsymbol{P}_{xx} - \boldsymbol{P}_{xz}\boldsymbol{P}_{zz}^{-1}\boldsymbol{P}_{zx})\det(\boldsymbol{P}_{zz}) \end{aligned} \tag{C.54}$$

From Eq. (C.34), it can be derived that

$$\begin{aligned} p(\boldsymbol{x}|\boldsymbol{z}) &= \frac{p(\boldsymbol{x}, \boldsymbol{z})}{p(\boldsymbol{z})} \\ &= \frac{1}{\det(2\pi(\boldsymbol{P}_{xx} - \boldsymbol{P}_{xz}\boldsymbol{P}_{zz}^{-1}\boldsymbol{P}_{zx}))^{1/2}} \frac{\exp\left[-\frac{1}{2}(\boldsymbol{y} - \bar{\boldsymbol{y}})^{\mathrm{T}}\boldsymbol{P}_{yy}^{-1}(\boldsymbol{y} - \bar{\boldsymbol{y}})\right]}{\exp\left[-\frac{1}{2}(\boldsymbol{z} - \bar{\boldsymbol{z}})^{\mathrm{T}}\boldsymbol{P}_{zz}^{-1}(\boldsymbol{z} - \bar{\boldsymbol{z}})\right]} \end{aligned} \tag{C.55}$$

The exponential term of the numerator in Eq. (C.55) is

$$-\frac{1}{2}(\boldsymbol{y} - \bar{\boldsymbol{y}})^{\mathrm{T}}\boldsymbol{P}_{yy}^{-1}(\boldsymbol{y} - \bar{\boldsymbol{y}}) =$$

$$-\frac{1}{2}\begin{bmatrix}(\boldsymbol{x}-\bar{\boldsymbol{x}})-\boldsymbol{P}_{xz}\boldsymbol{P}_{zz}^{-1}(\boldsymbol{z}-\bar{\boldsymbol{z}})\\ \boldsymbol{z}-\bar{\boldsymbol{z}}\end{bmatrix}^{\mathrm{T}}\begin{bmatrix}\boldsymbol{P}_{xx}-\boldsymbol{P}_{xz}\boldsymbol{P}_{zz}^{-1}\boldsymbol{P}_{zx} & \mathbf{0}\\ \mathbf{0} & \boldsymbol{P}_{zz}\end{bmatrix}^{-1}$$
$$\begin{bmatrix}(\boldsymbol{x}-\bar{\boldsymbol{x}})-\boldsymbol{P}_{xz}\boldsymbol{P}_{zz}^{-1}(\boldsymbol{z}-\bar{\boldsymbol{z}})\\ \boldsymbol{z}-\bar{\boldsymbol{z}}\end{bmatrix} \tag{C.56}$$

Its substitution into Eq. (C.55) can obtain

$$p(\boldsymbol{x}|\boldsymbol{z})=\frac{1}{\det(2\pi(\boldsymbol{P}_{xx}-\boldsymbol{P}_{xz}\boldsymbol{P}_{zz}^{-1}\boldsymbol{P}_{zx}))^{1/2}}\cdot$$
$$\exp\Big[-\frac{1}{2}(\boldsymbol{x}-\overline{\boldsymbol{x}}-\boldsymbol{P}_{xz}\boldsymbol{P}_{zz}^{-1}(\boldsymbol{z}-\overline{\boldsymbol{z}}))^{\mathrm{T}}(\boldsymbol{P}_{xx}-\boldsymbol{P}_{xz}\boldsymbol{P}_{zz}^{-1}\boldsymbol{P}_{zx})^{-1}$$
$$(\boldsymbol{x}-\overline{\boldsymbol{x}}-\boldsymbol{P}_{xz}\boldsymbol{P}_{zz}^{-1}(\boldsymbol{z}-\overline{\boldsymbol{z}}))\Big] \tag{C.57}$$

Obviously, $p(\boldsymbol{x}|\boldsymbol{z})$ satisfies the condition of Gaussian PDF, so

$$\begin{aligned}E\{\boldsymbol{x}|\boldsymbol{z}\}&=\bar{\boldsymbol{x}}+\boldsymbol{P}_{xz}\boldsymbol{P}_{zz}^{-1}(\boldsymbol{z}-\bar{\boldsymbol{z}})\\ \boldsymbol{P}_{xx|z}&\triangleq E\{(\boldsymbol{x}-E\{\boldsymbol{x}|\boldsymbol{z}\})(\boldsymbol{x}-E\{\boldsymbol{x}|\boldsymbol{z}\})^{\mathrm{T}}|\boldsymbol{z}\}\\ &=\boldsymbol{P}_{xx}-\boldsymbol{P}_{xz}\boldsymbol{P}_{zz}^{-1}\boldsymbol{P}_{zx}\end{aligned} \tag{C.58}$$

which completes the proof.

Corollary *If the positions of* $\boldsymbol{x}$ *and* $\boldsymbol{z}$ *in* Eq. (C.48) *are exchanged, then* $p(\boldsymbol{z}|\boldsymbol{x})$ *is given by:*

$$p(\boldsymbol{z}|\boldsymbol{x})=\mathcal{N}(\boldsymbol{z};\bar{\boldsymbol{z}}_{z|x},\boldsymbol{P}_{zz|x}) \tag{C.59}$$

where

$$\begin{aligned}\bar{\boldsymbol{z}}_{z|x}&\triangleq E\{\boldsymbol{z}|\boldsymbol{x}\}=\bar{\boldsymbol{z}}+\boldsymbol{P}_{zx}\boldsymbol{P}_{xx}^{-1}(\boldsymbol{x}-\bar{\boldsymbol{x}})\\ \boldsymbol{P}_{zz|x}&\triangleq E\{(\boldsymbol{z}-E\{\boldsymbol{z}|\boldsymbol{x}\})(\boldsymbol{z}-E\{\boldsymbol{z}|\boldsymbol{x}\})^{\mathrm{T}}|\boldsymbol{x}\}\\ &=\boldsymbol{P}_{zz}-\boldsymbol{P}_{zx}\boldsymbol{P}_{xx}^{-1}\boldsymbol{P}_{xz}\end{aligned} \tag{C.60}$$

Theorem C.2 *For the linear measurement equation*

$$\boldsymbol{z}=\boldsymbol{H}\boldsymbol{x}+\boldsymbol{v} \tag{C.61}$$

If $\boldsymbol{x}$ and $\boldsymbol{v}$ are Gaussian random variables, and

$$\begin{aligned}E\{\boldsymbol{v}\}=\mathbf{0},\ E\{\boldsymbol{v}\boldsymbol{v}^{\mathrm{T}}\}=\boldsymbol{R}\\ E\{\boldsymbol{x}\boldsymbol{v}^{\mathrm{T}}\}=\mathbf{0}\end{aligned} \tag{C.62}$$

then

$$p(z|\boldsymbol{x}) = \mathcal{N}(z; \boldsymbol{Hx}, \boldsymbol{R}) \tag{C.63}$$

Proof Because $[z^{\mathrm{T}}, \boldsymbol{x}^{\mathrm{T}}]^{\mathrm{T}}$ is the linear combination of Gaussian variables $\boldsymbol{x}$ and $\boldsymbol{v}$, $\boldsymbol{x}$ and z satisfy the joint Gaussian distribution. Calculate $\boldsymbol{P}_{zx}$, $\boldsymbol{P}_{xx}$ as

$$\begin{aligned} \boldsymbol{P}_{zx} &= E\{(z - \bar{z})(\boldsymbol{x} - \bar{\boldsymbol{x}})^{\mathrm{T}}\} = \boldsymbol{H}\boldsymbol{P}_{xx} \\ \boldsymbol{P}_{zz} &= E\{(z - \bar{z})(z - \bar{z})^{\mathrm{T}}\} = \boldsymbol{H}\boldsymbol{P}_{xx}\boldsymbol{H} + \boldsymbol{R} \end{aligned} \tag{C.64}$$

Substituting the above equation into Eq. (C.60) completes the proof

Corollary *For* Eq. (C.61) *and the condition in* (C.62), *denote* $\boldsymbol{y} \triangleq [\boldsymbol{x}^{\mathrm{T}}, z^{\mathrm{T}}]^{\mathrm{T}}$. Then

$$p(\boldsymbol{y}) = \mathcal{N}\left(\begin{bmatrix} \boldsymbol{x} \\ z \end{bmatrix}; \begin{bmatrix} \bar{\boldsymbol{x}} \\ \boldsymbol{H}\bar{\boldsymbol{x}} \end{bmatrix}, \begin{bmatrix} \boldsymbol{P}_{xx} & \boldsymbol{P}_{xx}\boldsymbol{H}^{\mathrm{T}} \\ \boldsymbol{H}\boldsymbol{P}_{xx} & \boldsymbol{H}\boldsymbol{P}_{xx}\boldsymbol{H}^{\mathrm{T}} + \boldsymbol{R} \end{bmatrix}\right) \tag{C.65}$$

Appendix D
Constrained Optimization

For the constrained optimization problem:

$$\min\ f(\boldsymbol{x}), \quad \text{s.t. } g(\boldsymbol{x}) = \mathbf{0} \tag{D.1}$$

Define the Lagrange function as

$$J(\boldsymbol{x}, \lambda) = f(\boldsymbol{x}) + \lambda g(\boldsymbol{x}) \tag{D.2}$$

Then $\boldsymbol{x}^*$ is a local minimum point, if and only if the following conditions are hold

1. $\nabla J_x(\boldsymbol{x}^*, \lambda^*) = \mathbf{0}$ and $\nabla J_\lambda(\boldsymbol{x}^*, \lambda^*) = 0$ (necessary condition)
2. $z^{\mathrm{T}} \nabla^2_{\boldsymbol{xx}} J(\boldsymbol{x}^*, \lambda^*) z \geq 0 \quad \forall z$ s.t. $\nabla_{\boldsymbol{x}} g(\boldsymbol{x}^*)^{\mathrm{T}} z = 0$ (sufficient condition)

where $\nabla^2_{\boldsymbol{xx}}$ represents the second partial derivative with respect to $\boldsymbol{x}$.

D. Wang et al., *Spacecraft Autonomous Navigation Technologies Based on Multi-source Information Fusion*, Space Science and Technologies, https://doi.org/10.1007/978-981-15-4879-6

Appendix E
Coordinate Transformation of Optical Imaging Sensor

As shown in Fig. E.1, denote the focal length of imaging sensor (camera) as f, the imaging resolution as $p \times p$ pixels, and the total field angle as 2θ. Then, the field angle of each pixel is

$$\frac{2\theta}{p} \tag{E.1}$$

The Cartesian coordinate length corresponding to each pixel is

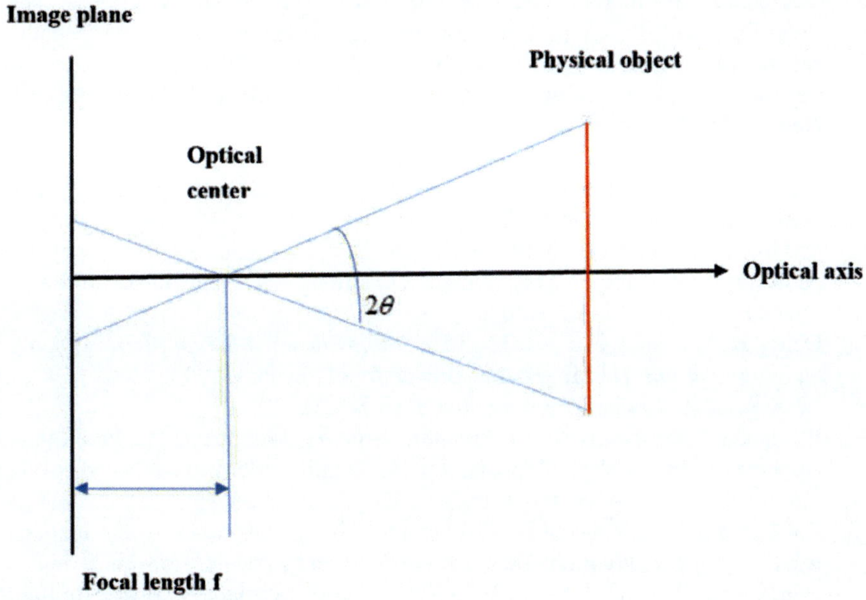

Fig. E.1 Illustration of imaging

D. Wang et al., *Spacecraft Autonomous Navigation Technologies Based on Multi-source Information Fusion*, Space Science and Technologies, https://doi.org/10.1007/978-981-15-4879-6

$$\frac{2f \tan\theta}{p} \tag{E.2}$$

Equations (E.1) and (E.2) are often used for optical navigation simulation. The corresponding angle unit or Cartesian coordinates can be obtained by using the pixel noise. For example, for the optical sensor of the "Deep Space 1," the focal length is 677 mm, the pixel matrix of its imaging element is 1024 × 1024, and the total field angle is 13.4 mrad. Then, the field angle of each pixel is

$$\frac{13.4 \times 10^{-3}}{1024} \simeq 13.1\ \mu\text{rad} \tag{E.3}$$

The Cartesian coordinate length corresponding to each pixel is

$$\frac{2f \tan\theta}{p} = \frac{2 \times 0.677 \times \tan(13.4 \times 10^{-3}/2)}{1024} = 8.8593 \times 10^{-6}\ \text{m} \tag{E.4}$$

References of Appendices A–E

1. Standish, E. 1995. *JPL planetary and lunar ephemerides DE405/LE405*. JPL IOM 312.F-98-048.
2. Seidelmann, P.K. 2006. *Explanatory supplement to the astronomical almanac.* University Science Books.
3. Wikipedia Contributors. 2020. *Minute and second of arc.* Wikipedia, The Free Encyclopedia. https://en.wikipedia.org/w/index.php?title=Minute_and_second_of_arc&oldid=934654747. Accessed at 14 Jan 2020.
4. Crassidis, J.L., and J.L. Junkins. 2011. *Optimal estimation of dynamic systems.* New York: CRC Press.
5. Magnus, J.R., and H. Neudecker. 1988. *Matrix differential calculus with applications in statistics and econometrics.* Wiley series in probability and mathematical statistics. New York: Wiley.
6. Wikipedia Contributors. 2020. *Matrix calculus.* Wikipedia, The Free Encyclopedia. https://en.wikipedia.org/w/index.php?title=Matrix_calculus&oldid=930941643. Accessed at 15 Jan 2020.
7. Deng, Z., P. Zhang, and W. Qi. 2016. *Robust fusion Kalman filtering theory with applications.* Harbin: Harbin University of Technology Press.
8. Li, X.R. 2016. *Review of probability (class notes).*
9. Wikipedia Contributors. 2020. *Probability theory.* Wikipedia, The Free Encyclopedia. https://en.wikipedia.org/w/index.php?title=Probability_theory&oldid=930545712. Accessed at 15 Jan 2020.
10. Bar-Shalom, Y., X.R. Li, and T. Kirubarajan. 2004. *Estimation with applications to tracking and navigation: Theory algorithms and software.* New York: Wiley.
11. Han, C., H. Zhu, and Z. Duan. 2009. *Multi-source information fusion.* Beijing: Tsinghua University Press.

Appendix F
Mathematical Terms

General Specifications

Italics	Scalar, for example x
Lowercase bold italics	Vector, for example $\boldsymbol{x}$
Capital bold italics	Matrix, for example $\boldsymbol{A}$
$\mathbb{R}$	Real number filed
$\triangleq$	Same definition.

Mathematical Symbols

arg min	The variable value that minimizes the objective function
arg max	The variable value that maximizes the objective function
adj(·)	Adjoint matrix, for example, $\text{adj}(\boldsymbol{A})$ is the adjoint matrix of $\boldsymbol{A}$
cov(·, ·)	Thc cross-covariance of two random variables (or vectors), which can also be denoted as $C_{..}$ for random variables or $P_{..}$ for random vectors
const	Constant
$C(m, n)$	Number of the combinations composed of n elements taken from m elements
$\boldsymbol{C}_i^j$	Attitude transformation matrix from the coordinate system i to the coordinate system j
$\delta(\cdot)$	Continuous delta function
δ_{ij}	Discrete delta function
diag(·)	Diagonal matrix
det()	Determinant operator
$E\{\}$	Mean value. For example, $E\{x\}$ is the mean of x, also denoted as $\bar{x}$
exp{}	Exponential function
$F_x(x)$	Distribution function of the random variable x, denoted as $F(x)$; it can also be used for random vectors

D. Wang et al., *Spacecraft Autonomous Navigation Technologies Based on Multi-source Information Fusion*, Space Science and Technologies, https://doi.org/10.1007/978-981-15-4879-6

$F_{x,y}(x, y)$	Joint distribution function of the random variables x and y, denoted as $F(x, y)$; it can also be used for random vectors
$\nabla^2_{\boldsymbol{xx}}$	The second partial derivative with respect to $\boldsymbol{x}$, also denoted as $\frac{\partial^2}{\partial \boldsymbol{x}^2}$
$\boldsymbol{I}$	Identity matrix
$\mathcal{I}$	Fisher information matrix
$\lambda_i(\cdot)$	The ith eigenvalue of a matrix. For example, $\lambda_i(\boldsymbol{A})$ is the ith eigenvalue of $\boldsymbol{A}$
$\mathcal{N}(\boldsymbol{x}; \bar{\boldsymbol{x}}, \boldsymbol{P}_x)$	The probability density function of the Gaussian random vector $\boldsymbol{x}$ with the mean of $\bar{\boldsymbol{x}}$ and the variance of $\boldsymbol{P}_x$
$\frac{\partial f}{\partial \boldsymbol{x}}$	The partial derivative of the function f with respect to $\boldsymbol{x}$, also denoted as $\nabla_{\boldsymbol{x}} f$
$P(A)$	Occurrence probability of the event A
$p_x(x)$	Probability density function
$\text{rank}(\cdot)$	Rank of a matrix
$\boldsymbol{R}_x$	Autocorrelation matrix of the random vector $\boldsymbol{x}$
$\boldsymbol{R}_{xy}$	Cross-correlation matrix of the random vector $\boldsymbol{x}$ and $\boldsymbol{y}$
σ	Standard deviation
s.t.	Making the later condition true
$\text{Tr}[\cdot]$	Trace of square matrix. For example, $\text{Tr}[\boldsymbol{A}]$ is the trace of $\boldsymbol{A}$
$\text{var}(x)$	(Auto)variance of the random variable x, also denoted as σ_x^2 or C_x
$\text{var}(\boldsymbol{x})$	Autocovariance matrix of the random variable x, also denoted as $\boldsymbol{P}_x$ or $\text{cov}(\boldsymbol{x}, \boldsymbol{x})$. Without causing confusion, it can be simplified as $\boldsymbol{P}$
$z_{1:n}$	$\{z_1, z_2, \ldots, z_n\}$
$\boldsymbol{0}$	Zero matrix or zero vector
$\Vert\cdot\Vert$	2-norm of vector
$\emptyset$	Null or empty set
$(A\vert B)$	The conditional statistical properties of A under the condition of B occurrence. For example, $E\{\boldsymbol{x}\vert z\}$ is conditional mean.

Operators

$\otimes$	Kronecker operator
$\text{vec}(\cdot)$	vec operator
$[\boldsymbol{x}\times]$	Cross-product operator of the 3D vector x.

Superscripts

T	Matrix transposition
(i)	The variable related to the ith node (or local filter) in the information fusion system
(ij)	Co-action quantity of the ith, jth node (or local filter) in the information fusion system.